Reading Retail

Reading Retail

A geographical perspective on retailing and consumption spaces

Neil Wrigley
Professor of Geography
University of Southampton

Michelle Lowe
Senior Lecturer in Geography
University of Southampton

ARNOLD

A member of the Hodder Headline Group
LONDON
Co-published in the United States of America by
Oxford University Press Inc., New York

First published in Great Britain in 2002 by
Arnold, a member of the Hodder Headline Group,
338 Euston Road, London NW1 3BH

http://www.arnoldpublishers.com

Co-published in the United States of America by
Oxford University Press Inc.,
198 Madison Avenue, New York, NY10016

British Library Cataloguing in Publication Data
A catalogue record for this book is available from the British Library

Library of Congress Cataloging-in-Publication Data
A catalog record for this book is available from the Library of Congress

ISBN 0 340 70661 9 (hb)
ISBN 0 340 70660 0 (pb)

1 2 3 4 5 6 7 8 9 10

Production Editor: Rada Radojicic
Production Controller: Bryan Eccleshall

Typeset in 10 on 13 pt Palatino by Phoenix Photosetting, Chatham, Kent
Printed and bound in Great Britain by MPG Books Ltd, Bodmin, Cornwall

For Amelia and Theodore

Contents

PART 3

MAKING AND RE-MAKING THE GEOGRAPHIES OF RETAIL CAPITAL 129

Preface

The concept for this book has been in our minds for quite some time. In 1996, in *Retailing, consumption and capital*, we had set out a manifesto for a 'new retail geography'. What could be simpler then, we thought, than a major text, using the organizational frameworks and style of our undergraduate courses at the University of Southampton, to flesh out that manifesto and to capture the intensity of the debates beginning to unfold in the late 1990s on the geographies of retailing and consumption spaces.

But how could we achieve that in the straitjacket of the conventional academic text? After all, our courses were built around encouraging our students to develop a sense of *engagement* with the swirling and rapidly developing debates in the field – providing a framework and a route map into study of the geographies of retailing and consumption, but asking our students to share with us the sense of a research literature unfolding before them as they were taught. The last thing we wanted to achieve was a freezing of that dynamic in a conventional 'magisterial' textbook account. Rather, we wanted a format which would encourage our undergraduates to 'read retail' through the lens of 'a new retail geography' perspective, with a sense of stimulation and liberation – to gain an active sense of the range, intensity and research literature sources of a rapidly evolving series of debates to which we, our academic collaborators and our research students were attempting to contribute.

And so was born the 'hybrid' text which we lay out below: part organizational framework, part text, part 'reader' – a packaging of our vision of, and approaches to, the study of the geographies of retailing and consumption spaces – illuminated through the eyes and words, not only of ourselves, but the many academics who have been as fascinated as we have during the past decade with subjects that have suddenly found themselves at the very heart of contemporary theoretical debate in the social sciences and humanities.

The 68 short and edited 'readings' in the book include many by those UK and US academics with whom we have engaged over that period. The work of others is discussed in detail in the text. To all of those represented in this way, or who have taken part in, and continue to contribute to, what our students consistently tell us is one of the most interesting and challenging areas of study we extend our thanks. To Laura McKelvie, previously Geography Editor at

Arnold who was brave enough to go along with our vision of this hybrid text, and to the current Geography Editor at Arnold, Liz Gooster, who has seen it to completion, we also express our gratitude. We also extend warm thanks to our secretaries, Adam Corres and Denise Pompa, who produced the typescript, to Bob Smith of the Cartographic Unit, Department of Geography, University of Southampton, for the excellence of his diagrams and to Andrew Currah for his help with the Subject Index. Finally, to all our colleagues in other universities who have been pressing us to finish the book for the past two years so that they could use it with their students, we are grateful for your patience. All we can say in mitigation is that the arrival of Theo two years ago, and the absence of sleep which followed, certainly changed the nature of our retail and consumption experiences.

Neil Wrigley and Michelle Lowe
Winchester
Summer 2001

Acknowledgements

The authors wish to express their thanks to the following for permission to reproduce material from the works listed below.

John Wiley and Sons, Inc. for Reading 1.2 by Leigh Sparks in *Agribusiness* (1997) Vol. 13, 153–67; ITPS Ltd for Reading 1.3 by Daniel Miller in *Material Cultures: Why Some Things Matter* (1998) UCL Press: London, 169–87; Blackwell Publishers for Reading 1.5 from Ron Johnston, Derek Gregory and David Smith (editors) *The Dictionary of Human Geography* 3rd Edition (1994) Blackwell: Oxford, 533–35; Pion Ltd for Reading 2.2 by Alan Hallsworth in *Environment and Planning A (*1991), Vol. 23, 1217–23; University of North Carolina School of Law for Reading 2.7 by D. Gordon Smith in *North Carolina Law Review* (1996) Vol. 74, 101–203; Royal Geographical Society (with the Institute of British Geographers) for Reading 3.2 by Louise Crewe and Eileen Davenport in *Transactions of the Institute of British Geographers* (1992) NS Vol. 17, 183–97; Edward Arnold Ltd for Readings 3.5 and 13.5 by Deborah Leslie and Suzanne Reimer in *Progress in Human Geography* (1999) Vol. 22, 401–20; Harvard University Press and the President and Fellows of Harvard College for Reading 4.2 by Morris A. Adelman in *A&P: A Study in Price-Cost Behaviour* (1959) Harvard University Press: Cambridge MA, 59–64; Frank Cass Ltd for Reading 4.3 by John Fernie in *Service Industries Journal* (1997) Vol. 17, 383–96; Taylor and Francis Ltd for Reading 4.4 by David L.G. Smith and Leigh Sparks in *International Review of Retail, Distribution and Consumer Research* (1993) Vol. 3, 35–64, http://www.tandf.co.uk/journals; Pion Ltd for Reading 6.1 by Nicholas Blomley in *Environment and Planning D: Society and Space* (1986) Vol. 4, 183–200; Royal Geographical Society (with the Institute of British Geographers) for Reading 6.3 by Michelle Harrison, Andrew Flynn and Terry Marsden in *Transactions of the Institute of British Geographers* (1997) NS Vol. 22, 473–87; ITPS Ltd for Reading 7.3 by Cliff Guy and J. Dennis Lord in Rosemary Bromley and Colin Thomas (editors) *Retail Change: Contemporary Issues* (1993) UCL Press: London, 88–108; ITPS Ltd for Reading 9.1 by John Hannigan in *Fantasy City: Pleasure and Profit in the Postmodern Metropolis* (1998) Routledge: London and New York, 90–92; ITPS Ltd for Reading 10.2 by Frank Mort in *Cultures of Consumption: Masculinities and Social Space in Late Twentieth-Century Britain* (1996) Routledge: London, 157–63; Sage Publications Ltd for Reading 11.1 by Mica Nava in *International Journal of Cultural Studies* (1998) Vol. 1,

163–96; ITPS Ltd for Reading 11.6 by Sean Nixon in *Hard Looks: Masculinities, Spectatorship and Contemporary Consumption* (1996) UCL Press: London, 522–56; ITPS Ltd for Reading 12.5 by Daniel Miller, Peter Jackson, Nigel Thrift, Michael Rowlands and Beverley Holbrook in *Shopping, Place and Identity* Routledge: London, viii–xi; Pion Ltd for Reading 12.6 by Rob Shields in *Environment and Planning D: Society and Space* (1989) Vol. 7, 147–64; Pearson Education Australia Pty Ltd for Reading 12.7 by John Goss in Kay Anderson and Fay Gale (editors) *Inventing Places: Studies in Cultural Geography* (1992) Longman Cheshire: Melbourne, 158–77 © 1992; ITPS Ltd for Reading 13.4 by Alison Clarke in Roger Silverstone (editor) *Visions of Suburbia* Routledge: London, 132–60.

In addition we are indebted to the following authors from whose work we have taken shorter extracts:

Michael D. Smith (Reading 1.1); Nicholas Blomley (Readings 1.4, 9.3 and 11.3); Cliff Guy (Reading 2.6); Jo Foord, Sophie Bowlby and Christine Tillsley (Reading 3.1); Christine Doel (Reading 3.3); Alexandra Hughes (Reading 3.4); Rachel Bowlby (Reading 4.1); Paul Freathy and Leigh Sparks (Reading 5.1); Steve Smith (Reading 5.2); Susan Porter Benson (Reading 5.3); Paul du Gay (Reading 5.5); Terry Marsden, Michelle Harrison and Andrew Flynn (Reading 6.4); Richard Longstreth (Reading 7.1); Ken Jones and Jim Simmons (Reading 7.2); Jon Goss (Readings 7.4, 9.4); Marilyn Lavin (Reading 7.5); David Brooks (Reading 7.6); Steve Wood (Reading 8.1); John Fernie and Steve Arnold (Reading 8.3); Bill Lancaster (Reading 9.2); Margaret Crawford (Reading 9.5); John F. Sherry Jr. (Reading 9.6); Mike Featherstone (Reading 10.1); John Fernie, Christopher Moore, Alex Lawrie and Alan Hallsworth (Reading 10.3); Sharon Zukin (Reading 10.4); Nicky Gregson and Louise Crewe (Reading 10.5); Mona Domosh (Reading 11.2); Robyn Dowling (Reading 11.4); Frank Mort (Reading 11.5); Wilbur Kowinski (Reading 12.1); Jeffrey Hopkins (Reading 12.2); Gwyn Rowley and OXIRM (Reading 12.3); Thomas J. Schlereth (Reading 13.1); Jonathan Reynolds (Reading 13.2); Robert Di. Romualdo (Reading 13.3).

PART I

Introduction

1
Reading retail: purpose and organization of the book

The strains of 'Tesco, Tesco, Tesco . . .' drifting across the African bush in the closing scenes of the BBC's well-known television documentary *Mange Tout* captures for many geographers the sense of a world at the beginning of the twenty-first century in which the geographies of production are being as actively shaped by multi-national retail capital as the familiar landscapes of consumption – the shopping malls, superstores and retail streetscapes – which crowd the daily experiences of their predominantly western, urban lives. But how are they to understand this world of retail capital and its production and consumption spaces? How are they to 'read' retail?

In this book we offer a guide. Essentially our approach is to build on and expand our well-known, but more abstract, argument (Wrigley and Lowe, 1996) that within human geography a reconstructed sub-discipline of retail geography emerged during the 1990s – a 'new retail geography', characterized above all else by theoretical engagement and by a shared appreciation that retail geography is potentially one of the most interesting and challenging areas of study 'given the subtlety and importance of retail capital, consumption and space' (Blomley, 1996: 256). The 'new retail geography' which we and our co-authors sketched out in the mid-1990s was not only one in which retail capital and its transformation were seen as vital and relevant topics for research demanding urgent attention (Ducatel and Blomley, 1990: 225), but also a subject whose central problematic and potential lay in the fact that the arbitrary categories of 'economy' and 'culture' required constant shattering in its study. That is to say, a subject which to be worthy of its name needed to take *both* its economic and cultural geographies seriously, combining 'an exploration of the economic structures of retail capital with an analysis of the cultural logic of retailing' (Blomley, 1996: 239). As a result, we argued that the 'new retail geography' found itself at the very heart of contemporary debate within critical human geography, presenting opportunities which should be seized and challenges which should be addressed.

But this book is not about such abstract argument. Rather it is an attempt to show a largely undergraduate audience how it is possible to 'read retail' through the lens of a 'new retail geography' perspective, allowing them in the process to gain an active sense of the range, intensity and research literature sources of the swirling and rapidly developing series of debates which focus on the geographies of retailing and consumption spaces. Not only, therefore, do we adopt organizational frameworks which have proved useful in our undergraduate courses on these topics at the University of Southampton, but we also use a wide range of short 'readings' selected from the work of both well-known and lesser-known authors in the field. The 'readings' are provided both to offer depth and perspective on key issues and to encourage readers to develop a sense of engagement with the theoretical debates and research literature shaping the field. Together our organizational framework, text and selected 'readings' offer our readers a route map into the study of the geographies of retailing and consumption spaces. Our intention in choosing and packaging the material in the way we have is not to strait-jacket our readers. Indeed, it is the very opposite. As we go on to demonstrate below, multiple readings of particular topics are the very essence of academic writing and debate in the field. Our hope is that our choices and approaches will serve to stimulate and liberate – encouraging readers to explore and develop still further the potential of subjects which we believe are among the most interesting and challenging in human geography – subjects which lie at the very heart of contemporary theoretical debate within both the discipline and social science more generally.

Developing a sense of the debate: reading and re-reading retail

But how can our readers develop a sense of the debates shaping the study of the geographies of retailing and consumption spaces? Crucial to that development we suggest must be an appreciation of the nature of academic writing which, by its very nature, invites critical review from a variety of perspectives. In this book we essentially use three approaches to sharpen that appreciation. First, we show how single issues in retailing and consumption will inevitably be approached from a variety of angles – in particular, we contrast 'economic' and predominantly 'cultural' readings of the same topic. Second, we show how single pieces of academic writing will often be constructed to offer multiple readings of a topic. Third, we demonstrate how a single piece of academic writing may be read and re-read over time from widely differing perspectives, or in markedly different ways at the same time by a variety of audiences.

Example 1 – The empire filters back: Starbucks coffee

Let us begin then with a consideration of Michael D. Smith's fascinating paper in *Urban Geography* (1996) on the Starbucks Coffee Company – the 'bean and beverage' retail and mail order enterprise, first established in Seattle in 1971, which grew phenomenally following its transition into a coffee bar format in 1987. In that paper, Smith offers two readings of Starbucks. The first attempts to situate Starbucks 'within a broader historical and contemporary geography of coffee production closely linked to various phases in the history of Euroamerican colonialism and imperialism' – a reading informed, in part, by recent critiques of consumption and post-colonial theory and 'offered to suggest how consumption and production, the symbolic and the real, the local and the global are all indissolubly bound up one with the other'. The second considers key aspects of the 'styling' and 'placing' of Starbucks to 'offer a reading of Starbucks as the site of a specific set of local consumption practices'.

Smith's first reading is rooted in the widely held perception that coffee, one of the most important tropical commodities in international agricultural trade, 'reflects unambiguously the relationships of dependency that connect the poor underdeveloped nations of the South to the rich industrialized countries of the North'. Having considered aspects of this dependency Smith asks the question: 'So where in this historico-geographical matrix of coffee relations . . . does Starbucks fit?' His answer goes well beyond the sheer economics (however fascinating) of a vertically integrated retail business which, in 1990, paid around $2 a pound for Columbian coffee but sold it for $8 a pound in its stores as beans or, as brewed coffee, for the liquid equivalent of $20 a pound. In Smith's view: 'What makes Starbucks particularly interesting is that it trades not only in the 300-year-old market for this tropical commodity, but in an equally enduring if less tangible symbolic economy of images and representations that are the cultural correlates of Euroamerican domination.' As a result he shows how 'Starbucks . . . appropriates, refashions and redistributes elements from the history and contemporary economic geography of coffee production to produce its own "cognitive map" of coffee'. He argues that 'this is not merely an embellishment of the Starbucks marketing strategy, but is, on the contrary, an integral component of the Starbucks product, not only in the form of the commodity itself but in the symbolic meaning that is invested in that commodity and in the act of consuming it'. As a result, 'it is for precisely this reason that . . . production and consumption are inseparable in the case of Starbucks, which has succeeded in creating what we might term a "cultural geography of production"'.

Smith's second reading of Starbucks focuses on the 'styling' and 'placing' of the firm – the flair for design, positioning, service and promotion which established it as a highly successful retailer in what he terms a largely 'privileged, urban(e) milieu of consumption' in North America. Reading 1.1 captures the spirit of this second reading.

Reading 1.1 – 'Styling' and 'placing' Starbucks

Starbucks outlets are integrally connected to those 'landscapes of leisure' where people with disposable income, not to mention cultural capital go to consume, display themselves, and watch others; as such they are part and parcel of the cultural geography associated with the 'recolonization' of North American inner cities by the yuppie shock troops of a putatively post-industrial capitalism. Starbucks' store design thus relies upon its links not

> only with the city, but with those gentrified neighbourhoods and districts whose cachet it has absorbed, even as the chain expands into suburban streetscapes, airport arcades, and shopping malls . . .
>
> Starbucks' retail design could be said to revolve around an enticement to 'eat the street'. All Starbucks outlets are equipped with floor-to-ceiling windows, behind which are lined counters and rows of elevated stools providing a perfect and legitimized vantage point for the individual voyeur. Patrons are meant to watch when at Starbucks, to take in the streetscape as they sip their coffee . . .
>
> The symbolic appeal of Starbucks coffee cannot be separated from the manner in which it is served, since service is so central to the Starbucks model. The preparation of specialized coffee drinks – cafe latte, espresso – is integral to the act of consumption. There is thus a performative element in Starbucks, an aestheticization of the commodity, as the 'baristas' transform the formerly mundane act of serving coffee into the theatrics of consumption.
>
> **Extracted from: Michael D. Smith (1996): The empire filters back: consumption, production and the politics of Starbucks Coffee. *Urban Geography*, 17, 502–24.**

Example 2 – Questioning the globalization of Coca-Cola

In contrast to Michael D. Smith's explicit multiple readings of Starbucks within a single journal article, our second example focuses on a case in which two pieces of academic writing approach a single topic from rather different perspectives – one a predominantly 'economic' reading, the other a predominantly 'cultural' reading. The topic itself has interesting links to the themes explored by Smith – Coca-Cola after all is a quintessential example of what Smith in that paper terms the 'symbolic economy of images ... the cultural correlates' of American domination of certain aspects of global trade. Despite their rather different perspectives, both writers accept this, beginning their papers with phrases such as 'the iconic status of Coca-Cola is taken for granted' (Sparks, 1997: 153) and 'Coca-Cola is one of three or four commodities which have obtained the status ... of meta symbol' (Miller, 1998a: 170). However, the orientation of their readings of Coca-Cola differs considerably.

The first reading by Leigh Sparks – geographer turned business school professor – is distinctly 'economic' in orientation and is taken from a special issue of the US agricultural economics journal *Agribusiness* (1997) in which economists and geographers debated issues of retailer–manufacturer relations and the convergence of food manufacturing and distribution systems (Cotterill, 1997). It focuses on the way in which Coca-Cola – a seemingly inviolate global brand – was challenged during the 1990s by the growing power of retailers and the development of retailer 'own-label' products. The broader issues of retailer–manufacturer relations, shifts in power within those relations, and the strategic significance of retailer own-label products within those shifts will be dealt with in detail in Chapter 3. Here we focus simply and briefly on an 'economic' reading of one particular facet of the Coca-Cola and globalization story. The reading raises interesting issues about growing retailer power in a market which has traditionally been assumed not only to be manufacturer-dominated, but in essence merely a battlefield for global hegemony between two major corporations, Coca-Cola and PepsiCo.

Reading 1.2 – Restructuring the cola market

Retailers seeking to expand their retailer brand offer . . . have initially exploited weaker or newer market segments and avoided manufacturer-dominated sectors such as the cola market . . . The last five years or so, however, have seen even the soft drink market challenged seriously by retailer brands, both in North America and Europe. In North America, many retailers (including Wal-Mart) have followed Loblaw's Canadian lead by introducing 'President's Choice' or some similar retailer brand variant. As the title suggests, this is not the typical down-market view of retailer brands found often in North America, but is rather a statement of trust, reputation, and high quality. This change has had a dramatic effect on the perception and practice of brands and brand management. In the UK, J. Sainsbury challenged Coca-Cola head-on with its Classic Cola launch in 1994. So similar was Classic Cola in brand positioning terms (including the packaging and labelling) that Coca-Cola threatened to sue J. Sainsbury for 'passing off' unless the packaging and design were changed. In late 1994, Virgin launched its own branded cola, initially available only in selected retailers but again a threat to the established order. Even in highly brand-conscious Japan, retailer brand cola (Classic Selection Cola) has been launched by Ito-Yokado (including Seven-Eleven Japan). The common element to all these examples is the Cott Corporation, a Canadian producer of retailer brands, mainly soft drinks. From 1989 it has transformed itself and thrown down a challenge to the leading cola corporations – when customers purchase Coca-Cola are they paying a brand premium or a brand tax, and are they willing to continue paying this in the light of enhanced (price and quality) competition? . . .

Initially Coca-Cola and PepsiCo saw retailer brands and Cott as a minor irritant, but as time passed, local markets were picked off in deals and big players such as Wal-Mart and KMart added to Cott's portfolio, then reaction became critical. This has taken the form of brand reinforcement and pricing reductions to meet the threat and has led (in part) to a stabilisation of retailer brand penetration . . . The free ride that Cott had has been ended . . . Whatever the outcome [of this fightback, however], Cott has arguably destabilized brand markets worldwide . . . [Its] challenge [to the Coke/Pepsi duopoly] illustrates the desire of retailers to gain control of supply chains to extract value [and raises important issues] relating to actual and latent retail power . . . regulatory frameworks conditioning the ability of retailers to develop and exercise such power . . . the ways in which strategy and branding are interlinked and delivered by operational practices . . . relationship building and partnerships in vertical channels . . . [and] the restructuring underway in manufacturer–retailer–consumer relationships.

Extracted from: Leigh Sparks (1997): From coca-colonization to copy-cotting: the Cott Corporation and retailer brand soft drinks in the UK and the US. *Agribusiness*, 13, 153–67.

The second reading is by Daniel Miller (1998a), an anthropologist and well-known writer on theories of consumption (Miller, 1995, 1998b, 2001) often in collaboration with geographers such as Peter Jackson and Nigel Thrift (see Miller et al., 1998). Miller's cultural–ethnographic reading of the production and consumption of Coca-Cola in Trinidad has at its heart an attempt to introduce vitally important nuances into conventional interpretations of archetypical global products such as Coca-Cola and to 'assert the scholarship of contextualization' (Miller, 1998a: 185). Rather than focusing on what he views as rather sterile debates about global homogenization focused on meta-symbols such as Coca-Cola, or discussions of corporate strategies for so-called 'global localization', Miller (1998a: 170) favours the slow building up of a stance towards western capitalism which is 'informed and complex' and which rests 'on the comparative ethnography of practice'. His concern (Miller, 1998a: 18) is with how different groups use 'commodities to create a much more subtle and discriminatory process of incorporation and rejection than allowed for in simple models of Americanization or globalization'. Reading 1.3 captures some of the flavour of Miller's examination of the localization of both production and consumption of Coca-Cola in Trinidad.

Reading 1.3 – Contextualizing Coca-Cola

Coca-Cola could be argued to be a remarkably unsuitable candidate for this role as the key globalized corporation ... The enforced restitution of Classic Coke (following the company's decision to change the composition of the drink in the 1980s as a response to the increasing popularity of Pepsi) was surely one of the most explicit examples of consumer resistance to the will of a giant corporation we have on record ... The second reason why Coca-Cola is not typical of globalization is that from its inception it was based on a system of franchising. The company developed through the strategy of agreeing with local bottling plants that they would have exclusivity for a particular region and then simply selling the concentrate to that bottler. It is only in the last few years that Coke and Pepsi have begun centralizing the bottling system and then only within the USA. There are of course obligations by the bottlers to the company ... [but] the franchise system allows for a considerable degree of local flexibility ...

To understand the details of marketing Coke in Trinidad demands knowledge of ... local, contingent and often contradictory concerns that make up the way capitalism operates locally, together with the way these affect the relationship between local imperatives and the demands emanating from the global strategists based in Atlanta. Often the net result was that Coke representatives in Trinidad were extremely uncertain as to the best marketing strategies to pursue even when it came to choosing between entirely opposing possibilities such as emphasizing its American or its Trinidadian identity ...

The companies produce soft drinks. The public consume sweet drinks. This semantic distinction is symptomatic of the surprising gulf between the two localized contexts ... From the point of view of consumers, the key conceptual categories are not the flavours and colas constantly referred to by the producers. In ordinary discourse much more important are the 'black' sweet drink and the 'red' sweet drink ... This distinction between

> drinks relates in part to the general discourse of ethnicity that pervades Trinidadian conversation and social interaction . . . [there is a] dominant association of these drinks, red with Indian, black with African [the dominant population groups of the country are split 40 per cent ex-African, originally mainly though by no means entirely ex-slave, and 40 per cent ex-South Asian, almost entirely originally ex-indentured labourers]. This does not, however, reflect consumption. Indeed marketing research shows that if anything a higher proportion of Indians drink colas . . . Many Indians explicitly identify with Coke and its modern image . . . In examining the connotations of such drinks we are not therefore exploring some coded version of actual populations . . . the complex context of consumption often frustrate[s] the producers who are looking for a consistency in the population that they can commoditize.
>
> **Extracted from: Daniel Miller (1998a): Coca-Cola: a black sweet drink from Trinidad. In Miller, D. (ed.) *Material cultures: why some things matter*. London: UCL Press, 169–87.**

Example 3 – From pornography to post-modern road-map

Our final example demonstrates how a single piece of writing may, over time, be read and re-read from widely differing perspectives. The piece in question is the novel *Au bonheur des dames* published in 1883 by the French writer Émile Zola. Set in Paris, it documents the rise of the late nineteenth-century department store – that quintessential retail form and 'landscape of consumption' which we consider in greater detail in Chapter 11. Zola grounded his novel in painstaking observation of the great Parisian department stores of the era – in particular the Bon Marché – and it contains a particularly fine appreciation of the logic of capital accumulation pioneered within those stores which were, at one and the same time, part factory, part palace and cathedral. In particular, the book is infused with insights into how store design and layout, incessant spatial reconfiguration of in-store space, display, manipulation of customer circulation patterns, and disorientation of shoppers to induce consumption, were used within those stores to create a magical dream-like consumer paradise. It also captures with considerable sensitivity facets of the rise of consumer society in the world cities of that era, and some of the relationships between consumer culture, consumption spaces and urban form in those cities – (see also Domosh, 1996a, 1996b, on New York and Boston) – Zola's aim being to craft what he terms *'un poème de la vie moderne'*.

However, the book is essentially a novel in which consumption and sexuality are interwoven. It offers in lyrical and sensuous prose the story of Octave Mouret, an aggressive and unscrupulous merchant who builds a retail empire – in the process pursuing members of his female staff and maintaining several mistresses – and Denise Baudu, a young woman from the provinces, who climbs the career ladder within that empire, resisting over a considerable period Mouret's advances, until finally agreeing to marry him. At the time of its publication it was regarded and marketed as a 'racy', perhaps even somewhat pornographic piece. A century later, however, it was a novel which was rediscovered by growing numbers of feminist writers and cultural historians, and was 'increasing *de rigueur* in any historical analysis of consumption and gender' (Blomley, 1996: 248).

Reading 1.4 is by Nicholas Blomley – one of the pioneers of the 'new retail geography' of the 1990s (see Ducatel and Blomley, 1990) based at Simon Fraser University in Canada. Blomley's objective in his essay was to read *Au bonheur des dames* to provide insight into the cultural geographies of retailing, in particular issues of femininity, masculinity, power and retail space. In the process, however, he demonstrates how Zola's novel has progressively been reinterpreted – read and re-read – over the space of a hundred years, and speculates on the potential of Zola's account for 'a reconstructed retail geography'.

Reading 1.4 – Re-reading Zola: signpost towards a 'new retail geography'

In this chapter I want to use *Au bonheur* (translated as 'The ladies' paradise') as a signpost towards a 'new retail geography' . . . Zola is usually seen simply as a novelist. A novelist he is, and a very accomplished one at that, but he is also, I want to argue, a consummate retail geographer, alert to questions of space, capitalism and patriarchy. As expressed most clearly in *Au bonheur des dames*, he offers a number of valuable insights that I think could be of central importance to a new retail geography. These include the social construction of masculinity and femininity; the intersection of space and power; sexuality and consumption; and the nature of retailing as a public and private space . . .

Like many of Zola's works, it resists closure, playing on a series of ambivalences. While it offers a dispassionate account of modern commerce, it also provides lush, lyrical and sensuous prose. While lauding the aspirations of Mouret, it also carefully and sympathetically documents the store's annihilation of the traditional retail landscape, especially the neighbouring shop owned by Denise Baudu's uncle. While the novel can be read as an endorsement of certain patriarchal claims, ending as it does with marriage and the triumph of a woman who wins, as Zola notes, 'purely with her femininity', it also carefully documents the manner in which Mouret himself cynically and cold-heartedly manipulates and conquers his women customers: the store, he exclaims, is built upon 'women's flesh and blood' (Zola, 1992: 68).

The history of the novel over the last century reveals these multi-valences and ambiguities. The novel was, apparently, the first of Zola's to be translated into English. It has been suggested that this novel was chosen because of the growing notoriety of the author in Britain. With the subtitle 'A realistic novel' it was marketed as a 'racy', even pornographic piece. A more accurate (and hence more explicit) translation was issued in 1886 by a firm specializing in sensational fiction. Henry Viztelly, the publisher, was later imprisoned for three months for publishing Zola's *La terre*, deemed an 'obscene libel' by the English courts.

As interest in Zola waned, so the novel disappeared from view, such that I had a hard job finding an English translation in 1990. An extensive library search finally unearthed a copy, reissued in 1976. A limited edition of the 1886 translation had been reprinted, curiously enough, by the Ohio-based retail consultancy Management Horizons. They deemed this a 'Service to Retailing', 'assuring the continued prosperity' of the company's clients. Reflecting the spatial preoccupations of retail capital, the book was characterized as primer on store design, or 'a truly unique and distinctive work on how to create store excitement. The principles and techniques described in Ladies' Paradise are as pertinent today as they were one hundred years ago. Ladies' Paradise, therefore, is a pragmatic guide which can be used in our time to increase the merchandising effectiveness of almost any retail store'.

In 1992, the Viztelly translation again resurfaced with a reprinting by the fashionable University of California Press, with an introduction by the equally fashionable Kristin Ross, author of *The emergence of social space: Rimbaud and the Paris Commune*. She notes the relevance of the novel in very different terms: 'with the convergence of a wide variety of theoretical interests – those of Marxists, postmodernists and feminists, cultural theorists and social historians – intent on charting the transition from an economy based on production to one based on consumption, Zola's novel has again come to the forefront as an indispensable document in the history of . . . "the society of the spectacle" (Zola, 1992: iii)'.

So from pornography to pragmatic marketing guide, and on to post-modern road-map, the novel has been positioned in very different ways over time. This seems suggestive: given its subtlety and resistance to closure, the novel can indeed be read in all of these ways. It is precisely for this reason that it lends itself to a reconstructed retail geography. Not only does it offer a detailed economic statement – as Management Horizons realized – but also it provides a cultural account of consumption and sexuality, as perceived by the first and most recent characterizations. What all three miss, however, is the degree to which the novel is richly geographic. I hope to explore its economic and cultural geographies.

Extracted from: Nicholas Blomley (1996): 'I'd like to dress her all over': masculinity, power and retail space. Chapter 13 in Wrigley, N. and Lowe, M.S. (eds) *Retailing, consumption and capital: towards the new retail geography*. Harlow: Addison Wesley Longman, 238–56.

Constructions of retail geography: placing the book and its organization

Our readers should by now have gathered an impression of the purpose, orientation and style of the book. Our aim, as noted above, is to encourage them to develop an active sense of the wide-ranging debates and academic literatures which focus on the geographies of retailing and consumption spaces. Our book, by construction then, is no magisterial text offering a single route to truth and knowledge. Rather it celebrates multiple readings and re-readings and challenges our readers to explore and, above all, engage with the writers and debates currently shaping the field. But clearly, via its organizational framework, its interpretations, and its selection of 'readings', the book reflects our choices, our positioning as human geographers, and our sense of the reconstructed sub-discipline of retail geography which emerged during the 1990s. To what extent can this broader orientation be justified? What are its consequences in terms of the topics which we include in the book? Conversely, what does it mean in terms of the material which is neglected in the book but which some may feel lies at the heart of a geographical perspective on retailing? In simple terms, why do we offer space in the book to, say, Blomley's consideration of the potential of Zola's retail geography and to the views of cultural theorists and anthropologists who have engaged with geographers in their consideration of consumption spaces (Jackson et al., 2000), but not to central place theory or to retail location modelling, geodemographics, market area analysis, and similar concerns of what has been termed 'orthodox retail geography'?

These are difficult and contentious questions. In part our choices reflect the fact that excellent books are available which cover these more traditional approaches to the geography of retailing – ranging from Brian Berry's classical books of the 1960s (Berry, 1963, 1967), through Jones and Simmons' (1990) definitive account of locational approaches to analysing the retail environment, to Clarke and Birkin's valuable description of retail modelling and market area analysis linked to GIS techniques (see Longley and Clarke, 1995; Birkin et al., 1996). But, essentially, our organizational structure and selections reflect our perception of the reconstructed nature of the subject which emerged during the 1990s, and in this context we direct our readers to two sources.

The first is the well-known *Dictionary of human geography* edited by Ron Johnston, Derek Gregory and David Smith. In its third edition published in 1994 (a fourth edition appeared in 2000 and our readers should also consult that new edition) a strong sense of an ongoing redirection and reconfiguration of retail geography flows through the *Dictionary*'s definition and description of the field. Reading 1.5, written by Nicholas Blomley, attempts to capture that sense in a brief extract from an entry in the *Dictionary*.

Reading 1.5 – Defining and questioning 'retail geography', 1994

Conventionally defined as the study of interrelations between the spatial patterns of retail location and organization on the one hand, and the geography of retail consumer behaviour on the other ... A well-established subdiscipline, mainstream retail geography has certain general characteristics. Broadly speaking, neoclassical economic principles predominate, with considerable emphasis placed upon the structuring role of individual consumer decisions ... With strong links to marketing, [conventional] retail geography is also applied in its emphases ... relatively limited attention [is] given to the retail capital.

The mainstream literature has many powerful insights ... It is notable, however, that retail geography appears to have largely ignored – and been ignored by – the many turbulent theoretical debates of the wider discipline ... In part this reflects an institutional resistance to theoretical change within the subdiscipline . . . However, a widely scattered yet growing literature makes it possible to delineate an alternative model of retail geography. Such an alternative, although far from unified, offers very different interpretations of space and the retail economy, and claims to recover the full potential of retail geography from its overly focused orthodoxy.

Such an alternative reading of retail geography might begin with the analysis of Ducatel and Blomley (1990) who seek to retheorize retail capital itself . . . situat[ing] it within a larger system of production, distribution and consumption . . . Such an analysis helps in directing attention to the restructuring of retail capital, a process which seems to have intensified in the past decade. Characteristic corporate tactics include changes in internal labour markets and work organization, the importance of new forms of sales and warehousing technology, and shifts in capital concentration and deployment. Moreover, spatial changes seem of vital importance as witnessed by the construction of re-configured 'consumption spaces' (such as the 'mega-mall'), and shifts in both the intra-urban and international location of retail capital.

A retheorization of retail capital also directs attention to the importance of the wider social, cultural and economic context that is both expressed in, and shaped by, retailing and consumption. This expression and mediation appears to be spatial at several levels, and at several scales. One woefully underexplored topic, in this regard, is the way in which gender roles are formed in the 'retail spaces' such as the department store and the suburban retail landscape.

On this alternative reading, the question of the roles, functions and encodings of space take on a new urgency. Not only is space a critical concern for an understanding of the process of retail restructuring, but it also bears directly on the manner in which the relations between retail capital and other spheres of social, cultural and economic life are played out over time. However, rather than being seen as a passive surface upon which consumer decisions are inscribed, space is understood as strategic; as open to manipulation and production.

Extracted from: Retailing, geography of. In Johnston, R.J., Gregory, D. and Smith, D.M., eds. (1994) *The dictionary of human geography*, third edition. Oxford, Blackwell, 533–5.

The second source is our own well-known essay 'Towards the new retail geography' (Lowe and Wrigley, 1996). Written in the mid-1990s, it considers in much greater detail many of the themes and questions outlined in the *Dictionary*'s delineation of 'an alternative reading of retail geography'. The essay begins by demonstrating how the early 1990s was a period when retail geography began, at last,

to take its economic geographies seriously and began to explore such issues, as: the geography of retail restructuring; the interface between retail and financial capital; the complex and contradictory relations of retailing with the regulatory state; the interrelations of corporate strategy, corporate culture and market structures in the industry; the social relations of 'production' in retailing; the structure of channel relations (notably retailer–supplier relations); the spatial switching of retail capital; and the configuration of retail spaces to induce consumption.

However,

> no sooner had a theoretically embedded new economic geography of retailing, focused upon retail capital and its transformation and upon the extent to which 'the transformation of the retail sector and of consumption work represents a second wave of industrial restructuring' (Christopherson, 1996) begun to be embraced, than there were calls for a cultural perspective on retailing and consumption spaces to be adopted ... as the theoretical high ground of critical human geography [began to] experience a rapid and significant 'cultural turn' (Lowe and Wrigley, 1996: 4).

As a result, the objective of our essay, and the edited volume of which it formed the introduction, was to show how a reconstructed retail geography which took both its economic and cultural geographies seriously could be, and indeed was in the process of being, developed. Our view was that:

> retailing, conceived either in terms of the traditional maxim of 'location, location, location', or in terms of the unique position which retail capital occupies between production and consumption, mirroring its struggles with productive capital in its relations with consumers, or alternatively in terms of being uniquely suited to analysis of the links between femininity, masculinity, place and consumption, is an inherently *geographical* phenomenon.

We argued therefore that it was of 'central importance both to the broader progress of critical human geography and to the wider study of circulation and consumption in the social sciences, to have a vibrant reconstructed retail geography positioned at the cutting edges of theoretical debate' (Lowe and Wrigley, 1996: 5). As a result, we identified and outlined six key themes in the emergent economic and cultural geographies of retailing and began the process of articulating an organizational framework within which such a reconstructed retail geography might be debated and taught (see also Crewe, 2000).

Organization of the book and the wider agenda

Essentially in this book, and in the teaching courses around which it is based, we adopt and expand that earlier framework, focusing in particular on economic geographies of retail capital and retail restructuring and cultural geographies of consumption spaces and places. Above all, we attempt to combine explorations of the economic structures of retail capital with an analysis of the cultural logic of retailing. In the next section (Part 2) of the book we begin that task – setting the scene via a consideration of five of the key issues which we believe must underline any attempt to understand the processes of capital accumulation, competitive strategy and corporate restructuring that actively transform the retail industry. In particular, we consider the rise of corporate retail power and its transformation, the reconfiguration of corporate structures in retailing and of retailer–supply chain interfaces, organizational/technological transformations of the industry, changing employment relations, and issues of retail regulation and governance.

Our intention in Part 2 of the book is to lay the groundwork for 'a more sophisticated understanding of the *reciprocal* nature of the relations between space and corporate retail activity' (Clarke, 1996: 292), and it is that understanding which we pursue much more explicitly in Part 3 of the book. Here we focus on what, adopting a phrase of Storper and Walker (1989), we term the 'inconstant geography' and spatial switching of retail capital, together with issues relating to the configuration, manipulation and contestation of retail space. We show how corporate retail capital 'actively explores and penetrates specific spaces at a number of scales' (Ducatel and Blomley, 1990: 225), the way it manipulates spatial layout and design to induce consumption, and the way it often remakes an inherited spatial configuration via a process of capital switching (i.e. via a reconfiguration of the locus of profit extraction), but sometimes gets locked into (and finds it difficult to abandon) existing geographies. In contrast with 'orthodox' retail geography, we show how space in the new economic geographies of retailing is interpreted in a far more dynamic way – being conceived as the product of social activity, in which the active creation and recreation of markets, the 'grounding' of capital, the relations between spatial configuration, spatial discipline and control, and so on, are viewed as central issues in both the capital imperatives facing corporate retailers and in the contested retailer–consumer relation. In Blomley's (1996: 239) terms, retail spaces, rather than being viewed as passive surfaces, 'are increasingly being cast as actively produced, represented and contested'.

In turn, this interpretation leads us, in Part 4 of the book, to a consideration of cultural geographies of consumption spaces and places. In particular, we focus on four sites which lie at the heart of our readers' everyday experiences of consumption at the beginning of the twenty-first century – that is to say, the street, the store, the mall, and the home. We focus, in particular, on the gendered and contested nature of these consumption spaces, on geographies of display, on the production and representation of new landscapes of consumption, on issues of place and identity, on the peopling of consumption spaces and, above all, on the constant remaking and remoulding of such spaces.

Readers will have noticed, perhaps, that our approach to the wider issues of consumption studies, as they are now understood in the social sciences and humanities following a decade of explosive growth in research in the field, is, as a result, a rather focused one. It is the consumption spaces of retail capital which provide our primary concern in this book – and sometimes the fringes of those spaces (Gregson et al., 1997; Crewe and Gregson, 1998). In consequence, we stray only occasionally into the broader issues of contemporary consumer culture and consumption which have provided the focus for cultural theorists and cultural historians. Our readers must look elsewhere, therefore, for geographical writing on these broader topics. For example, to Peter Jackson (1991, 1994, 1996) on the semiotics of advertising and the cultural politics of identity, to Mona Domosh (1998) on wider aspects of consumer culture in nineteenth-century American cities, to Paul Glennie and Nigel Thrift (1996a) on consumption and identity in early modern England (as part of an attempt to understand 'what is modern about modern consumption'), to Ian Cook and Philip Crang (1996) on the globalization of culinary culture, to Philip Crang (1996) for consideration of displacement, consumption and identity, to Nicky Gregson and Louise Crewe (1998) for work on second-hand consumption, and to David Clarke (1996) for work on what he terms the 'hyperspaces of the consumer society'. A useful starting point, they will find, is the attempt by Peter Jackson and Nigel Thrift (writing with Miller, Holbrook and Rowlands – see Miller et al., 1998) to provide a brief history of the study of consumption, as a preface to their more detailed consideration of shopping, place and identity – work which we will return to in Chapter 12 in our discussions of the shopping mall.

Conversely, our book offers readers, we would argue, the opportunity to engage with a far more comprehensive, more informed, and hopefully more subtle interpretation of the role of retail capital in contemporary consumption than was frequently the case in the work of some of the cultural theorists of the 1990s. Like so many economic geographers during that decade (see Crang, 1997; Gertler, 1997; Sayer, 1997 and related papers in Section 1 of Lee and Wills, 1997), our response to the wider 'cultural turn' in human geography, when 'suddenly culture and the cultural are absolutely everywhere ... [and] culture – in a variety of substantive and conceptual guises – is increasingly key to economic geography's research agendas' (Crang, 1997, 4), has been to seek, within our own substantive focus on the geographies of retailing and consumption spaces, an overall perspective in which the economic, the social and the cultural are viewed as interdependent 'interarticulated, enmeshed and mutually constituted' (Pred, 1996). Like Schoenberger (1994, 1997), Gertler (1995, 1997), Thrift and Olds (1996), Barnes (1995), and the contributors to Lee and Wills (1997), we have sought economic geographies that embrace forms of cultural analysis, but which remain sensitive, as Andrew Sayer (1997: 25) puts it, to the fact that 'to give the impression that economic logic has become subordinated to culture is to produce an idealized picture of an often brutal economically dominated world'. We are essentially at one with Philip Crang (1997: 14) when he concludes that economic geography's value lies in 'its emphasis on understanding differentiated spaces, places and practices of production, circulation and consumption, and the forms of surplus extracted within and between them'. Our perspective on retailing and consumption, like

Crang's (1997: 15) on economic geography in the era of the 'cultural turn' more widely, rests on the view that 'these spaces, places, and practices are never purely economic, and nor are the surpluses they produce', but it is 'the commitment to study these vital economic moments – production (in its broadest sense), circulation and consumption – and their regulation, by whatever means necessary, wherever they take place, and whatever materials are produced, circulated and consumed in them' that is essential.

In outlining this motivation and agenda for our reading of retailing and consumption spaces, we are conscious that some of our readers may feel that we have moved a long way away from what they might regard as the heart of a geographical perspective on retailing. However, as we have shown in this introductory chapter, the past decade has been a period when, above all else, the reluctance to engage in the theoretical debates of the wider discipline and related subjects, characteristic of the 'orthodox' retail geography of the previous 20 years and highlighted in the *Dictionary of human geography*'s summary of the field in 1994 (see Reading 1.5), has well and truly disappeared. Our reading of retail capital and its production and consumption spaces reflects that sea change. We challenge our readers to explore with us some of the key features of that reconfigured perspective; to engage with the writers and debates shaping the field; to 'read' retail with us and to see it, at first through our eyes but increasingly we hope through their own, as one of the most fascinating, demanding and challenging areas of study in contemporary human geography.

PART 2

Setting the scene: retail dynamics

2
Reconfiguration of corporate structures in retailing

The sudden disappearance of the long-established F.W. Woolworth's variety stores from 'Main Street' USA in 1997, the bankruptcy of Macy's – New York City's and the world's largest department store – in 1991, and the increasingly common usage of phrases such as the 'Wal-Martization of America', provide some simple yet graphic illustrations from just one economy of the dynamic nature of the transformation of retail capital. In this section of the book that dynamic, and the processes of retail restructuring which power it, provide our focus. In particular, we consider five key issues: the concentration of capital and reconfiguration of corporate structures in retailing, the character and shifting power balance of retailer–supplier relations, organizational and technological transformations of the industry, changing retail employment relations, and retail regulation and governance.

Our starting point in this chapter is the first of these issues – the increasing level of retail concentration which has been such a marked feature of many western economies since the end of the Second World War, particularly over the past 25 years. Such concentration, as we shall see, has been accompanied by the growth of large retail corporations whose concern has been both to create and maintain what might be termed their 'competitive space'. As retail geography began, in the early 1990s, to take its economic geographies more seriously, so the strategic market behaviour of these large retail corporations and the corporate restructuring of entire sectors of the industry increasingly became the focus of attention. In this chapter we guide our readers through some of those issues and that literature, noting how they raise important questions about the transformation of retail firms as organizational entities (their financial structures, management and governance) and about the creation and implementation of corporate strategy. As a result, we will see that it focuses attention on a range of issues which are central to wider debate within economic geography.

Concentration of capital and the rise of the retail corporation

Corporate retail, either in the form of the department store 'cathedrals of consumption' which emerged in Paris, New York, London and other world cities in the late nineteenth century (see Chapter 11), or in the less flamboyant form of food store chains such as the Kroger Grocery and Baking Company, the Jewel Tea Company, and the Great Atlantic and Pacific Tea Company (A&P) which began to develop in the USA during the same period, has enjoyed an upward trajectory throughout much of the twentieth century. Indeed, as early as the 1930s in the USA, that growth had already sparked intense anti-chainstore sentiment and the passage of restrictive regulation (the Robinson–Patman Act, 1936), as corporate retail growth coincided with, and became unpopularly equated with, significantly increased small business failure rates during the Depression years (see Wrigley, 1992). Nevertheless, prior to the Second World War and despite certain well-known exceptions such as Woolworth's and the supermarket chains A&P, Safeway and Kroger in the USA, or the Burton's menswear chain which had 595 branches in Britain and Ireland by 1939 (Alexander, 1997), retailing in most western economies remained an industry dominated by many hundreds of thousands of small-scale firms. These were either single-outlet independents or very small corporate chains, reflecting the fundamental ease of entry, in terms of low initial capital requirements, into the industry. It was essentially the post-war period, and especially the final quarter of the twentieth century, in which retailing was transformed via a strong trend towards the concentration of capital into an industry increasingly dominated by 'big capital' in the form of large corporations. Indeed, the final years of the century were characterized by the development of global empires by some of those firms (see Chapter 9).

The scale of these trends can be seen in official retail census figures. For example, in the USA in the immediate post-Second World War period, US Bureau of the Census figures show retailing still dominated by small independent firms operating just a single store. In 1948, 70 per cent of total US retail sales were accounted for by such firms and only 18 per cent by the larger chains operating more than ten stores. By the early 1980s, however, a significant reversal had taken place – the share of total US retail sales accounted for by the single-store independent firms had fallen to just 48 per cent whilst that of the larger chains had risen to over 40 per cent. Similar trends had also occurred in Britain, so that, by 1984, 58 per cent of total retail sales had been captured by equivalent chains.

But something much more significant had also occurred, and began during the 1980s to attract increasing amounts of attention by writers from a number of disciplinary perspectives (see, for example, our own contributions, e.g. Wrigley, 1987). That was the rise of the retail corporation – the mega-chain – and the increasing importance such firms had begun to assume in several western economies. In the USA and Britain, for example, retail census figures show that the relatively small number of very large chains (defined conservatively as those operating more than 100 stores), which comprised just a tiny proportion – less than 1 per cent – of all retail firms, had dramatically increased their share of total retail sales. In the USA, their share had almost tripled from 12 per cent in 1948 to 30 per cent by 1982, whilst in Britain it had more than doubled in the same period to 42 per cent by 1984. Moreover, these overall figures concealed much greater levels of dominance which the mega-chains had begun to achieve in particular sectors of the industry.

The 1980s was a decade in which these trends were strongly reinforced in many western economies. In Britain, for example, a small group of food retail corporations emerged, progressively tightened their grip on the market, and by the end of the decade totally dominated that sector of retail trade. In the process, not only did these firms become major employers of labour – the J. Sainsbury Group, for example, almost tripled the size of its labour force during this period to reach 100,000 employees in 1989/90 – but they also became both hugely profitable and politically significant entities in UK economic life. In the USA, despite strong financial forces encouraging deconglomeration in the mid- to late 1980s (see Reading 2.4), overall retail concentration levels continued to increase. Moreover, these were the years in which the archetypal retail corporation Wal-Mart emerged – spreading rapidly from its Arkansas heartland (and origins as a single-outlet firm in 1962), first across the US south (Laulajainen, 1987; Graff and Ashton, 1994) and then nationally, increasing its sales and profits more than seventeen-fold during the decade and growing to over 1100 stores.

In countries such as Britain, in which the manufacturing base of the economy had been decimated (most dramatically during the recession years of 1979–82), but in which economic prosperity and a consumer-led boom characterized the middle years of the decade, a strong sense of an increasingly service-sector-based economy shaped by retail capital and an overwhelming 'retail revolution' began to emerge. Indeed, the political economist Robin Murray, in a number of well-known articles published in the late 1980s (see Murray, 1989), argued that retailing, and its emerging retail corporations, had become a pivotal industrial sector – representative of what he termed the 'economy of innovation' and the shift to a 'post-Fordist' age. His view that 'in Britain, the groundwork for the new system' – a system which geographers such as Harvey (1989a), Gertler (1988), Schoenberger (1988) and Sayer (1989) termed 'flexible accumulation' or 'post-Fordism' – 'was laid not in manufacturing but in *retailing*' (Murray, 1989: 42) was influential and widely accepted.

Reading 2.1 is an extract from the book *Consuming passion: the rise of retail culture*, published in 1989 and written in a highly accessible style by Carl Gardner, then editor of *Interior Design* magazine, and Julie Sheppard. The book successfully captures both the general spirit of Britain's 'retail revolution' of the 1980s, and the particular role which the emerging retail corporations played within it.

Reading 2.1 – Recalling the 1980s – Britain's 'retail revolution'

In little more than twenty years, retailing has moved from being a dull, business backwater to become one of the most important, dynamic sectors of the British economy. It has had an irreversible impact on our towns and cities and, for many people, transformed shopping itself into a pleasurable 'leisure experience'. In the process it has increasingly offered consumers an ever-changing array of products and services – and new environments in which to buy them. The resulting 'retail culture' is everywhere – it has colonised huge areas of our social life outside the traditional high street, from sporting venues to arts centres, from railway termini to museums. Many see it as the epitome of Thatcher's Britain, breeding acquisitive individualism and destroying the nation's traditional manufacturing base. Others see it as a sign of 'new times', a potential way forward for an ailing economy . . .

The 1980s saw an unprecedented wave of mergers and take-overs in the retail industry. Between 1982 and 1987, the height of the retail boom, £13.6 billion was transacted as 360 retail businesses changed hands. Even in a buoyant market, the scope for retail expansion through organic growth was restricted – the easiest way to sustain it was by merger and acquisition. No sector was left untouched by merger mania – nearly a quarter of businesses changed hands in variety stores and food, a fifth of department stores and 10 per cent in the furniture trade. This spate of activity exacerbated existing tendencies within the industry towards fewer and fewer firms controlling larger and larger shares of the market . . . .

Many well-known high street names changed ownership in this period, some more than once. Dixons bought Curry's. Paternoster bought Woolworth's which in turn bought Comet. Burton's bought Debenhams. Habitat merged with Mothercare, and then Habitat Mothercare merged with British Home Stores and became Storehouse. Asda merged with MFI in 1985, and then demerged at a loss two years later. Dee Corporation who already owned Gateway bought International Stores from BAT, and then Fine Fare. Tesco sold Victor Value to Bejam. Argyll bought Safeway. Next bought Grattan. Ladbroke bought Homecharm from Texas Homecare and then sold it to Harris Queensway. Ratners bought H. Samuel. By the end of this confusing merry-go-round, the already powerful were even more powerful, and high street favourites had disappeared altogether.

Now, the top ten retailers control 26 per cent of all retail sales, with multiples controlling an enormous 80 per cent, compared with 33 per cent in 1960. In certain retail sectors a few large multiples control a substantial market share. In 1984, in the middle of the retail boom, the five largest accounted for 44 per cent of footwear sales, 42 per cent of domestic electrical and gas appliances, 36 per cent of men's and boys' wear, 31 per cent of carpets and furniture . . . But even these figures conceal the extent of power exerted by some of the biggest retailers. M&S, for example, controls 25 per cent of the market for men's socks and trousers, 33 per cent for men's underwear, 34 per cent of lingerie and 20 per cent of women's swimwear.

British retailing is uniquely the province of big business. In this process of increasing concentration, some popular high street names have been swallowed up and renamed by their new owners – others continue operating under their old trading names but are now part of large retail conglomerates. Presto, the northern supermarket chain, disappeared [in England] when Argyll bought Safeway . . . Fine Fare got similar treatment at the hands of the Dee Corporation when the company decided to continue the Fine Fare operation under the Gateway fascia. Conversely, Debenhams, Principles, Top Shop, Dorothy Perkins,

> Harvey Nichols and Hamleys have all retained their identities but are all part of the Burton Group.
>
> **Extracted from: Carl Gardner and Julie Sheppard (1989): *Consuming passion: the rise of retail culture*. London: Unwin Hyman, 24–6.**

Readers will have noted that Gardner and Sheppard not only highlight increasing concentration and the growing significance of retail corporations in the 1980s, but also the routes via which that retail concentration was occurring – mergers and acquisitions on the one hand, and 'organic' (new store development driven) expansion on the other. In addition, they outline an important consequence of that concentration – the existence within the large retail corporations of multiple trading names (fascias). As retail corporations grew during this period they faced an increasingly segmented ('demassified') consumer market. As a result, they became increasingly sophisticated at both responding to and actively orchestrating that market – providing precisely targeted retail offers to particular segments. In some sectors of the industry it became extremely difficult for consumers to see the controlling corporation behind the multiple fascias.

So let us move on now to consider these and related issues of corporate reconfiguration in greater detail, updating our discussions into the 1990s in the process. To do that, however, it is important that we shift focus from the aggregate industry level to that of particular retail sectors. As we will see, aggregate figures and commentary on the retail industry as a whole tend to conceal more dramatic stories of the transformation of retail capital at the individual sector level. Interrogation of these stories via what Clark (1998) has termed the 'close dialogue' offered by case studies of corporate restructuring has been used by geographers to understand the strategic market behaviour of the large retail corporations and the restructuring of retail sectors and markets.

Concentration and the retail corporation in the 1990s

We now consider four case studies taken from Britain and the USA, covering the department store, drug store and food retail sectors. These studies serve to highlight and illustrate important facets of corporate reconfiguration in the industry. Having outlined these examples, we then move on to discuss some conceptual themes raised by these cases.

Example 1: The road to oligopoly – US drug store retailing

One sector of the US retail industry where the trend towards concentration and the emergence of mega-chains has been most dramatically in evidence since the 1980s is the drug store industry. The 'greying' of the US population since the 1980s has resulted in drug stores gradually increasing their share of total US retail sales. But set against this trend have been intense pressures exerted on drug store profit margins from two directions. First, from managed health care providers (the HMOs – health management organizations) who increasingly attempted during the 1990s to hold down the escalating costs of the US healthcare delivery process. Second, from retail competitors – in particular, the food retailers who shifted their primary formats to food/drug combination superstores during the period, the discount store general merchandisers (Wal-Mart, etc.), and to a limited extent mail order retailers of maintenance drugs. Competitive advantage in the drug store industry shifted irreversibly to the larger chains who could exploit their scale economies and their buying power leverage over suppliers to mitigate these pressures. It was these chains also which had the capital resources to invest in both the state-of-the-art computer technology that became increasingly essential in the sector to provide the efficiency gains necessary to offset profit margin pressure, and to replace traditional post-war strip-centre and mall stores with the larger (10–12,000 sq. ft.), free-standing, corner-location stores with drive-through pharmacies, which became the critical competitive vehicle of the 1990s in the drug store industry.

The result was a wave of mergers and acquisitions in the sector which swept across the USA in the 1990s as smaller regional chains and independents, whose competitive position was becoming increasingly vulnerable, were consolidated into a small number of major firms. In turn, in the mid-1990s, these major firms began to acquire each other. For example, CVS and Revco, previously the fourth and fifth largest chains, merged to form the second largest chain; Rite Aid, previously second largest, acquired Thrifty PayLess, previously seventh largest (Thrifty having acquired PayLess in 1994 to create at the time the West Coast's largest drug store chain); and J.C. Penney's drug store division, the eighth largest chain, acquired Eckerd previously the third largest. By 1997, in an industry whose total sales in 1996 had reached $91 billion, just five mega-chains dominated the market (Figure 2.1) accounting for almost 53 per cent of total US sales compared to a mere 28 per cent in 1990. Together these five chains operated approximately 13,000 stores across the USA, and the consensus view was that the industry was moving rapidly towards an oligopoly.

Indeed, that trend was reinforced, early in 1998, when CVS acquired Arbor, an important regional chain based in Michigan, raising its annual sales to around $15 billion from over 4000 stores, and edging past Walgreen's to become the largest drug store chain in the USA. At the same time, it was estimated that the remaining independent drug stores – in the order of 22,000 across the USA – were closing at a rate of approximately 2000 per year (Merrill Lynch, 1997b).

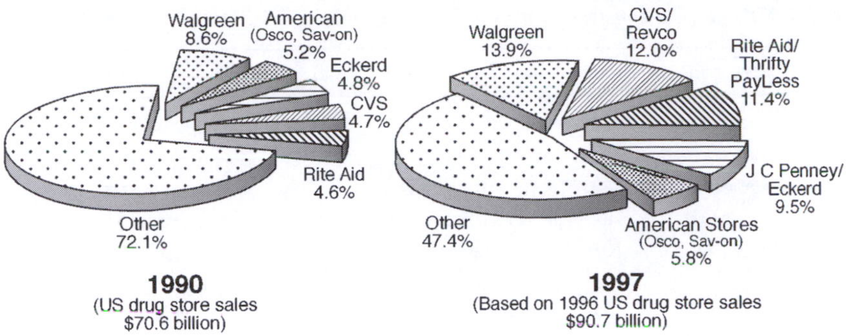

Figure 2.1: *Rapidly increasing concentration in US drug store retailing in the 1990s*
Source: Merrill Lynch (1997a, 1997b).

Example 2: What became of the great American department stores?

Our second example concerns the US department store industry, a historically important sector of US retailing which has progressively lost share of US general merchandise, apparel and furniture (GAF) sales to the discount store retailers (Wal-Mart, Kmart, Target, etc.), though that loss began to stabilize in the mid-1990s as certain demographic trends (the 'greying' of the US population) began to favour the sector. Within this relatively slow-growing 'mature' sector, the advantages of scale have become progressively more compelling as smaller, long-established firms found themselves increasingly disadvantaged against the buying power and operating expense leverage enjoyed by the larger chains. As in the drug store sector, the result has been significant consolidation, with many well-known companies such as Macy's, Younkers, Herberger's and Carson Pirie Scott being swallowed up during the 1990s by a small group of highly active mega-chain consolidators – notably Federated Department Stores, May Department Stores, Dillard's and Proffitt's. Table 2.1 summarizes this trend by documenting the many well-known US department store fascias (Macy's, Bloomingdale's, Lord and Taylor, Filene's, Hecht's) which by 1997 were merely divisions of Federated, May or Proffitt's.

Using the US Census Bureau's figures on total sales in conventional US department stores – a market valued at $58.3 billion in 1997 – it can be seen (Table 2.2) that by the end of that year the wave of acquisitions by the mega-chains resulted in 74 per cent of the market being controlled by the top five firms. Moreover, as the table indicates, the trend to consolidation was further strengthened in 1998 with the merger of Dillard's and Mercantile Stores and Proffitt's acquisition of Saks – two developments which raised the share of the market controlled by the top five firms to over 80 per cent.

Amongst the four major consolidators within the industry during the 1990s, Proffitt's and Federated Department Stores provide particularly interesting

Table 2.1: *The department store divisions of Federated, May and Proffitt's in 1997*

Firm	Store name	No.	Firm	Store name	No.	Firm	Store name	No.
Federated	Bloomingdales	23	May	Lord & Taylor	63	Proffitts	Proffitts	19
	The Bon Marche	42		Hecht's	49		McRae's	31
	Burdines	49		Strawbridge's	22		Younkers	50
	Macy's East	87		Foley's	55		Parisian	40
	Macy's West	101		Robinsons-May	55		Herberger's	37
	Rich's	20		Kaufmann's	47		Carson Pirie Scott	28
	Lazarus	47		Filene's	40		Bergner's	13
	Goldsmith's	6		Famous-Barr	18		Boston Store	11
	Stern's	25		L.S. Ayres	12			
				Mercier & Frank	8			
	Totals	400			369			229

Source: adapted from J.P. Morgan Securities, New York (1998).

Table 2.2: *Leading firms in the US department store industry, 1997 and 1998*

Rank	Firm	1997 Sales ($ mill)	1997 Market Share (%)	Projected Share (%)	1998 Rank
1	Federated	15,668	26.9		
2	May	12,352	21.2		
3	Dillard's	6,632	11.4	16.5	3
4	Nordstrom	4,852	8.3		
5	Proffitt's (Saks Inc)	3,545	6.1	9.2	4
6	Dayton Hudson	3,162	5.4		
7	Mercantile Stores	3,055	5.2	Acquired	
8	Belk	2,042	3.5		
9	Neiman Marcus	1,951	3.3		
10	Saks Holdings	1,835	3.1	Acquired	
	Total US conventional department store sales 1997: $58,335 million				

Source: adaped from J P Morgan Securities, New York (1998), with modifications by authors.

case studies. Alabama-based Proffitt's grew exceptionally rapidly from just a small 15-store regional chain in 1992 to become the fourth largest in the industry with over 300 stores by 1998 – in the process absorbing many smaller regional chains in the midwest and south-east and, ultimately, in a $3 billion deal, Saks Fifth Avenue – changing its name in the process to Saks Inc. Federated, on the other hand, provided one of the most fascinating stories of the intense period of financial re-engineering and leveraged buyouts and acquisitions which gripped corporate America in the late 1980s (see Example 4 below for further details).

Federated, originally founded as a holding company by several family-owned regional department store chains (including Filene's of Boston) and which, by the mid-1980s, included Bloomingdale's, Bullocks, Burdines and Rich's, plus (rather unusually) Ralph's the Los Angeles-based food store chain, was acquired in 1988 at a cost of more than $8 billion by the Campeau Corporation controlled by Robert Campeau a successful Canadian real estate developer. Previously, in 1986, Campeau had gained control of Allied Stores in a high-yield ('junk') bond-financed deal worth over $4 billion and he used a similar financing structure to acquire Federated – in the process defeating a rival bid from Macy's. To pay down some of the debt incurred in the acquisition of Allied and Federated, Campeau sold off more than 25 of the chains owned by the two holding companies (including Brooks Brothers, Filene's and Ralph's). However, as we will see in Reading 2.2, this proved insufficient to prevent Campeau collapsing into bankruptcy early in 1990 – the largest bankruptcy to that point in US history. Campeau was ousted as chairman and chief executive of the company later that year and, under the protection of the bankruptcy court, a Chapter 11 restructuring of Federated began. Unprofitable stores were closed, non-core divisions were divested, and operating profits reduced significantly. In 1992, Federated emerged from bankruptcy as a large and well-managed chain and subsequently began to make strategic acquisitions – particularly Macy's (which itself had collapsed into bankruptcy in 1991) and Broadway in Southern California (which it combined, along with Bullocks into the Macy's division). Between 1992 and 1997, it doubled its share of US department store sales, significantly improving its operating profit margins in the process.

Reading 2.2 by Alan Hallsworth, a British geographer, extracted from an article published in 1991 in the leading urban and regional research journal, *Environment and Planning A*, provides insight into the acquisition of Federated by Campeau and its subsequent bankruptcy.

Reading 2.2 – Campeau and the bankruptcy of Federated

Robert Campeau, founder and head of Campeau Corporation, came to prominence in the Canadian capital of Ottawa. There he established himself as a housebuilder and his fortunes rose as the population of Ottawa expanded. Eventually, Campeau diversified his activities into the lucrative sphere of shopping-mall development ... By 1986, Campeau became interested in the US market, even though Canadian developers had experienced mixed fortunes there ... The difference with Campeau was that the timing was perfect for highly-leveraged deals ...

In seeking mall investments in the USA, Campeau ... began to look at Allied Stores Corporation. Though owners of such famous retail names as Brooks Brothers and Ann Taylor, Allied primarily interested Campeau in respect of its mall property holdings. The overall Allied group, it emerged, was showing poor profits and the temptation soon arose to take over the whole operation – including the malls ... In October 1986, Campeau bought Allied for $3.6 billion, plus assumed debt and fees – $4.4 billion in all. This, at the height of the junk bond boom, was financed by $3 billion in bank loans from a consortium led by Citibank: $875 million in bridging finance from First Boston and only $300 million in actual 'equity' . . . The immediate aftermath of the Allied takeover exemplified the ethos of junk as Campeau trimmed Allied to pay off debt. In 1987, sixteen of Allied's smaller units were sold for a total of around $1.16 bill . . . [but] he still held key Allied divisions such as Brooks Brothers . . .

By the middle of 1987, Campeau's apparent success led him to look at Federated Stores which had an asset base that included Bloomingdales . . . Again the timing seemed perfect, as the stock market crash of October 1987 had depressed Federated shares still further . . . [they] were trading as low as $36. Had Campeau been able to acquire Federated at that price there is little doubt that it would have been a sound 'investment' for property and potential cash-flow reasons . . . Federated held some of the longest-established names in US department store retailing – and a long-established food chain, Ralph's . . .

Having acquired very few shares in Federated, he launched a takeover bid at $47 per share that valued the company at $4.2 billion. The mere prospect of a bid pushed the shares higher, with arbitragers also acquiring shares, and exaggerated the 'windfall profits' that had already ensued to Federated shareholders. It is believed that Campeau intended to go no further than $60 a share [but he was being raced for control of Federated by Macy's which felt itself in a vulnerable position] . . . In the end, by April 1988, Campeau had acquired Federated at $73.5 a share – a total of $6.6 billion for the shares alone. Debt and fees added $2.2 billion, to give a total debt of $8.8 billion . . . To raise finance for the Federated deal Campeau arranged to sell Brooks Brothers to Marks & Spencer for $750 million . . . It did not take long for the backers of the Allied deal to conclude that the assets in which they had invested were to be sold to fund the Federated shareholders . . . [and] as the financial process rumbled on 'confidence' fell . . .

As soon as the Federated deal was made, Campeau began to divest assets in order to cut short-term debt. He disposed of I Magnin and Bullocks to Macy's for $1.1 billion, and Filene's and Foley's to Mays for $1.5 billion . . . Ralph's became a separate spin-off subsidiary . . . [However] problems arose. The retail market was flat and the junk bond market was overflowing . . . As Allied bonds fell in value, Federated junk became harder to market. First Boston were attempting to sell over $1 billion of Federated junk bonds at around 14 per cent interest and finding no takers . . . [In addition] another jewel in the Allied crown – Ann Taylor – sold for less than expected. Campeau [essentially] had high-interest debts and an asset base that was rapidly declining back to earlier valuations . . .

Campeau's ploy was to try to move away from the faltering junk market and into more solidly-based funding. His idea was to find financial partners who would offer low-interest finance – secured on the group's real estate . . . [as a result] he spurned conventional refinancing options . . . and seemed finally to lose the crucial 'confidence' of the market . . . [Despite additional involvement by Olympia and York – the Canadian property developers – and a pledge to sell Bloomingdales – that crucial 'confidence' could not be restored and the share value of Campeau Corporation fell substantially] . . . On 15 January 1990, Allied–Federated filed for Chapter 11 bankruptcy in Cincinnati, OH, after a deadline was set on $2.3 billion of payments . . . Campeau personally defaulted on a loan pledged by the National Bank of Canada against many of his shares in Campeau Corporation . . . [Campeau was ousted as Chairman and Chief Executive of Campeau Corporation in

August 1990, meanwhile] . . . Citicorp and Chemical Bank paid up $700 million to keep operations going under Chapter 11 restructuring.

Extracted from: Alan Hallsworth (1991): The Campeau takeovers – the arbitrage economy in action. *Environment and Planning A*, 23, 1217–23.

Example 3: Store wars and after – UK food retailing

In contrast to the previous examples, our third case study concerns a sector in which retail concentration and the rise of the mega-chain has attracted considerable academic attention. During the 1980s, as we noted above, Britain experienced the emergence of a small group of food retail corporations whose turnover, employment levels, profitability, and sheer market and political power came to rival the largest industrial firms in any sector of its economy. The growth of these mega-chains – Sainsbury, Tesco, Argyll (now Safeway), Asda, and Gateway (now Somerfield) – was powered by both merger and acquisition and

Table 2.3: *Key events in the growth of Argyll (Safeway) in the 1980s*

Date	Event	Rationale
1981	Acquires the Lo-Cost, Mojo and Cordon Bleu freezer centre chains from Oriel Foods Ltd	
1982	Acquires Allied Suppliers, operators of the Liptons, Presto, Templeton and Galbraith chains. With Allied's 2–3 per cent of UK market, Argyll becomes major food retailer	Market share and leverage over suppliers
1984	Acquires Hintons chain in north-east England	Entry into new geographical market, enhanced UK market share
1985–86	Argyll's large stores operated under Presto name, smaller under Lo-Cost name. But Presto has limited consumer franchise in England compared with Sainsbury, Tesco, Asda	
1987	Acquires UK operations of long-established, high-quality US food retailer Safeway for £681 million. Adds approx. 2.5 per cent market share and an 'off-the-peg' corporate identity	Market share and corporate identity, plus enhanced leverage over suppliers
1987–90	Consolidates in England into two trading formats. Safeway for large stores, Lo-Cost for small discount stores.[a] Major new store investment programme under Safeway name	New store development vital in 'store wars' era. Safeway corporate identity provides viable competitive vehicle

[a] Lo-Cost stores finally divested in 1994.

by 'organic' (new store development driven) growth. For example, Table 2.3 outlines key events in the growth of the Argyll Group (Safeway) from its origins as a tiny shell company in the late 1970s to a position as the third largest UK food retailer with sales of almost £4.6 billion per annum by 1989/90. Similarly, Guy (1994a: Table 7.4) lists the key acquisitions driving the rise of Gateway (Somerfield) during the same period. In contrast, the growth of Sainsbury – the leading UK food retailer at the end of the 1980s – like that of Wal-Mart in the USA, was entirely by organic growth within the UK. Throughout the 1980s, Sainsbury's pre-tax profits rose at never less than 20 per cent per annum (indeed they averaged over 26 per cent per annum) as a 'virtuous circle' of heavy capital investment in new store development, IT systems, distribution/logistics, and supply-chain management (see Chapters 3 and 4) produced faster stock/inventory turnover and higher sales per square foot in its stores than those of its rivals. In turn, this resulted in greater opportunities to lower operating costs and raise operating profit margins to higher levels than its rivals, which as a result produced more retained profit which could be ploughed back into growing the company, and allowed differentially favourable access to capital when necessary for additional investment (see Wrigley, 1991).

By the late 1980s, the critical arena of competition between the major food retailers in the UK had increasingly become the new store development process. Larger, edge-of-city superstores, operating at lower costs and with negligible catchment area competition from rival stores of the same vintage became the vital 'driver' of the increasing sales and profit levels of the major firms. Competition between these firms for the most attractive development sites became intense and the late 1980s and early 1990s were widely characterized as an era of *store wars* (Wrigley, 1991, 1993a, 1994) in which an increasingly frantic store-building boom forced up prices in a distorted land market. As two of the top five firms (Gateway and Asda) began to suffer from severe financial difficulties related to debt burdens assumed during the leveraged buyout of Gateway in 1989, the top three – Sainsbury, Tesco and Argyll (Safeway) – began to separate out in terms of growth, profitability and capital investment levels. Indeed, by 1991/92, this 'big three' had raised their joint annual capital investment levels (largely in new store development) to £2.3 billion per annum, whilst their operating profit margins, which had increased consistently during the 1980s (Table 2.4) had risen to levels which were quite unusual in international terms.

During the 1990s, the upward trend in concentration continued in a sector which had already moved significantly in that direction by the late 1980s. Using the most conservative measure of sector concentration – that of the Institute of Grocery Distribution, which is based on a very broad definition of the total market for food retailers (including sales from specialized foodstores, chemists, off-licences, confectioners, tobacconists and newsagents) and which, as a result, produces significantly lower concentration figures than other measures based

Table 2.4: *Trends in operating profit margins (%) of 'big three' UK food retailers 1985–92*

Firma	1984/85	1985/86	1986/87	1987/88	1988/89	1989/90	1990/91	1991/92
Sainsbury	5.25	5.53	6.09	6.54	7.31	7.61	8.32	8.71
Tesco	2.72	3.10	4.10	5.21	5.86	6.18	6.62	7.09
Safeway	3.57	3.89	4.34	4.69	5.19	5.94	6.74	7.49

[a] Figures based on turnover exclusive of VAT. Takes account of only major fascia of firm, not secondary fascias/operations such as Lo-Cost/Presto (Argyll), Savacentre/Homebase/Shaw's (Sainsbury).
Source: adapted from Henderson Crosthwaite (1992).

on narrower definitions of the market (see Wrigley and Clarke, 1999) – the top four firms are shown to have increased their combined market share by over 8 per cent between 1990 and mid-1996, to reach 42 per cent of a market worth by that stage £81 billion (approx. $130 billion). By mid-1998, as a consequence of further merger and acquisition activity (the merger of Somerfield and KwikSave – the previous fifth and sixth largest firms), the most conservative measure of sector concentration suggested that the five largest UK food retailers controlled almost 55 per cent of national trade in this broadly defined total market. Less conservative measures placed that share at over 65 per cent.

However, despite such trends, the major story in this sector of UK retailing during the 1990s was not the continuation of concentration, but certain threats and sectoral shifts which have resulted directly from that concentration. The major firms experienced threats directly related to the dominance they achieved during the 'store wars' era and, in turn, their strategic market response altered the competitive environment and the relative positioning of firms within the sector.

Reading 2.3, by Neil Wrigley, which should be supplemented by Burt and Sparks (1994) and Wrigley (1996) and updated by a consideration of Wrigley (1998a) and Wrigley and Clarke (1999), discusses a period, 1993/94, of deep crisis in the industry when, suddenly, the dominant strategy of accumulation during the store wars era – growth via out-of-town superstore development – began to be questioned, the barriers to market entry erected during that era began to appear vulnerable, and the major food retailers faced intense threats to their profitability.

Reading 2.3 – The vulnerabilities of market dominance

1991 was the annus mirabilis of UK food retailing. Profits and margins of the 'big three' reached record levels and, at the very bottom of a services/property-led recession in the UK, the big three raised £1.4 billion of new capital in a wave of public rights issues to fund their seemingly ever-accelerating expansion programmes. At the same time, however, a number of retail analysts began to express their growing sense of unease about the long-term viability of such investment strategies and about the continuation of the 'golden age' . . . By early 1992, the general sense of unease was beginning to take a more tangible form. In particular, worries began to focus upon two issues: first, the wider price-competition ripple effects and possible destabilization of profit margins that were beginning to result from the entry into the UK of the continental-European 'deep' discounters; second, the considerable disparity that had emerged between the trajectories of the general UK property market and the food-retail property market, and the alarming overvaluation of the food retailers' property portfolios which that disparity implied.

The rise and competitive impacts of the deep discounters – The corporate strategies adopted by the major food retailers in the late 1980s, focused as they were on hugely increased investment and differential access to capital, left a gap in the barriers to market entry that was exploitable by low-margin, low-capital-intensity, limited-range 'deep' discounters. Into that potential market gap, during 1990, rode the European discounters Aldi and Netto, sensing the emergence, as the UK moved into a rapidly deepening recession, of opportunities for aggressive low-margin operations. They were attracted to the UK not only by the logic of their own long-term corporate expansion trajectories, but also by the high ROCE (return on capital employed) achievable in the discount sector in the UK, by general 'margin envy', and by the relatively limited UK competition in the sector . . .

At first, the impact of the deep discounters was less than had been feared . . . The big three were able to project the view that Aldi and Netto had entered a subsector of the market that was totally separated from that which they occupied, and that price-competition effects were unlikely to cross market-segment boundaries. The difficulty with this argument, however, was that it failed to take account of the position of the vulnerable, debt-encumbered, second-tier food retailers (Asda and Gateway), and the significant role that they were beginning to play in amplifying the primary impact of Aldi and Netto's entry . . . Unable to compete with the ever-increasing capital investment programmes of the big three, the second-tier food retailers found themselves threatened by an expanding discount sector . . . Their response was to refocus corporate strategy on price-competitiveness . . . and to reconfigure certain stores that were at the point of obsolescence, underperforming, and/or facing locally intense competitive pressure from the deep discounters into a discount superstore format. Underperforming Gateway superstores were converted at relatively limited cost into a low-price, low-margin discount format under the label Food Giant, [whilst] Asda began to experiment with a similar format under the label Dales . . . In this way, the second-tier retailers began to develop a more flexible competitive response to the market entry of the deep discounters, and achieved a vital leveraging of low-cost assets.

By mid-1993, a network of discount superstores had emerged and the expansion of Aldi and Netto had also gathered pace. Moreover, the successful expansion of the deep discount sector had triggered in its turn the entry of other mainland-European discount operators . . . Overall, the forecast was that the discounters (including Kwik Save) would capture 15.5 per cent of the total UK market by 1995 . . . However, at the level of particular regions and urban areas, the impact of the discounters was far more significant . . . As a result, severe local price wars began to be reported, as the big three were forced to introduce special 'price fighter' brands to compete with the discounters, and were faced with the erosion of the local spatial monopolies so vital to their profitability in the 1980s as the discounters sought 'parasitic' locations adjacent to their established superstores.

By the end of 1993, the wider chain reaction to the market entry of the deep discounters could no longer be dismissed and the big three could no longer attempt to sustain a position aloof from the reality of rapidly changing competitive conditions. Between September and December 1993 they all publicly acknowledged the threat and announced new forms of long-term downward repositioning on price ... By early 1994, the question was not whether the price-competition ripple effects ... were beginning to impact on the profit margin structures of the industry in general, and the big three in particular, but to what extent they prefaced a one-off relatively orderly downward correction in margins or a chaotic and continuing downward slide ...

Worries over property valuations – After 25 years in which the property market in the UK had provided a seemingly low-risk escalator to the accumulation of household and corporate wealth, commercial and domestic property values fell precipitously in both real and absolute terms during the recession of 1989/90 to 1992. By late 1992 average domestic property values in southern England had fallen 30 per cent from the peaks that they had reached in 1988. Curiously, during these years, the UK food-retail property market seemed to have managed to isolate itself from the malaise affecting the rest of the property market. The major UK food retailers, locked into the logic of the store wars, continued to bid up the prices paid for superstore development sites. Perversely, freehold site costs were driven up by around 30 per cent in the three years to 1992, with the average site costs of a new Tesco superstore rising from £15 million to £22 million. Moreover ... a great deal of that new store expansion was taking place ... in the very areas of the country that were suffering the greatest falls in value in the general property market.

Not surprisingly, by mid-1992 ... increasing concern began to be voiced about the incredible disparity that had emerged ... Essentially, there were three facets to [these] arguments.

(1) The asset portfolios (superstores and development sites) of the major UK food retailers had become significantly overvalued, and their traditional policy of non-depreciation of freehold (and long-leasehold) buildings and land could no longer be justified in face of the much reduced 'residual value' assumptions that flowed from the UK property crisis.

(2) The accountancy practices of the major food retailers – specifically the capitalization of interest paid on the finance used to fund the development of new superstores – appeared to have exacerbated the overvaluation of the superstores constructed since the mid-1980s.

(3) Correction of (1) and (2) ... was likely to reduce the pre-tax profits of the major food retailers significantly ...

At first, the major food retailers made a determined attempt to ignore these arguments ... [but] by the end of 1993, acknowledgement by the big three of the ... changing nature of competition in UK food retailing was followed by their belated acceptance of the validity of the arguments concerning property overvaluation and depreciation. In December 1993, Argyll became the first member of the big three to announce that it was to start depreciating its store values ... [at a cost] of £40 million p.a. ... Argyll was then followed by Tesco, which in January 1994 announced that it was to begin depreciating its buildings in a similar fashion at a cost of £36 million p.a. ... the value of its land [at a cost of £32 million p.a.] and accept that it had paid, in the late 1980s/early 1990s, a 'premium' for much of the land on which its superstores were built that was well in excess of any conceivable 'alternative use' value that might be realized on exit from those sites . . . Finally, Sainsbury was forced to follow Argyll and Tesco and begin depreciating its buildings, thus incurring an annual depreciation charge of £40 million. Unlike Tesco, however, Sainsbury took the view that the fall in the value of land (superstore sites) that it had purchased in the late 1980s/early 1990s should be reflected, not via an annual amortization, but via a major one-off write-down of £365 million . . . The only question remaining in early 1994 was whether or not the property write-downs and the new policies on depreciation went far enough.

Extracted from: Neil Wrigley (1994): After the store wars: towards a new era of competition in UK food retailing?. *Journal of Retailing and Consumer Services*, 1, 5–20.

Stung into competitive response by the pressures described in Reading 2.3, the major food retailers began to reposition their offer, introducing price-fighter 'Value' and 'Economy' own-label ranges to match the discounters, and adjusting downwards gross profit margins more generally – particularly on so-called 'known-value' items. In addition, and led by Tesco in early 1995, they began to introduce customer loyalty/reward cards – successfully leveraging the competitive advantages offered by their advanced IT systems (see Chapter 4) into invigorated marketing strategies. In the process, the consumer dynamics which had driven the growth of the discounters in the early 1990s began to change. Aided by falling store building costs which widened the range of opportunities for profitable development – facilitating the development of new smaller store formats and allowing the major firms to target the smaller towns which they had by-passed during the store wars era – returns on new store investment also began to improve (Wrigley, 1998a).

One important consequence of the successful repositioning of the major food retailers was the triggering of an intense rationalization of the discount food retail sector which had enjoyed such rapid growth during the early 1990s. Discount operators, which were highly leveraged, which could not call upon the financial backing of a cash-rich parent company committed to long-term presence in the market, which had expanded too rapidly in an insufficiently cost-controlled manner or, conversely, which had not expanded enough to obtain a critical threshold presence in the market before the major food retailers repositioned their offer, found it impossible to sustain ongoing losses. Sparks (1996a) provides a valuable insight into one of these cases – Shoprite, a firm which had entered the market in 1990, expanded to over 100 stores in Scotland and northern England, but which crashed into financial crisis and failure in late 1994 – whilst Wrigley and Clarke (1999) consider the shake-out in the entire discount sector. In particular, they discuss the demise of Kwik Save, the UK's largest discount food retailer (see Sparks, 1990) which found itself progressively squeezed in the difficult middle ground between the deep discounters on the one hand and the repositioned major food retailers on the other. Between 1994 and 1998 Kwik Save faced an increasingly desperate struggle to arrest an accelerating decline in its like-for-like sales, operating profits and market capitalization, as the key features of its competitive differentiation melted away.

Another important consequence of this period of repositioning and of the differential strategic market responses by the major firms, was certain shifts in relative rankings in the sector. In particular, Tesco, which proved more adept at responding to the challenges of the competitive threats of the early 1990s, which gained significant 'first mover' advantages from the introduction of customer loyalty cards, and which made an important strategic acquisition – the regional chain Wm Low in Scotland (Sparks, 1996b) – replaced Sainsbury as the leading firm in the sector and began to entrench itself more and more firmly in that position. Meanwhile Asda, which implemented substantial capital restructuring and a well-executed renewal programme between 1992 and 1995, rejoined the top tier of UK food retailers and enjoyed buoyant sales increases from its low price positioning within that tier – the 'big three' of the early 1990s became the 'big four' once again.

Example 4: LBOs and deleveraging – US food retailing

Our final example considers a sector which during the 1980s, despite a continuing, albeit weak, trend towards concentration at the aggregate level (Connor and Schiek, 1997: Table 10.5), quite remarkably saw no concentration whatsoever in terms of the share of the market controlled by the top ten firms. Why then did this sector of US retailing stand out from the prevailing trend elsewhere in the industry, and from the very rapid consolidation of food retailing in other western economies during that decade, and what subsequently happened during the 1990s?

As a result of almost 50 years (1930s to early 1980s) of regulation in the USA hostile to the development of 'big' retail capital in the food industry and to market share being concentrated into the hands of a small number of major chains operating multi-regionally and enjoying considerable purchasing leverage (Wrigley, 1992, 1999b), a food retail sector had emerged by the early 1980s which was much less concentrated than might have been expected, and in which there was far less shift in power from food manufacturers to retailers than in countries such as Britain (see Chapter 3 and Hughes, 1996a). As a result, it was a sector which had largely become structured, in spatial terms, as a set of regionally dominant chains. From the early 1980s, however, when the Reagan administrations began to slacken the regulatory constraints under which big business operated, and when the enforcement of US antitrust laws dropped dramatically (Mueller and Paterson, 1986: 403), conditions became conducive to rapid consolidation within US food retailing. Strangely that did not occur.

In simple terms, consolidation at the level of the major firms did not occur in US food retailing during the 1980s because relaxation of investment regulation and the development of new financial instruments and markets – specifically the high-yield ('junk') bond market – created immensely strong countervailing tendencies of *deconglomeration* and *deconcentration* within the US economy (see Wrigley, 1999b). US food retailing became caught up in an intense wave of leveraged buyouts (LBOs) and leveraged recapitalizations which swept across corporate America during the mid- to late 1980s, effectively restructuring and rebuilding firms and industries (Yago, 1991). The US industries most susceptible to restructuring in this way were those with stable demand and strong cash flows but urgently needing to increase efficiency and productivity. Food retailing was quintessentially of this type and the result was that, between 1985 and 1988, 19 of the 50 largest firms, accounting for almost 25 per cent of US supermarket sales, undertook LBOs or leveraged recapitalizations (Chevalier, 1995a). Corporate debt levels rose sharply – doubling between 1985/86 and 1988/89 (Cotterill, 1993) – as the scale of the LBOs and recapitalizations involving the top 20 firms ranged from $4–5 billion in the case of Safeway and Kroger (Denis, 1994) to $1–2 billion in the case of Stop & Shop and Supermarkets General.

Reading 2.4 consists of extracts from two papers by Neil Wrigley. The first part provides detail on the period of leveraged restructuring and deconglomeration in US food retailing during the late 1980s. The second considers the period of deleveraging in the sector during the early 1990s. This provided preconditions for a long-delayed period of consolidation in the late 1990s that began, at last, to move US food retailing closer to the levels of concentration found in other western economies.

Reading 2.4 – How financial re-engineering held back consolidation

The LBO wave of the mid- to late 1980s was a period when, in Jensen's terms (1986, 1993), capital markets were used to provide the mechanism for corporate control. 'Free-cash flow' theories were used to support the view that leveraging firms to very high levels – substantially increasing their levels of debt ... was an effective way to overcome 'corporate control failure', to open the management of large companies to monitoring and discipline from capital markets, and to force them to take the hard decisions which would eliminate excess capacity and maximise firm value[1] ...

Initially prompted by high-yield-security-financed hostile takeover threats – for example, both Safeway and Kroger, the leading chains in the industry, were threatened by takeover bids led by the Haft family – a wave of LBOs and leveraged recapitalizations swept through the US food retail industry between 1985 and 1988. During these high-leverage transactions, firms essentially borrowed against future cash flows to make payments to their shareholders – that is to say future profits were capitalized and built into the financial base of the firm through increased debt. As a result, debt to capitalization ratios across the industry rose abruptly.[2]

The LBO food retailers, committed to servicing their huge debt burdens, were forced to divest assets and cut capital expenditure programmes[1] ... Safeway, for example, rapidly divested $2.4 billion of assets in the two years following its LBO in 1986 – over 1000 stores and several entire divisions including those in the UK, Oklahoma, Salt Lake City, Dallas, El Paso, Kansas City, Little Rock, Houston and Southern California ... Such assets divestments ... prompted, and were facilitated by, a wave of spin-off LBOs in which the divisional management of the high-leverage chains, together with financial sponsor groups, were able to create, via buyouts, stand-alone regional chains. The creation of Homeland Stores from the Oklahoma division of Safeway, via a spin-off to existing management and investment group Clayton, Dubilier and Rice provides a typical example[2] ... The result was the emergence of a distinct 'geography of divestiture' as LBO firms divested marginal, peripheral and/or readily saleable parts of their business, with important implications for the degree of market concentration and the nature of competition in many regional/local markets. In turn, leveraged restructuring created important interactions between the LBO firms and the decisions of rival firms. The 'liquidity constrained' position of the LBO retailers created opportunities for predation via market entry and expansion, with important consequences for the nature of competition, pricing, and the probability of market exit and entry (see Chevalier, 1995a, 1995b, Wrigley 1998c).[1]

During the early 1990s, the major high-leverage food retailers were able, progressively, to retire debt, refinance to lower coupon debt, and reduce short-term borrowing – in the process lowering debt-to-capitalization ratios and improving cash-flow-coverage of interest expenses multiples ... In the case of Safeway, which had returned to public company status in 1990 following asset sales which had reduced its LBO debt from $5.7 to $3.1 billion, the early 1990s saw it retire $565 million of LBO debt in 1991, refinance $1 billion of public subordinated debt in 1992, retire a further $292 million of senior subordinated debt in 1994, and regain 'investment grade' status on its senior unsecured debt in 1995. The result was that, at an aggregate level, the US food retail industry gradually unwound from the highly leveraged position it had reached in the late 1980s.

During this period of deleveraging, the economics of the US food retail industry began to change quite dramatically. Competitive advantage began to shift to larger firms possessing critical mass and capital. In particular, the major multi-regional firms were able to reverse the diseconomies of scale which had dogged their fragmented divisional-level operating structures of the 1980s by developing the strong IT systems architecture necessary to allow centralized administration and control of distribution/logistics. By exploiting potential synergies in procurement, distribution and administration using these systems, by wresting increasing levels of control of the buying process and the supply chain from the food manufacturers, and by substantially increasing their capital expenditure programmes – encouraged by rising rates of return on investment – the major firms began to achieve differentially rapid sales/profit growth.

With competitive advantage shifting in this fashion ... the benefits conferred by scale of operation and from industry consolidation became increasingly apparent by the mid-1990s. However, in what by that point had become a low-inflation environment, the inherent limitations of increasingly well-capitalized major firms seeking growth simply via organic (new store development-driven) expansion became increasingly exposed. Instead, those firms and their European competitors began to turn to merger and acquisition, and to the synergistic cost savings available from such mergers, as the best means of growth available to them. As a result, by 1996/97, the industry was widely perceived as being 'poised on the cusp of dramatic consolidation' (J.P. Morgan, 1997). This perception soon became reality as a wave of acquisitions of regional chains by the larger firms swept through the industry. Ahold acquired Stop & Shop and Giant Food Inc.; Safeway acquired Vons, Carr Gottstein, and Dominick's; Food Lion acquired Kash n'Karry; Albertsons acquired Buttrey, Seessels and part of Bruno's; and Fred Meyer acquired Smith's/Smitty's, Ralphs/Food-4-Less and Quality Food Centres/Hughes. In turn, the leading firms then began to acquire each other, Albertson's announcing an $11.7 billion merger with American Stores in August 1998, and Kroger announcing a $13 billion merger with Fred Meyer in October 1998. During this wave, the valuations placed on acquisition targets – the so-called acquisition multiples – broke decisively out of their historic range and rose to new heights.

In spatial terms, the acquisition-driven consolidation of the industry had several important consequences. First, the early stages of the process saw the emergence of two major chains on the east and west coasts of the USA. Both ranked amongst the top five firms in the industry by mid-1998 but had not featured in the top ten firms at the end of the 1980s. In the east Ahold had created a US chain (as part of its rapidly expanding global business) of over a thousand stores with revenues of $19 billion p.a., stretching from Bi-Lo in the south, through Giant Food Inc. in Washington, DC/Baltimore, to Stop & Shop in New England. In the west and mountain states, a series of mergers in which the Yucaipa Companies – a California-based private investment group specializing in the acquisition and management of food retail chains – played a key role, saw the emergence of a radically expanded version of the existing retailer, Fred Meyer, into a chain of over eight hundred stores with revenues of $15 billion p.a. The chain, stretching from Seattle to Los Angeles and Phoenix, consolidated the existing Fred Meyer, Ralphs, Food-4-Less, Smith's, Smitty's, Quality Food Centers and Hughes businesses and operated in seven of the ten fastest growing states and five of the fastest growing cities in the USA.

Second, and beginning in mid-1998 as the leading firms began to acquire each other, embryonic national food retailers began to emerge in the USA. Kroger's merger with Fred Meyer, for example, is set to create a combined company operating over 2200 supermarkets in 31 states and over 800 convenience stores in 15 states – effectively the whole of the US with the exception of the northeast states, Minnesota, Wisconsin and Hawaii. Similarly, Albertson's merger with American Stores will create a company with almost 2500 food and drug stores in 37 states, lacking a presence only in parts of the midwest and the southeast.[2]

Extracted from: [1] Neil Wrigley (1998b): European retail giants and the post-LBO reconfiguration of US food retailing. *International Review of Retail, Distribution and Consumer Research*, 8, 127–46; [2] Neil Wrigley (1999b): Market rules and spatial outcomes: insights from the corporate restructuring of US food retailing. *Geographical Analysis*, 31, 288–309.

By the end of the decade, as Wrigley (2001a) discusses in a later paper, the share of US supermarket sales controlled by the top four food retailers had risen to over 36 per cent – a dramatic rise from the level of 23 per cent which it had stood at just three years earlier. And with that process ongoing – with Safeway acquiring Randall's in Texas and Genuardi in Philadelphia, Ahold acquiring Bruno's in Alabama as part of Grand Union in the north east, and

Delhaize acquiring Hannaford Brothers to add to its existing east-coast chains of Food Lion and Kash n'Karry – a further increase to almost 40 per cent in the market share of the four leading US food retailers appeared inevitable.

Some conceptual themes

The four cases above have been outlined in some detail for two reasons. First, because they provide important context for our discussions in subsequent chapters. As a result, we will revisit them on several occasions. Second, because they have served to introduce – in what we hope has been a relatively painless fashion – a large number of themes of considerable importance to the economic geographies of retailing. These have ranged from the transformations of the financial structures of firms and the role of individual entrepreneurs and corporate strategists, through shifts in power in the relations between retailers and manufacturers and the opportunities and vulnerabilities of market dominance, to the regulation of corporate retail. Chapters 3 to 6 take up many of these issues in much greater detail. Here, however, we choose to highlight just three interrelated themes which serve to illustrate some of the broader conceptual debates concerning corporate reconfiguration in the industry, and which provide, once again, examples of the multiple readings and re-readings of issues actively shaping the field.

Corporate lock-in, sunk costs and market exit

In a well-known paper published in *Environment and Planning A* in 1994, subsequently expanded in her book *The cultural crisis of the firm* (1997), Erica Schoenberger, a US geographer at Johns Hopkins University, introduced into the mainstream of debate in economic geography issues relating to power, identity and knowledge in the firm. Her work stresses the importance of what she terms the 'social asset structures' of managers – specifically their private benefits of control – and 'the way in which corporate identity and corporate culture frame the kinds of knowledge that can be produced and utilized by the firm in the creation and implementation of competitive strategies' (Schoenberger, 1994: 449 – see also Alexander, 1997 and Shackleton, 1998a for further discussion in a retail context). In particular, she attempts to understand why it is that there are so many well-documented examples of firms that become locked into inappropriate corporate strategies, delaying essential corporate reconfiguration, even though those firms realize the need for transformation and, indeed, often understand the structure of the transformation required.

Corporate lock-in – the inability to initiate changes in strategy and drive through essential corporate restructuring in the face of what may be rapidly changing competitive conditions – is a vital factor in any consideration of the reconfiguration of corporate structures in retailing. Indeed, we have already touched upon such issues in Example 3 and Reading 2.3 above which considered the threats encountered by the UK food retailers in the early 1990s and their competitive repositioning to meet such challenges. As Schoenberger indicates, corporate lock-in is intimately bound up with the issue of what are termed 'sunk costs' (see Clark and Wrigley, 1995, 1997a, 1997b for debate within economic geography) – defined as those costs to a firm which are 'irrevocably committed to a particular use, and therefore not recoverable in the case of exit' (Mata, 1991: 52). As Schoenberger (1994: 438) observes: 'firms flinch in the face of high sunk costs'. As a result, sunk costs often impede, and certainly transform the nature of, corporate restructuring and provide barriers to market exit. Conversely, as has long been recognized in economic theory (Caves and Porter, 1976, 1977; Eaton and Lipsey, 1980, 1981), sunk costs also provide important barriers to market entry – particularly in markets in which incumbent firms exercise significant market power (see Clark and Wrigley, 1997b: 340).

Corporate lock-in, the social asset structures of managers, sunk costs, and market exit are therefore interrelated issues which must be considered in any reading of the processes of corporate reconfiguration in retailing. As Michael Jensen (1993: 847–8) – the leading financial economist – has observed:

> Exit problems appear to be particularly severe in companies that for long periods enjoyed rapid growth, commanding market positions, and high cash flow and profits. In these situations, the culture of the organization and the mindset of managers seem to make it extremely difficult for adjustment to take place until long after the problems have become severe and in some cases even unsolvable ... Even when managers do acknowledge the requirement for exit, it is often difficult for them to accept and initiate the shutdown decision. For the managers who must implement these decisions, shutting [divisions] or liquidating the firm causes personal pain, creates uncertainty, and interrupts or side-tracks careers. Rather than confronting this pain, managers generally resist such actions.

In Reading 2.5 we continue our discussion, begun in Reading 2.3, of the threats faced by the UK food retailers in the early 1990s – specifically those relating to overvaluation of their property assets. But here, in an extract from a further paper by Neil Wrigley, we consider a reading which places emphasis on the role of corporate lock-in and sunk costs in both the production and management of those threats.

Reading 2.5 – Grounding retail capital and managing sunk costs

At the heart of the accumulation process in retailing is the constant need to 'ground' retail capital. However, that fixing in place (however short-term) brings associated vulnerabilities. It is the manner in which these tensions and contradictions are played out which forms a central component of the corporate strategies of major retailers. Yet, surprisingly, and in marked contrast to investigations of tensions at the interface between retail capital and productive capital, little has been written in the fields of retail geography and planning about these issues . . .

The neglect is all the more curious given the fact that during the 'store wars' of the 1980s the major UK food retailers became locked into corporate strategies of accumulation in which strategic capital investment and an ability to ground that capital via new store development programmes became the all-consuming engine of corporate growth. In turn, that resulted in what Guy (1996: 1575) has described as 'one of the largest construction programmes ever to have taken place in Britain'. One interpretation of that massive level of investment was that it was an attempt to raise barriers to market entry to such an extent that potential entrants would be faced, if their entry were to fail as a result of retaliatory strategic or tactical responses by the incumbents, with a level of non-recoverable sunk costs which would be quite unacceptable. To the extent that the 'big three' UK food retailers still appear unchallengeable in terms of their own strictly defined market position, that investment in the raising of entry barriers was to a large degree successful. However, it is also well known that excessive entry-discouraging investment involves considerable risks for incumbent firms, and that it may increase the probability of ruinous loss by increasing the fixity of costs and by lowering the average salvage value of the firms' assets (Clark and Wrigley, 1995). Faced, therefore, with a new form of competition (the deep discounters) who chose not (or were not able) to accept the sunk cost risks of challenging the incumbent major food retailers on their own terms . . . and by a re-evaluation of the role of store expansion programmes within the context of rapidly changing competitive conditions and the belated impact of the UK property crisis, the major food retailers were seen to have locked into local paths of capital accumulation that became dangerously self-reinforcing and narrow. The questions which must be asked therefore are:

1. Why did the major food retailers persist with such competitive strategies in the early 1990s when it was becoming increasingly obvious to commentators outside the industry that a change in the strategy of accumulation that had characterized the era of the store wars was becoming necessary?
2. What was the role of sunk costs in this corporate lock-in?
3. How can firms characterized by potentially significant sunk costs manage those costs to ensure their long-term viability and growth?

The first (and tangentially the second) of these questions is increasingly being posed and answered more generally within economic geography. Schoenberger (1994), for example, has attempted to understand why it is that there are so many well-documented examples of firms that become locked into inappropriate corporate strategies, delaying essential corporate restructuring, even though those firms realize the need for transformation and often know quite a lot about the dimensions of the transformation required. She has also posed the question of 'why firms flinch in the face of high sunk costs yet simultaneously invest gigantic sums in disastrously misguided initiatives' (1994: 438). Her answers to these questions centre on the way that top managers of these firms seek to defend their own social asset structures, as a result foreclosing certain kinds of corporate strategies however appropriate and vital. Schoenberger's view is that economic geographers must focus on how 'the need to transform the firm in order to remain competitive may encounter its most serious obstacle in the social being, position, and perceptions of the people that run it' (1994: 448), and on 'the ways in which corporate identity and corporate culture frame the kinds of knowledge that can be produced and utilized by the firm in the creation and implementation of competitive strategies' (1994: 449).

Clark and Wrigley (1997a) address similar themes in an attempt to answer the third of these questions. They argue that the management of sunk costs is a profoundly social process in which there will inevitably be 'principal–agent' type problems in firms characterized by the separation of ownership and control, and contested bilateral relationships between managers and the firms' internal stakeholders (such as workers) which must be taken into consideration. The management of sunk costs within the firm (more specifically the retail manager's capacity to exploit the strategic/competitive options implicit in a portfolio of stores differentiated by age and exchange/use value) is embedded within the more general underlying problem addressed by Schoenberger – namely how do managers achieve their goals and defend their social asset structures relative to the goals and interests of the firms' shareholders and stakeholders? Clark and Wrigley outline, in the context of manufacturing firms, a range of possible institutional options for stabilizing conflict between management and stakeholders and resolving problems of information flow, co-operation and commitment in firms characterized by significant sunk costs. What emerges from that analysis is that, in order to manage its sunk costs, a firm may have to differentiate spatially its institutional structure and differentiate (between production sites) its contracts with local stakeholders. It is of interest to ask, therefore, questions such as: to what extent are attempts to manage sunk costs in British food retailing leading to spatially differentiated labour contracts and work practices within firms, e.g. between the repositioned and non-repositioned store types/labels within a single firm?

More generally, Clark and Wrigley (1997a) suggest that how sunk costs are managed in relation to competitors and in relation to the interests of corporate managers and their stakeholders is a vital determinant of a firm's long-term growth. The sudden and dramatic problems of property overvaluation, non-recoverable initial investment, and the implications of sunk costs for corporate strategy, which engulfed the UK food retailers in 1993/94 confirm this and serve to highlight the fact that many aspects of the 'grounding' of retail capital have remained surprisingly unproblematized in retail geography. Clearly, there is a great deal yet to be written about the realization of value in firms characterized by significant sunk costs but, given the centrality and potential vulnerabilities that follow from the constant need to 'ground' retail capital, it is a task which cannot be ignored.

Extracted from: Neil Wrigley (1996): Sunk costs and corporate restructuring: British food retailing and the property crisis. Chapter 6 in Wrigley, N. and Lowe, M.S. (eds) *Retailing, consumption and capital: towards the new retail geography*. Harlow: Addison Wesley Longman, 116–36.

A central feature then of the story of the property crisis in UK food retailing, introduced in Reading 2.3 and considered in more general terms in Reading 2.5, is the way in which previously regarded assets came to be seen during a period of crisis as, in a certain sense, potential sunk costs – the 'premiums' paid during the 'store wars' era for development land being increasingly regarded as non-recoverable and in need of 'writing off'. But how generally applicable is this reading of retailers' land and property 'assets', what role have such assets in corporate strategy and restructuring, and what can it tell us, in particular, about issues of market exit in retailing?

These are matters which Cliff Guy, a city and regional planner at the University of Wales, Cardiff, and author of *The retail development process* (1994a) has considered in two papers. The first, published in *Environment and Planning A* (1997) considers in detail sunk cost dimensions of retailers' land and property investment, applying and extending several of the arguments outlined above. Guy's reading suggests that the crisis which afflicted the UK food retailers in the early 1990s was to a certain extent unusual and that, in regulated land markets, property assets may offer retailers certain operation flexibility and profitability advantages. However, he acknowledges that property 'assets' of retailers, because of their non-liquid nature, raise significant problems during periods of corporate reconfiguration that require market exit – either partially or totally. Quoting Weatherhead (1997: 9), Guy (1998a) notes that 'property can seem like a ball and chain, slowing down a business and making quick response to changing markets almost impossible'.

In his second paper, Guy (1998a) takes up, more specifically, these issues of exit, and the extent to which property 'assets' can become sunk costs in such circumstances, by considering a number of cases of attempted market exit by retailers in the UK which involve a property asset dimension. Reading 2.6 extracts from that paper by considering just two examples – one related to an attempted exit from a freehold property portfolio, the other related to exit from a rental property portfolio.

Reading 2.6 – Sunk costs and exit: Littlewoods and British Shoe Corporation

A large retail company can rationalise its operations in several ways . . . however, a frequent reaction to problems such as decreasing market share or profitability is to consider closing outlets, either those for which the retail offer is no longer considered appropriate, or those which appear to underperform . . .

In retailing, freehold property is always seen as an asset and is included in the balance sheet, ideally at 'open market value'. Exiting from a property should enable the owner to realise the asset . . . [and], in the right circumstances, selling the property can realise substantial financial gains over the original cost of purchase. So there is a potential debate over whether freehold property assets can be seen as sunk costs at all. Success in selling freehold properties is, however, very much dependent upon the nature and location of the properties concerned, and the state of the property market at the time of sale. A recent example in the UK has been the attempt by Littlewoods to sell 135 variety stores . . . amounting to 7 million sq. ft. of retail space . . .

Following years of declining sales and profits from its high street operation, in March 1997, Littlewoods put the whole chain up for sale, at a price rumoured to be around £550 million. [After rejecting lower offers from the retailer Kingfisher and a venture capital company] . . . the difficulties inherent in selling 135 large stores led to the company changing its strategy and continuing to use most of the outlets for retail purposes [just 19 stores were sold to Marks & Spencer] ... The general conclusion is that a large set of freehold properties is likely to bear an element of sunk costs. A company is very unlikely to be able to obtain full open market value on a substantial number of properties if they are to be sold simultaneously in a sluggish market ... Sunk costs will arise ... and affect the company's profit-and-loss account [either because the firm will be forced into] a 'fire sale' in which the price received will be well below the open market value of the stores ... [or into] continued retailing which might not be profitable but would allow a more controlled release of properties into the market. Littlewoods appear to have adopted the second strategy, after several months of indecision.

[In the case of a rental property retail portfolio] despite the supposed advantages of flexibility for this type of ownership ... difficulties can be encountered in disposing of stores ... A prime example has been the long-running attempt by the British Shoe Corporation (a subsidiary of Sears plc) to close its high street shoe retailing operation ...

Disposal of rented stores is rather different from disposal of freehold. The only elements which can be genuinely sold are the stock in the stores, and any 'goodwill' attached to the trading name ... In the case of BSC ... due to past poor performance, this was likely to be minimal. Disposal of the store meant in most cases assignment of the lease, usually at no cost to the 'purchaser', who would then take on responsibility for rent payments. The main 'purchaser' of BSC stores in 1995/96 was the rapidly expanding Facia company, which itself then went into receivership in 1996, apparently unable to pay the rent required on several hundred stores in expensive high street locations ... Responsibility for debts incurred on the stores which had been assigned to Facia then reverted to BSC, who in effect took over the running of these 380 shops until another 'purchaser' could be found.

The sunk costs involved in this operation are identifiable largely as a series of 'exceptional provisions' made by the parent group Sears to cover the actual costs of closure, and the loss of assets represented by short-lease properties in the balance sheet. These provisions included £16 million to cover the 'sale' of 245 outlets to Facia in 1995, £220 million in 1996 related to the disposal (again) of the ex-Facia shops; and a further £150 million in 1997 to cover the disposal of the remaining four chains of shoe shops.

Extracted from: Cliff Guy (1998a): Exit strategies and sunk costs: the implications for multiple retailers. Paper presented at the 5th International Conference on Retailing and Consumer Services, Baveno, Italy.

Capital structure decisions and corporate reconfiguration

Throughout our case studies the role of finance in the reconfiguration of corporate structures in retailing has been ever present. Having briefly considered vulnerabilities associated with the constant need to 'ground' retail capital, let us now turn to the issue of the capital structure of a firm and how that impacts on corporate reconfiguration. A telling example of this has already been provided in Reading 2.4. There we saw how the assumption of high levels of debt by US food retailers in the late 1980s constrained their ability to grow and consolidate, and how – via processes such as asset divestment – it significantly affected the spatial organization of individual firms. In addition, we noted how high levels of debt within the capital structure of the LBO food retailers placed them in a 'liquidity constrained' position and offered opportunities for predation by rival firms. But how are we to conceptualize such issues?

One possibility which some economic geographers have begun to explore involves engaging with certain aspects of the theoretical debates shaping the literatures of corporate finance and financial economics. For example, in Reading 2.4, the issue of the capital structure of the retail firm and how it affects the process of corporate restructuring can be considered within the framework of the so-called 'capital structure controversy' initiated by Modigliani and Miller (1958) – see Wrigley (1998c) for such an approach. As Robert Hutchinson, a professor of finance at the University of Ulster, and his co-authors, McCaffery, Hunter, Jackson and Cotter, demonstrate in a series of papers which develop a financial analysis of the UK retail industry (see Hutchinson and Hunter, 1995; McCaffery et al., 1997; Cotter and Hutchinson, 1998), the consensus position on that controversy is that the capital structure of the firm (its ratio of debt to equity) is of considerable importance to the operating performance of the firm and affects both the firm's capacity to fund growth and its vulnerability to unanticipated market downturns. In addition, it raises, as they demonstrate, important issues about conflicts of interest between managers and shareholders (see Reading 2.7), and shareholders and debt holders. Managers have incentives to expand the size and scope of the firm to satisfy their own interests (their private benefits of control) over those of shareholders and, in addition, enjoy privileged knowledge of the company's operations. They may, therefore, use any debt capital they raise for purposes other than originally stated (Reading 2.2 suggests this) – in effect shifting it to higher-risk investment. Suppliers of debt capital must seek to avoid such substitution. This can be achieved by detailed monitoring, but that is likely to be an extremely costly exercise. More usually it will take the form of attempting to impose restrictive covenants on the debt contract or seeking compensation by imposing a premium on the cost of the debt financing.

We have already seen in Reading 2.4 that the capital markets' traditional 'positive view of the sector's ability to service its debt commitments' (Hutchinson and Hunter, 1995: 76) was a vital element in the wave of highly leveraged corporate restructuring which swept through US food retailing in the late 1980s. Hutchinson and Hunter's research on UK retailing in the late 1980s/early 1990s confirms this strongly positive view and concludes that UK retailers were, as a result, able to 'sustain higher than average levels of debt, at lower cost of debt capital than expected' (see more generally Hallsworth and McClatchey, 1994). But what can be said about debt financing in retailing in more hostile environments than the late 1980s, and what does this tell us about the nature of corporate reconfiguration in the industry at the beginning of the twenty-first century? Clearly, an ability to 'read' the process of corporate reconfiguration in the retail industry requires an understanding of the willingness of financial markets to supply capital and of the financial instruments used. It also requires, as we will see below, an appreciation of how suppliers of that capital seek to control managers via the process of corporate governance.

Corporate strategists and their governance

Our discussion, in Example 2, of the role of Robert Campeau in the bankruptcy of Federated and the reconfiguration of the US department store industry more generally, raises a number of important issues. Those relating to the financial structures of retail firms, and broader matters concerning inappropriate corporate strategies and the way a firm often encounters its most serious obstacles to remaining competitive 'in the social being, position and perceptions of the people that run it' (Schoenberger, 1994: 448), have been noted above. Here we conclude the chapter by raising, briefly, concerns relating to what in Chapter 6 we refer to as the 'corporate governance' of the retail firm.

From what is termed an 'agency perspective' – see Jensen and Meckling (1976); Fama (1980), Fama and Jensen (1983a, 1983b) – it has been suggested that corporate governance deals with the ways in which suppliers of finance to corporations – the people who sink capital – assure themselves of getting back a return on their investment, and that its crucial question is 'how do suppliers of finance control managers?' (Schleifer and Vishny, 1997: 737). This is highly pertinent in the context of the problems the financial backers of Campeau encountered during Federated's slide into bankruptcy described in Reading 2.2. But it also raises very important issues concerning the way economic geographers conceptualize, more generally, the process of corporate restructuring. As one of us has argued elsewhere (Wrigley, 1999b), adopting a definition of restructuring as 'a strategy of corporate change that materially alters, sometimes with minor modifications and at other times fundamental transformations, the composition of the firm's asset portfolio and/or claims against those assets' (Weber, 1997) serves immediately to link concepts of corporate restructuring with issues of corporate governance. Unfortunately, forms of restructuring which involve transformations of firms as organizational entities (their capital structure, management and governance) have traditionally been underemphasized in economic geography (Clark and Wrigley, 1997b). This is a pity because, as we will see in the context of Reading 2.7, matters of corporate strategy creation and implementation are intrinsically bound up with claims on corporate assets and prospects. What, therefore, can a study of retail corporations tell us about the role of, and control of, corporate strategists within the process of corporate reconfiguration?

Reading 2.7 by D. Gordon Smith, a US law professor, provides background on the case of the removal of Joseph Antonini, the CEO of Kmart – the US discount store retailer – in March 1995. Smith uses this case to contextualize a valuable and wide-ranging discussion of corporate governance, institutional investor activism and managerial accountability.

Reading 2.7 – Antonini's removal from Kmart

By 1990, Kmart had overtaken Sears as the largest retailer in the USA. The person in charge when Kmart reached this retailing pinnacle was Joe Antonini, who began his career with Kmart as an assistant store manager in 1964 ... worked his way up to president of the apparel division in 1984 ... [and] was elevated to the offices of President and Chief Operating Officer in 1986 and named Chairman and Chief Executive Officer in 1987 ... When Antonini assumed the helm [Kmart faced serious problems relating to] inventory control, labour costs and store renewal ... [for example] the average age of the approximately 2200 Kmart discount stores was 15 years ... only 10 per cent of stores were less than three years old while more than 40 per cent of stores at Target and Wal-Mart [its main competitors] were less than three years old ...

Upon assuming the top post, Antonini seized on the vision of creating a combination discount and speciality retailer by developing and acquiring other speciality retailers ... when Antonini came to power Kmart owned three speciality retailers in the US – Waldenbooks, PayLess drug stores, and Builders Square – [during his tenure it added OfficeMax, The Sports Authority, Borders, Makro and the Pace Warehouse Clubs, plus international operations in central Europe, Mexico and Singapore] ... Antonini's vision for Kmart apparently would have dotted the world with Kmart Power Centres – shopping centres featuring Kmart discount stores surrounded by Kmart-owned speciality stores.

Antonini's efforts to automate and renew Kmart's discount stores and expand and improve Kmart's speciality stores required significant amounts of capital ... In January 1994 Antonini announced a plan to raise capital by creating and issuing separate classes of common stock [known as 'targeted stock'] whose terms would be tied to [the performance of] four of Kmart's speciality retail subsidiaries: The Sports Authority, Builders Square, Borders, and OfficeMax. ... shareholders were asked to authorize the [proposal] ...

[In June 1994, Kmart's shareholders defeated the proposal] ... The most vocal opponents claimed that Antonini was distracted by speciality retailing and that, instead of selling targeted stock, Kmart should sell its speciality retailing subsidiaries outright and focus on Kmart's core business, discount retailing ... [indeed] from the time Antonini took control in 1987, Kmart's market share of total discount retail sales had dropped from 34.5 per cent to 22.7 per cent while Wal-Mart's had risen from 20.1 per cent to 41.6 per cent ... The shareholder vote was quickly recognized by observers as one of the most astonishing investor rebellions in corporate history ... [as one of the leading investor activists stated at the time] ... 'We wanted to force the company to rethink their direction ... Either you replace the strategy or the strategist' ...

In response to the shareholder vote, Antonini agreed to adopt the strategy advocated by the dissident shareholders and announced plans to sell the speciality retail subsidiaries ... [however] the vote had given directors a clear mandate to shake up top management ... In January 1995 Kmart's board removed Antonini as Chairman [and on 21 March] unanimously agreed to ask Antonini to resign as CEO.

Extracted from: D. Gordon Smith (1996): Corporate governance and managerial incompetence: lessons from Kmart. *North Carolina Law Review*, 74, 101–203.

As Smith suggests, one view of the events at Kmart holds that this is corporate governance at its best. Institutional investors dissatisfied with the performance of a company in which they had invested took aggressive action to change the company's strategic course and to replace the 'strategist' – the CEO. However, as Smith demonstrates, in the broader paper from which Reading 2.7 is extracted, there are other interpretations which focus on the appropriate roles of the board of directors and shareholders in the corporate governance system. At this stage, we leave such issues to Chapter 6, and conclude simply by echoing Schoenberger's (1994: 448) stricture that we must remember that corporations are 'run by real people', and that in considering the matters of corporate reconfiguration that have provided the focus for this chapter, we must remain as sensitive to issues relating to the corporate strategist and the governance of firms – specifically how suppliers of finance seek to control managers – as much as corporate strategy and corporate restructuring in the abstract.

3

Reconfiguration of retailer–supply chain interfaces

In a surprise move in July 1998, the Office of Fair Trading (OFT) in the UK announced its first full investigation since the early 1980s (published as *Competition and retailing*, 1985) of the increasing power and market dominance of the major food retailers. Citing ongoing complaints from suppliers about the detrimental impact of retailer concentration and market dominance – referring to it as a 'perennial theme' – the investigation appears to have been prompted by pressure from food manufacturers and suppliers, in particular, complaints from the farming lobby. In essence, those complaints were that food retailers were not passing on to consumers the lower prices they were obtaining from producers of agricultural commodities, and that the increasing power derived from retail concentration was being exercised to the detriment of the consumer and to obtain excessive returns.

Although we leave the specific question of competition regulation in retailing to Chapter 6, and what subsequently happened in terms of this particular investigation to Chapter 8, here we take up a number of the issues raised by this case and focus on the changing character of retailer–supplier relations. In particular, we centre our discussion on four themes. First, how can we characterize retailer–supplier relations and to what extent has there been a shift in the balance of power within these relations in favour of retailers? Second, what is the strategic significance of retailer own-label products in that shifting power balance? Third, how can we take account of what Hughes (1999) terms 'the social and cultural energies which drive many of the changes taking place at the retailer–supplier interface'? Finally, what does it imply for the study of commodity chains more generally?

The changing character of retailer–supplier relations

During Britain's 'retail revolution' of the 1980s (see Reading 2.1), it was not only retail concentration, mergers and acquisitions, and the rise of the mega-chain which began increasingly to attract the attention of writers, but also the sense that retailers were rapidly gaining the ability to shape and control supply relations. The sense, as Robin Murray (1989: 43) put it, was of the emphasis shifting from 'the manufacturer's economies of scale to the retailer's economies of scope', of retailers developing innovatory 'information and supply systems which allow them to order supplies to coincide with demand',

and of the revolution in retailing demanding and reflecting 'new principles of production, a new pluralism of products and a new importance for innovation' (Murray, 1989: 44). Meanwhile, industrial economists of a somewhat more neo-classical bent were equally suggesting that:

> a fundamental shift in the balance of power in consumer goods distribution channels has taken place. Until the mid-1960s manufacturers held a pre-eminent position . . . They were the source of almost all product innovations and new-product developments, they controlled physical distribution to wholesalers and retailers, they were responsible for virtually all product advertising, they exerted a powerful influence on retailers' stocking and display of their products, and they controlled retailers' margins by setting retail prices . . . A major feature of the 'retailing revolution' in the UK . . . has been the replacement of manufacturers' dominance of distribution channels by that of the retail chains . . . Retailers have increasingly assumed control over a range of functions traditionally performed by manufacturers, including physical distribution, advertising, packaging, product design, and product development' . . . [and] the shift in power is reflected in the growth in the profitability of retailing companies relative to consumer-goods manufacturers (Grant, 1987: 43).

How then had this widely perceived shift in power come about, what were the implications of the increasing ability of the retail corporations to reshape and control supply relations, and how can the retailer–supplier relations which emerged out of the 'retail revolution' of the 1980s be characterized?

Shifts in the bargaining power of retailers and suppliers

In Michael Porter's (1980) well-known book *Competitive strategy*, it is suggested that any industry can be viewed as being situated within a constantly changing nexus of five broadly defined competitive forces – the bargaining power of suppliers, the bargaining power of buyers, the threat of new entrants to the industry, the threat of substitute products or services, and rivalry among existing competitors. In Porter's schema, each firm should seek within this nexus to find a position where it 'can best defend itself against these competitive forces or can influence them in its favour' (Porter, 1980: 4).

Figure 3.1 illustrates the application of Porter's general schema within the specific context of the competitive forces operating on food manufacturers. In somewhat simple terms, what occurred during Britain's 'retail revolution' of the 1980s (but, as we saw in Reading 2.4, somewhat later – in the mid-1990s – in the USA), in terms of the relations between food manufacturers and retailers, was that retail concentration – via merger and acquisition and organic growth – produced a handful of very large retail firms with enormously increased buying and bargaining power. Economists refer to markets dominated in this way by a small number of buyers as *oligopsonistic*. In addition, those food retailers increasingly began to develop substitute own-label products – retailer brands which directly challenged the existing food

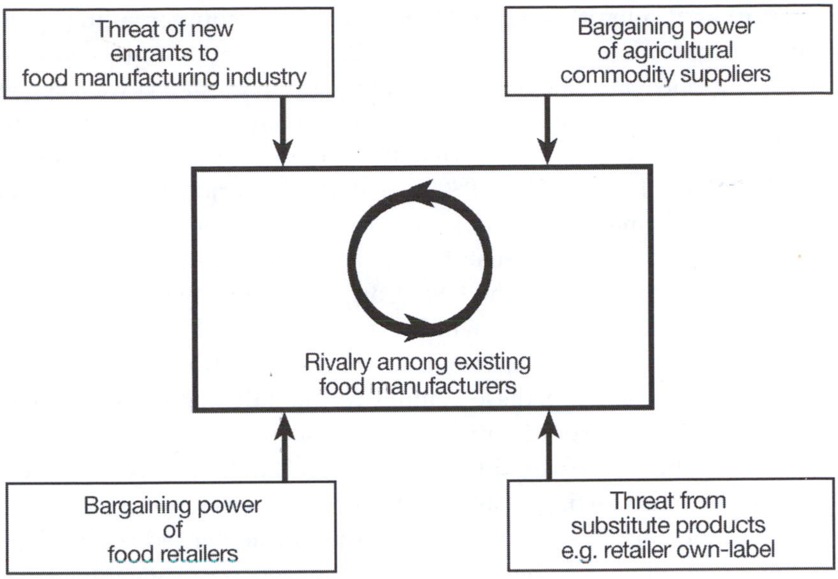

Figure 3.1: *Competitive forces operating on food manufacturers*
Source: Adapted from Hughes (1996b) after Porter (1980).

manufacturer brands. As a result, two of the competitive forces within Porter's schema shifted strongly in favour of the food retailers.

The roots of the bargaining power of the increasingly large food retail corporations lay in three factors. First, in their ability to deny, or credibly threaten to deny, to food manufacturers access to the retail markets vital to the sale of their branded products. This could occur via a threat of, or actual, 'delisting' of the manufacturers' brands or, alternatively, via a reduction in the allocation of (or quality of) selling space offered to those manufacturers' brands within the retailers' stores (Davies et al., 1985). Second, in the ability of the major retailers, as buyers of increasingly large volumes of any particular product, to demand specially negotiated and arguably 'discriminatory' price discounts – where the term 'discriminatory' implies discounts which are much greater than can be justified simply in terms of the large size of any order. Sometimes, in cases where retail merger and acquisition created much larger chains, this could involve demands for retrospective backdating of discounts to the period prior to the merger. Third, in the increasing ability of retailers to take control of their own distribution systems – taking supply of their stores out of the hands of the manufacturers and developing centralized, logistically efficient, quick-response systems (see Chapter 4). This had the effect of reducing retailer inventory holdings and the amount of capital tied up in those holdings, and of passing back to food manufacturers more and more of those costs.

However, as Galbraith's (1980: 118) thesis of 'countervailing power' recognizes, 'opportunity to exercise such [buying/bargaining] power exists only when suppliers are enjoying something that can be taken away, i.e. when they are enjoying the fruits of market power from which they can be separated'. As a result, what occurred in retailer–supplier relations during Britain's 'retail revolution' of the 1980s also depended critically on that central cell in Porter's schema in Figure 3.1 – the rivalry between the manufacturers themselves and – the upper left-hand cell – the threat of competition from new manufacturers. Indeed, as Grant (1987) argued at the time, the consumer goods manufacturing industries, which the increasingly large retailers of the 1980s dealt with, remained in general far more concentrated than any retail sector – effectively, therefore, these manufacturing industries were oligopolies in which price/output coordination against the retailers was theoretically possible. In practice, however, this did not occur. First, because the threat of competition to the existing manufacturers from new firms increased as retailers became larger – essentially because of lower entry barriers to supplying larger retailers rather than smaller. Second, because larger retailers were more able to utilize their buying/bargaining power to induce rivalry between the manufacturers – i.e. to breakdown their oligopoly coordination – particularly where excess manufacturing capacity existed. As a result the retailers were, in most cases, able to counteract the market power of their manufacturing suppliers to extract ever greater discounts and concessions, and reshape supply relations in their favour.

Reshaping the nature of supply relations

The sense that retailers were rapidly gaining the ability to reshape and control supply relations in the 1980s led academics from several disciplines into attempts to characterize and conceptualize the nature of the retailer–supplier relations which were emerging. Often these were positioned within wider theoretical debates on emerging forms of inter-firm organization in the 1980s – in particular the adoption more widely of Japanese-style industrial practices in western economies (see Florida and Kenney, 1991; McMillan, 1990; Morris and Imrie, 1992). Attention was focused on the extent to which long-term cooperative partnerships with suppliers – the non-adversarial, trust-based networks of relationships with suppliers characteristic of what Best (1990) more generally termed the 'new competition' – were becoming a more pronounced feature of retailer–supplier relations as the balance of power shifted progressively in favour of the larger retailers.

By the early 1990s there was a growing consensus, at least in Britain, that retailer–supplier relations had experienced important shifts in character. *Arms-length* relationships in which retailers simply placed orders for products with manufacturers in a purely market-based transactional manner, to the extent that they had ever truly existed, had clearly withered. In addition, in what Dawson and Shaw (1990) refer to as *administered* relationships – that is to say, supply relationships in which one party (the manufacturer or retailer) coordinates activity on the basis of leadership or power – it was increasingly the retailers who exercised that leadership, assuming control of and coordinating many of the supply-chain functions traditionally performed by the manufacturers. However, it was the rise of what Dawson and Shaw (1990) refer to as *associative* relationships involving a high level of collaboration and cooperation between retailers and manufacturers – value-adding 'partnerships', often long-term and centred on information-sharing networks – which attracted the greatest amount of attention. The sense was that an 'emphasis on interactive, flexible and stable supply networks was a key retailing strategy' (Foord et al., 1996: 68), and that the crucial 'question to be asked about the supplier–retailer relationship in the future is, will retailers, from a position of increased market power and under conditions favouring more stable relationships, use their growing influence to create truly associative relationships, or will they use their power to reinforce relationships of long-term domination?' (Dawson and Shaw, 1990: 36).

We now turn to two readings by British geographers who conducted research on the changing nature of retailer–supplier relations in the late 1980s and early 1990s, and whose work addresses these issues. The first is by Jo Foord, Sophie Bowlby and Christine Tillsley who studied two products, bread and women's hosiery (i.e. tights and stockings), and supply relations in these fields, adopting (Bowlby and Foord, 1995) the term *relational contracting* from Williamson (1985) and Hodgson (1988) to describe the transformed *administrative* and the emerging *associative* type of relationships outlined above. Reading 3.1 focuses on some of their observations on relational contracting between retailers and manufacturers concerning the supply of bread. Readers should be aware, however, that the term 'relational contracting' has been the subject of controversy, being criticized (Sayer and Walker, 1992: 128; Doel, 1996) as something of a 'chaotic concept' for, in this case, grouping together somewhat different types of retailer–supplier relationships. Nevertheless, this case study serves usefully to draw out some important features of the emerging supply relations of the early 1990s.

Reading 3.1 – Relational contracting: the case of bread

Several researchers have examined the changes taking place in the retail supply chain. From this varied empirical research it appears that, by the late 1980s, the dominant form of retail–supplier relationship in Britain was based on a system of 'relational contracting'. It should, however, be noted that relational contracting is by no means an entirely new form of retailer–supplier relationship . . . What is new is its increasing dominance across the retail sector; the form in which it is negotiated; and the implications of contemporary relational contracting for the division of the costs and revenues of realizing surplus value between retailer and manufacturer . . .

In our research on bread . . . we found that relational contracting dominated the form of interaction between all the major retailers and their suppliers . . . Bread is interesting because its short shelf-life and perishable nature has meant that it is a difficult product to integrate into retailers' new ordering and delivery systems. However, even bread is now being absorbed into the new centralized ordering mechanisms. [As a result it] can be used to summarise a number of features common to the new forms of retailer–supplier relationship.

First, *day-to-day management of contracts and product negotiations are largely centralized.* The perishable nature of bread has helped to maintain store-level ordering . . . However, even for this product there are now moves to centralize daily ordering. For most products all orders for fresh supplies are placed by retail headquarters. For own-label products retailers check on the quality of the merchandise and the production process both through occasional (and unannounced) factory visits and, sometimes, through inspecting samples of the product in their own quality control departments.

Second, the *application of compatible IT is critical* to the evolving structure of retailer–manufacturer interaction. In the bread supply chain couriers are still used in localized areas to deliver orders to manufacturers and telephone ordering is commonplace. However, automation of this process is increasing . . . Thus, retailers and suppliers are increasingly linked through compatible systems. Compatible automation increases centralized control, speed and accuracy through rapid communication of sales and stock information from retail outlets to the retail headquarters where orders to manufacturers for new production or 'call off' of existing stock are generated.

Third, *delivery* of products from manufacturer to retailers' warehouse is undertaken by both manufacturers and by third party specialist hauliers. However, specialist hauliers now dominate the final distribution of goods to individual stores. Delivery dates are specified and tightly monitored. Delivery windows at the warehouse or individual store are narrow and threats of sanctions are made by retailers to manufacturers if these are not observed.

Fourth, *product development and specification/price* are agreed through negotiation between retailer and manufacturer headquarters. These negotiations may be short in the case of a standardized product with which the manufacturer and retailer are both familiar . . . [However] in the development of a new 'value added' retailer-brand product, the manufacturer may become involved in market research, extensive product testing, evaluation and trialling products as a response to demands made by, and in some type of 'partnership' with, specific retailers. [Such products] could take up to two years to reach the shops.

Fifth, *establishing and maintaining the trading relationship* between individual firms requires investment of time and personnel . . . We found that this investment on the retail side was made through buying and merchandising teams often under the management of a senior director [whereas] manufacturing firms often created new posts for contract managers and/or for personnel with specific responsibility for individual retail customers. Retailers' desire to establish long-term 'partnerships' with groups of suppliers derives from the need for repeated purchase of reasonably large quantities of the same product; the need for consistency and reliability in terms of product specifications and delivery; and access to new product/process ideas and innovations. Search for new products and processes does initiate the search for new suppliers but developing these with a known and reliable supplier is less risky than total reliance on a new firm . . .

[Overall] the close relationships developed through relational contracting both *create* and *structure* the supply market in a reflexive manner . . . Long-term relational contracting involves costs as well as benefits. Maintaining and servicing the relationships can be costly in time of skilled personnel and may also involve specific capital investment. Thus there are pressures to minimize the number of such relationships for both retailers and manufacturers. [As a result] both try to maintain some choice of partners in order to avoid becoming too dependent on one another.

Extracted from: Jo Foord, Sophie Bowlby and Christine Tillsley (1996): The changing place of retailer–supplier relations in British retailing. Chapter 4 in Wrigley, N. and Lowe, M.S. (eds) *Retailing, consumption and capital: towards the new retail geography*. Harlow: Addison Wesley Longman, 68–89.

Two issues which emerge with consistency in Foord, Bowlby and Tillsley's account of retailer–supplier relations in the early 1990s relate to the concept of 'partnerships' and the avoidance of 'dependency'. In our second reading by Louise Crewe and Eileen Davenport, which considers the changing nature of supply relationships in British clothing retailing during the same period, these issues become the central concern. Crewe and Davenport (1992: 189) argue that, during the 1970s and early 1980s, two contrasting forms of supply relations had existed in this sector. The first was the 'arms-length' type, typified by the retailer C&A which was 'concerned only with the final product and not where or how the garment was produced', for whom 'continuity of the supply base was not regarded as important', and who 'had no desire to visit factories, supervise production, or encourage the adoption of new technology'. The second was the 'close-control' type, typified by Marks & Spencer which developed long-term relationships with its suppliers and became 'involved in every aspect of the manufacturing process, from choice of fabric and close attention to design, to colour, styling and finishing'. Crewe and Davenport suggest that by the early 1990s this marked contrast had begun to break down and that retailer–supplier relations in UK clothing retailing had begun to converge on what they term a 'preferred-supplier/hierarchical-control' model, as shown in Figure 3.2.

In this format, a dominant retailer sources from a complex multi-tiered supply chain, using the first tier of 'preferred' or favoured suppliers to police quality control within the chain, allowing the retailer to distance itself from the associated risks and responsibilities. In Reading 3.2, Crewe and Davenport pose important questions about dominance and dependency within such 'partnership' relationships. To what extent, for example, can they be considered 'associative' relations in the sense of Dawson and Shaw (1990)?

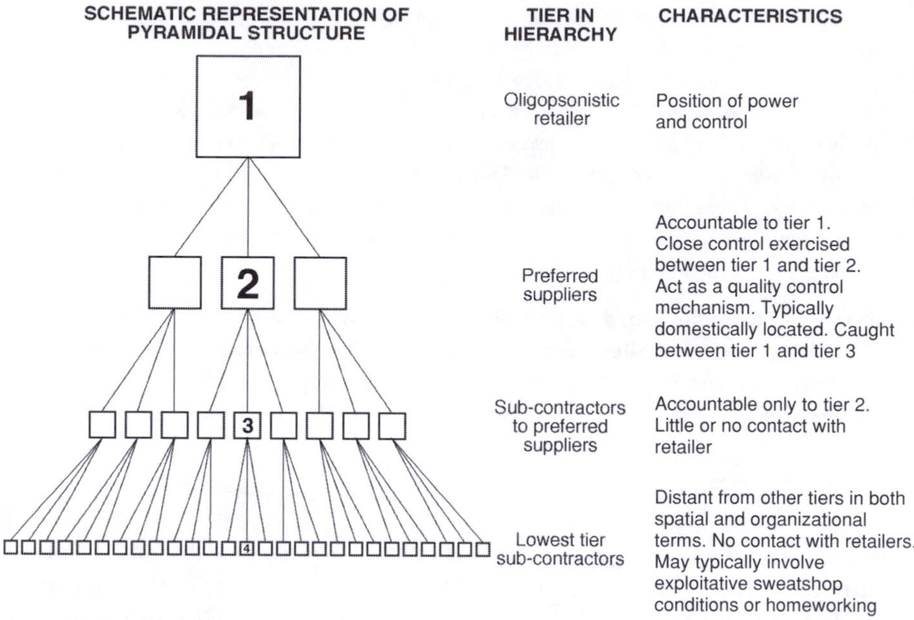
SCHEMATIC REPRESENTATION OF PYRAMIDAL STRUCTURE
TIER IN HIERARCHY
CHARACTERISTICS
1
2
3
4
Oligopsonistic retailer
Position of power and control
Preferred suppliers
Accountable to tier 1. Close control exercised between tier 1 and tier 2. Act as a quality control mechanism. Typically domestically located. Caught between tier 1 and tier 3
Sub-contractors to preferred suppliers
Accountable only to tier 2. Little or no contact with retailer
Lowest tier sub-contractors
Distant from other tiers in both spatial and organizational terms. No contact with retailers. May typically involve exploitative sweatshop conditions or homeworking

Figure 3.2: *The preferred-supplier/hierarchical-control model of retailer–supplier relations*
Source: Adapted from Crewe and Davenport (1992).

Reading 3.2 – Dominance, dependency and preferred suppliers: clothing retailing

Transformations within retailing are having profound implications on the organization of buyer–supplier relationships. Essentially, the motivation for such a restructuring of sourcing is the desire on the part of retailers to find new ways with which to maintain control over production without bearing the risks typically associated with a fickle and unpredictable fashion market . . .

It appears that convergence along the 'arms length'/'close control' spectrum is emerging. This convergence, we suggest, has resulted in the development of a single, pyramidal model. The pyramid is hierarchical in structure and comprises the dominant retailer along with multiple tiers of suppliers. We have called this the 'preferred supplier model' because, most importantly, it is predicated on the establishment of a first tier of 'preferred' suppliers . . . The further down the hierarchy one moves, the more tenuous and distant the relationship between retailer and supplier becomes, in both organizational and spatial terms. The characteristic feature of this pyramidal control structure is that of long term, close working relationships between retailers and their core of preferred suppliers . . . However, we raise a note of caution regarding the idea that this new sub-contracting pattern has trust as its basis . . . it is important not to confuse long term relationships with notions of partnership, alliance and mutual trust . . .

The emergence of this hierarchical control model has important implications for the British clothing industry . . . Simply stated, it is important to determine whether or not the strong power base enjoyed by retailers over suppliers is being challenged by the emergence of the new order . . . On the positive side . . . many clothing manufacturers are increasingly being consulted on design issues, rather than passively accepting retailers' designs and functioning powerlessly in a 'cut, make and trim' capacity . . . This movement towards design as a collaborative, interactive process may be one way of affording greater autonomy and control over production to clothing manufacturers. On the other hand, the expectation that the manufacturer takes responsibility for design may be less an equalising mechanism, and more a means of placing yet greater demands upon suppliers, as retailers relinquish yet another of their roles . . . Again on a positive note, the concentrated nature of the British clothing retail sector, and the move towards closer working relationships with preferred suppliers enables an element of stability to exist within the higher echelons of the sub-contracting system . . .

However, there is another side to this equation, the fact that suppliers are locked into a position of dependency over which they have little control. This relationship of dominance and dependency is perhaps most dramatically exemplified by suppliers to Marks & Spencer; with 90 per cent of S.R. Gent's output and 85 per cent of I.J. Dewhirst's output going to Marks & Spencer, it is easy to envisage many manufacturing operations losing their own identity, becoming pawns in the retailers' kingdom, and functioning according to the exacting standards dictated by the retail buyer . . . The unfortunate result is that at times of fluctuating market demand, directional and style changes in consumer demand, or when the economic screw tightens, the manufacturer may find its vital life support system removed, as the retailer seeks out alternative sources of supply in response to pressures to maintain market share and cut supply cost. The manufacturer becomes powerless under such circumstances, with retailer concentration creating a powerful barrier in the search for alternative markets. Such control is perhaps most effectively epitomized by Marks & Spencer; although they provide their customers with stable and historically huge contracts, in return they insist on scrupulous standards and wafer thin profit margins. They also demand exclusivity on different designs, different fabrics, and even differently coloured threads from competitors . . . Being a Marks & Spencer supplier guarantees nothing. The impact of retrenchment has been felt by many of [its] 600 suppliers . . . their security rests upon a very shaky foundation, a foundation which is only as strong as their last delivery . . .

Alongside close control over favoured suppliers, there exists a complicated chain of sub-contracting relationships further down the hierarchy, whereby favoured suppliers may themselves sub-contract production . . . the route by which the supplies reach the stores may well be highly complex, with the retailer having only the most tenuous link with the sub-contractors at the bottom of the pyramid. The implication of this [is that] preferred suppliers are finding themselves being locked into a double bind situation, between the exacting retailers at the top, and the risk and responsibilities over sub-contractors at the bottom, whose faults they must correct, and whose delivery dates they must oversee . . . To take Marks & Spencer as a case in point. Formerly a company with a well-established record of domestic sourcing, the group has recently opted for an aggressive involvement in international sourcing . . . In this way, the risk falls upon the favoured supplier to ensure that foreign goods are 'up to the Mark'.

Extracted from: Louise Crewe and Eileen Davenport (1992): The puppet show: changing buyer–supplier relationships within clothing retailing. *Transactions of the Institute of British Geographers*, NS17, 183–97.

It can be seen from Readings 3.1 and 3.2, that the picture of retailer–supplier relations emerging from research in the early 1990s was a complex and multi-layered one. Whilst there was little dispute that the focus of supply relations had shifted towards the 'partnership' end of the spectrum shown in Figure 3.3, the typical form of those 'partnerships, the exact nature of 'relational contracting', and what that implied in terms of dominance and dependency within such relationships was a matter of intense debate. Whilst Crewe and Davenport (1992: 196), for example, stress that the 'long term, obligational relationships which are emerging between buyers and suppliers must not be confused with the notions of mutual trust, partnership or alliance', and Ogbonna and Wilkinson (1998: 84) report major retailers publicly talking of developing 'partnerships' with dominant-brand manufacturers but privately stressing the need to 'erode their power', Foord, Bowlby and Tillsley (1996: 68) suggest, on the other hand, that the emerging retailer-led emphasis on interactive, flexible and stable supply networks 'was not uncontested by suppliers, nor was a large retail firm guaranteed superior power in all situations – manufacturers had counter-strategies which changed the supply relation and, in certain circumstances, limited retailers' control'.

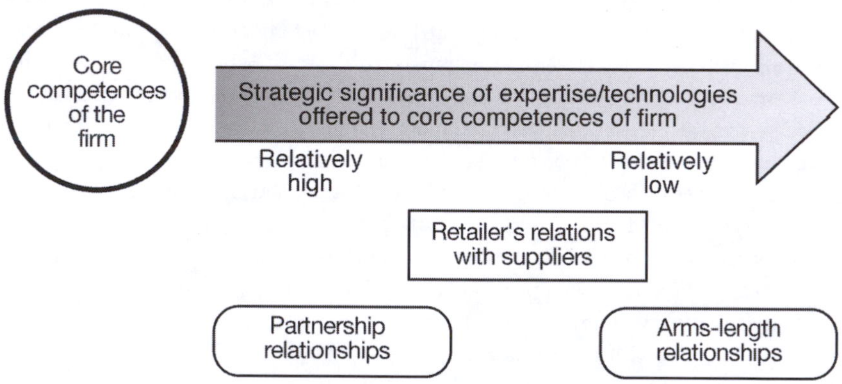

Figure 3.3: *Positioning types of retailer–supplier relationship with respect to core competences of retail firm*

Clearly the retailer–supplier relations which developed during the 1990s were characterized by highly complex power dynamics and by considerable ambiguity and contradiction – cooperation and competition often going hand-in-hand. In some respects, as Figure 3.3 suggests, the degree to which relations ever became truly 'associative' in the sense of Dawson and Shaw (1990) – that is to say, long-term value-adding partnerships, involving high levels of collaboration and cooperation, centred on information-sharing networks – depended on how strategically significant was the expertise and/or technologies offered by the supplier to what might be termed the 'core competences' of the retail firm (Reve, 1990). In other words, to the skills which enable that firm to sustain its competitive position within the market. In this respect, as Ogbonna and Wilkinson (1998: 84) argue, 'very different power dynamics appear to be at play between retailers and brand manufacturers, and retailers and own-label manufacturers'. As a result, it is on the strategic significance of retailer own-label products in the shifting power balances and changing character of retailer–supplier relations that we must now focus.

The strategic significance of retailer own-label products

We have already seen, within the context of Porter's general schema (Figure 3.1), how the rise of substitute own-label products was one of the competitive forces which shifted strongly in favour of retailers during Britain's 'retail revolution' of the 1980s. In the specific case of food retailing, for example, own-label products increased their share of the UK 'packaged grocery' market in step with the rising concentration levels of the major retailers, from 23 per cent at the beginning of the 1980s to 36 per cent of sales (or 48 per cent when perishables were included) a decade later. Moreover, the leading firms typically developed much higher levels of own-label sales than average and, as a group, increased those levels more rapidly (Table 3.1). In addition, in other sectors such as clothing retailing, where own-label products had traditionally held a larger market share, there were leading retailers such as Marks & Spencer whose products (including food) were entirely own-label.

However, it was not just the increase in the market share of own-label products which characterized Britain's 'retail revolution' of the 1980s; the nature of those products also changed. As the 1980s progressed, own-label products – particularly those of the leading food retailers – were successfully transformed away from the inferior-quality, cheap-substitute, 'generic', or sub-brand image which had been common in the 1960s and 1970s in the UK and which, as we will see, continued to characterize own-label products in some

Table 3.1: *Trends in own-label packaged grocery shares amongst the major UK food retailers, 1980–95*

	Own-label share (%)		
	1980	1992	1995
Sainsbury	54	55	56
Tesco	21	41	48
Safeway (Argyll)	28	35	44
Asda	5	32	41

Source: Wrigley (1998e).

sectors in other countries, particularly the USA, throughout the 1980s and early 1990s (Hughes, 1996a). In Britain, the image of own-label, in food retailing particularly, was successfully shifted towards high quality and innovation – own-label products effectively became repositioned as *retail brands* (see de Chernatony, 1989; Burt, 1992; Pellegrini, 1993; Laaksonen, 1994; Sparks, 1997). In essence, a link was established between those brands, the retailer's corporate identity, and consumer trust in the retailer's reputation, such that the retailer brand became a guarantee of quality and consistency. Reciprocally, innovation in retailer branding became a key element in the further enhancement of the retailer's consumer image.

As Terry Leahy (1987), now chief executive of the UK's leading food retailer, Tesco, outlined in the late 1980s, own-label products and the shift in positioning of those products into 'retailer brands' offered several strategic advantages to the food retailers. The most obvious lay in market development, product control and product innovation. Retailers could use own-label to fill gaps left in the range of products offered by the manufacturers, in that way spurring broader market development. In addition, they could develop own-label products quickly, under tighter control, and could take more risks with product innovation. The result was that during the 1980s the major UK food retailers were increasingly drawn into making significant investments in product specification, development, packaging and quality testing, and the developmental role of the retailer's own food technologists became ever more critical (Senker, 1986; Omar, 1995; Doel, 1995; Hughes, 1996b). New product innovation by the retailers was aided by the consumer preference/sales information – item by item, store by store, region by region – which began to flood out of their EPOS (electronic point of sale) scanner systems from the mid-1980s onwards (see Chapter 4). As a result, they could effectively take more risks than the food manufacturers, not only because they did not have to buy distribution and access to shelf space at the same marginal cost as the manufacturers, but also because their proprietary scanner data allowed them to position new own-label products with greater precision at the leading edge of changing consumer tastes.

At the cutting edge of these developments in UK food retailing was the emergence of the chilled ready-meals sector – a market whose value in the UK was worth over £300 million per annum by the early 1990s, but which barely existed in 1983, and which remained underdeveloped in many other countries (particularly the USA) by the mid-1990s. This was the first major food market segment in the UK to be totally dominated by own-label. As such, it was not only of 'immense symbolic significance . . . visibly epitomising the changing dynamics and shifting power balance within the food industry' (Doel, 1996: 55), but it was also significant in terms of demanding new types of supply network. Given that chilled ready meals were, in the mid-1980s, a completely novel, retailer-initiated, product sector, complex both to manufacture and distribute, there was simply no pre-existing brand manufacturer presence. That is to say, there was no manufacturer surplus capacity to appropriate as the basis of an own-label supply. As a result, the retailers – Marks & Spencer the pioneer of the sector, closely followed by Sainsbury, Tesco and the other major firms – were forced to develop their own ready-meals supply chains in the absence of organizational alternatives. As we will see in Reading 3.3 by Christine Doel, the chains that developed had some distinct characteristics. In particular, they were characterized by larger numbers of small and medium-sized specialist suppliers than was typical elsewhere in the industry, and the early growth of those suppliers was frequently underpinned by retailer intervention/commitment – that is to say, they were much closer to the 'associative' supply relationships envisaged by Dawson and Shaw (1990). In return, the retailers' requirements were exacting. The supply relations that evolved were, in consequence, dynamically interactive, based around the imperatives of continuous new product development as the food retailers pursued product-centred competitive advantage. They were also characterized by stringent confidentiality and exclusivity agreements. Finally, the exacting requirements of own-label chilled meals distribution created and/or strengthened the trend towards 'contracted-out' relationships between specialist logistics companies and the retailers (see Chapter 4 for further discussion).

Reading 3.3 by Christine Doel, a British geographer, draws on her research in the early 1990s on the relations between UK food retailers and their own-label suppliers (see also Doel, 1999). Like Crewe and Davenport, she confirms the shift away from arms-length strictly price-based relationships, noting that 'own-label supply networks created by proactive retailers and characterized by intense interaction are assuming progressively greater qualitative and strategic significance' (Doel, 1996: 62). However, it is her observations, based on semi-structured interviews with retailers and manufacturers, concerning the initiation and post-initiation phases of own-label supply by highly specialized suppliers with considerable strategic importance to the major retailers, which are particularly intriguing. Here a strong sense of truly 'associative' relationships emerges. But there are important caveats.

Reading 3.3 – Building own-label supply relations

The firms that currently supply multiple retailers with ready-meals are diverse in character. They involve divisions of several large public companies, but of greater overall significance is the large number of small specialist firms. Tracing the histories of these is instructive regarding actual processes of supply-chain initiation ...

[In the case of one such firm] initial expansion from the domestic kitchen to a small scale factory unit required a commitment from [major retailer E]. As the national account manager of the firm recalled:

> Basically what [retailer E] said was: 'We want your products. We will put an order to you in writing that you will supply us. Take that to the bank and show them that you have got an order from [retailer E]. I think they gave us a contract for something like twelve months . . .' that is how we started off.

The manner in which the sub-contract was initiated . . . demonstrates why 'although it can be argued that supermarkets ruin businesses because they suddenly pull out, at the same time they also actually create businesses'. In addition, it is indicative of the types of initiatives that have to be undertaken by the retail multiples in order to develop their own-label ready-meal supply chains in the absence of organizational alternatives . . . Direct intervention during the [supplier's] inception and early growth may well underpin a more dynamically interactive supply-chain structure . . .

The active role of retailers as effective innovators through the own-label network has been recognized, particularly in high-margin and thus strategically important product areas. In this context, the value of particular suppliers to the retailers must be immense as these will often possess unique capabilities and 'relation-specific' skills. Thus retailers will 'very often come to rely on [such suppliers] making contributions through their specialized competences, close attention to the lead firm's technology and business practices, and innovative capacity' (Sayer and Walker, 1992: 132). Moreover, these are generally also the own-label sub-contract links that were initiated by direct retailer intervention during the supplier's inception and early growth. As they are concentrated in new, innovative product areas, few alternative sources of supply exist and, in any case, the complexity of the products ensures that specifications are not easily transferable.

But, despite all of this, the retailer will [usually] not make a formal commitment to the supplier . . . From the retailer's perspective, the rationale for this is obvious. The lack of any formal commitment despite the strategic significance of highly specialized suppliers means, in principle, that the discipline provided by the threat of delisting remains even if, in practice, the likelihood of this is remote because of the absence of alternative sources of supply. In this situation, in return for even the possibility of continued patronage, the retailers' requirements are exacting. Product specifications are highly confidential and surrounded by legally binding contracts . . . The need for such complex legal arrangements in the context of retailer–manufacturer relations that are supposedly based on 'partnerships' and 'trust' does appear contradictory . . . [But] the concern of the retailers about product exclusivity is indicative of the intensely competitive environment in which they operate – one manifestation of this [being] how closely retailers monitor the interactions between their own-label suppliers and other retailers.

Extracted from: Christine Doel (1996): Market development and organizational change: the case of the food industry. Chapter 3 in Wrigley, N. and Lowe, S. (eds) *Retailing, consumption and capital: towards the new retail geography*. Harlow: Addison Wesley Longman, 48–67.

Finally, and as noted earlier, any assessment of the strategic significance of own-label products and own-label supply networks must be placed within the wider context of the extent of both retail concentration and the degree of shift experienced in the balance of power in distribution channels from manufacturers to retailers. In other countries, particularly in the USA (see Messinger and Narasimham, 1995; Cotterill, 1997), where the shift in bargaining power from manufacturers to retailers was less advanced by the early 1990s than in Britain, the level of own-label product sales in many retail sectors was considerably lower. In addition, consumer perception of own-label quality was much less positive, as the vital repositioning of own-label into 'retailer brand', with all that implied in terms of a guarantee of quality and consistency, was only just beginning to occur in some sectors. Hughes (1996a) and Sparks (1997) consider US food retailing from this perspective – pointing out how, by the early 1990s, retailer own-label sales in US food retailing represented only one-third of the equivalent UK levels, and showing that 'the processes by which US retailers' own-label supply relationships are initiated show a marked divergence from the UK model proposed by Doel' (Hughes, 1996a: 2207). In particular, own-label supply relations in US food retailing were often mediated by third-party brokerages – a type of supply-chain organization not conducive to the long-term growth of innovative own-label strategies. For that reason, a British food retailer, Sainsbury, on entering the US market, attempted to develop forms of own-label supply relations which were not tied to such brokerages but which were modelled instead on UK practices (Wrigley, 1997a).

'The ties that bind' – the embeddedness of supply relations

In concentrating in the previous sections on the complex power dynamics and somewhat abstract conceptual nature of retailer–supplier relations, the sense that those relations are, in practice, the product of networks of ongoing personal relationships – what Granovetter (1985) terms 'the ties that bind' – between individuals charged with managing the supply chain has been underemphasized. Indeed, Hughes (1999) argues more generally that, with the exception of the work of Christine Doel, accounts of retailer–supplier relations presented within the geographical literature 'have a tendency to undertheorise the social and cultural energies which drive many of the changes taking place at the retailer–supplier interface'. In turn, Doel (1999) emphasizes the centrality of 'unspoken behavioural norms and other facets of "community" in the structuring of economic relations', and stresses the importance of exploring 'informal constraints, conventions, and embeddedness' when considering social aspects of inter-firm relations and the character of the governance process at that interface.

Reading 3.4 by Alexandra Hughes, a British geographer, draws on her research on retailer–supplier relationships in Britain and the USA in the early 1990s (Hughes, 1996b). Here we concentrate simply on own-label supply chains, but Hughes also considers the socially organized competitive frameworks relating to the supply of manufacturer brands, considering both the multi-level organizational contact which structures the character of that supply and how it is embedded in networks of interpersonal relations.

Reading 3.4 – Social relations and the organization of own-label supply chains

Own-label supply relationships in the UK, whether they are 'administered' or 'associative' are ... characterized by an absence of written contracts ... and organized through multi-level contact[a] ...That is to say, these supply relationships are controlled through more than one channel of ongoing communication between the companies, where top level retail management meet with top level management of supplying companies and middle level management of retailers (buyers or category managers) meet with middle level management of suppliers (account managers). In addition, there is an active channel of communication between the technical departments of the two parties and there can also be cross-over in contact points[b] [for example when] account managers are present at director-level meetings or buyers meet with a supplier's technical staff ...This apparently hierarchical organization of inter-firm contact has been likened to both a 'marking system' and 'a game of chess'[a] ...

The day-to-day organization of agreements and negotiation concerning pricing, payments and ordering is controlled at the level of middle management ... It is therefore through the social relations between these people that power and negotiation in own-label supply chains operates[a] ... It is the continual interpretation by middle management of particular corporate strategies which shapes the material practices of buying and selling[b] ...

Buyers and account managers in UK own-label supply relationships tend to meet frequently, though the exact frequency of meetings varies according to the stage of a product's development and the companies involved. Generally, they can meet face-to-face as frequently as three times a week and as infrequently as once every four to six weeks. These meetings are also backed up by regular telephone calls.The location of the face-to-face meetings is normally the headquarters of the retailing company, though the retailer's regular factory visits can also provide an arena for discussion between the two parties.[a]

Several of the interviewees in my UK research reflected on the importance of close, interpersonal relations [in particular the development of trust within those relationships] in forging business links within supply situations. However, the link between trust at the interpersonal and inter-firm levels also presents problems, which stem from the competitive tension which lies within vertical inter-firm relationships. A sharing of too much information on the part of buyers and account managers may result in a loss of competitive advantage for the companies involved ... illustrating the contradiction that trust at the level of middle management can both engender *and* repress competitive advantage.[b] This would seem to be one reason that leading UK food retailers frequently switch their buyers between product categories so that they are not always dealing with the same account managers.[a]

Acknowledging the embeddedness of retailer–supplier power relations within verbal agreements and ongoing communication reinforces the centrality of ... a theorisation of

> social relations within supply chains[a] ... It is through situated engagement with these practices ... that effective ways of thinking critically about the creative construction of competitive spaces are opened up.[b]
>
> **Extracted from: [a] Alexandra Hughes (1996b): Changing food retailer–manufacturer power relations within national economies: a UK–USA comparison. Unpublished Ph.D. thesis, University of Southampton, 226–31; [b] Alexandra Hughes (1999): Constructing competitive spaces: on the corporate practice of British retailer–supplier relationships. *Environment and Planning A*, 31, 819–39.**

Commodity chains and retailer–supplier relations

What then does our discussion of the specifics of the retailer–supplier interface tell us about those broader issues of global networks of commodity production and retail supply with which we introduced Chapter 1? Tesco's sourcing of mange tout from Kenya is at once both an example of one form of the emerging retailer–supplier relations of the 1990s and, at the same time, a bridge to a wider analysis of the globalization of retailer supply chains and issues of commodity 'circuits' in production, distribution and consumption.

David Harvey's (1990: 427) well-known statement that:

> The grapes that sit upon the supermarket shelves are mute: we cannot see the fingerprints of exploitation upon them or tell immediately what part of the world they are from. We can, by further enquiry [however] lift the veil on this geographical and social ignorance,

although focused on the necessity to 'get behind the veil, the fetishism of the market and the commodity, in order to tell the full story of social reproduction', served during the 1990s to encourage a wave of research by geographers focused on commodity chains. That research broadens, in a vital way, the more specific retailer–supplier interface issues which have occupied our attention so far in this chapter.

By considering the flow of goods through such chains, the social and economic practices which shape that flow, and by tracing connections between different sites along the chains, geographers have increasingly linked global networks of commodity production and retail sourcing to local sites and practices of consumption (Crang and Jackson, 1998). In addition they have traced the ways in which the meanings of commodities are reworked along and across different sites and spaces in those chains (Jackson and Thrift, 1995; Crang, 1996; Leslie and Reimer, 1999). Much of this work has centred on global food chains (Arce and Marsden, 1993), with the work of Cook (1993, 1994) on the chain linking exotic fruit production in Jamaica to the supermarket shelves of the major UK food retailers, and Cook and Crang (1996) – see also Cook et al. (1998) – on what they term the 'biographies and geographies' of food being particularly well known. More recently, however, it has broadened to include other sectors such as home furnishings (Leslie and Reimer, 1999) and the cut-flower trade (Hughes, 2000).

Reading 3.5 by Deborah Leslie and Suzanne Reimer is taken from a wide-ranging *Progress in human geography* (1999) article on the spatiality of commodity chains. The article is underpinned by the authors' primary research on commodity chains in the home furnishing industries of Canada and the UK and we will revisit that research in more detail in Chapter 13 in relation to its contribution to our understanding of the changing configuration of 'home consumption'. Here, however, we focus on their summary of what they identify as three key strands in the commodity chain literature.

Reading 3.5 – Conceptualizing the commodity chain

There has been considerable interest in the buyer–supplier interface within the commodity chain. Both the dynamics and the regulation of buyer–supplier relationships have been examined in detail [and] that work has contributed greatly to our understanding of the mediating role of retailers within the economy, operating at the intersection of production and consumption ... As yet, however, there has been little exploration of multiple sites in the chain, [and] relatively few commentators have extended their analysis to final consumption ...The notion of the commodity chain [in contrast] attempts to trace the entire trajectory of a product from its conception and design, through production, retailing and final consumption ...

Although there is considerable overlap and convergence within the commodity chain literature, three distinct strands of work can be identified.The first tradition encompasses the global commodity chain (GCC) literature, much of it deriving from world systems theory [see Gereffi, 1994] ... GCC narratives tend to centre upon the global dynamics of production/consumption/retailing linkages ... For Appelbaum and Gereffi (1994: 43) a central question is:'where does the global commodity chain "touch down" geographically, why, and with what implications for the extraction and realisation of an economic surplus?' ...

The direction of enquiry within the GCC literature is often highly revealing. Most accounts tend to treat consumption as a starting point from which to trace relations back to the underlying exploitative reality of production ... points of distribution and consumption are merely noted at the outset before commencing the real task of unveiling production and extraction.

A second tradition of commodity chain analysis includes Fine and Leopold's (1993) explication of 'systems of provision' ... Fine and Leopold reject 'horizontal' analyses of consumption [and retail capital], arguing for a 'vertical' approach which ... pinpoints differences in the ways in which production and consumption are linked in different commodities ... [Their] approach points towards the possibility of a more balanced treatment of the relationship between production and consumption, in which the interplay between supply and demand factors is recognized ... [They] also acknowledge the sign value and symbolic meanings of commodities, recognize the constantly changing relationship between the meanings of commodities and the physical content of commodities ... [and] argue that the historical contingency of chains demands consideration ... A central shortcoming of much of the commodity chain literature is that the process of 'unveiling' sites is underpinned by a focus on production as the focus of 'reality' ... Fine and Leopold's system of provision analyses goes the furthest in avoiding this productionist pitfall. Nevertheless, at times, their attempts to understand particular systems of provision appear to be aimed at criticising demand side explanations and asserting the primary importance of production factors.

A third approach to the investigation of the commodity chain involves tracing commodity 'circuits' ... Unlike the notion of a chain, circuits have no beginning or end. Commodities interrelate with other commodities as they travel along a course. Thus while a consideration of the commodity chain implies tracing relationships of causality and constructing explanatory sequences, the notion of the circuit is less interested in these connections ... Cook and Crang have utilized the notion of circuits to consider the ways in which geographical knowledges of commodity systems are shaped and reshaped [and their] consideration of 'circuits of culinary culture' incorporates a healthy scepticism towards the origins of commodities: they recognize that origins are always constructed ... One concern raised by the notion of circuitry, however, is whether abandoning the language of the 'chain' means abandoning a language around which we can mobilise ... the conceptualisation of a 'virtually endless circuit of consumption' (Jackson and Thrift, 1995: 205) may involve the loss of an important stance: the foregrounding of exploitation. It is for this reason that we are hesitant to abandon the concept of the chain altogether ... However, we would agree with Cook, Crang and others that the goal of commodity chain analyses should not be to privilege one site on the chain as the locus of reality ... Our work on the home furnishings sector, an increasingly visible component of the retail landscape in both Canada and Britain [allows us] to provide a series of 'windows' onto particular aspects of commodity chain geographies.

Extracted from: Deborah Leslie and Suzanne Reimer (1999): Spatialising commodity chains. *Progress in Human Geography*, 22, 401–20.

Our focus in this chapter, then, has broadened considerably – from the specifics of the changing character of retailer–supplier relations in Britain's retail revolution of the 1980s, to attempts to position understanding of the reconfiguration of retailer–supply chain interfaces within much wider commodity chain or commodity circuits literatures. It is important, however, to remain sensitive to the interconnections between these themes. For example, the development and shifting nature of own-label products and supply relations, and the way major UK food retailers increasingly learned to manage own-label supply chains in the 1980s and early 1990s 'in order to derive a fluid and differentiated portfolio of competitive advantages' (Doel, 1999) resulted, as we will note in Chapter 8, in the application of similar supply-chain management techniques to global produce markets. In the next chapter we will see how the own-label supply chains which emerged during this period in the UK were, in turn, underpinned by a logistically refined 'demand pull' system of supply-chain management. Global commodity chains involving British retailers inevitably reflect, therefore, the disciplines of such management and are shaped in the image of the types of retailer–supplier relationships examined above.

4

Organizational and technological transformations in retail distribution

As we saw in Chapters 2 and 3, it was retail concentration, mergers and acquisitions, and the sense that retailers were rapidly gaining the ability to reshape and control supply relations, that attracted increasing amounts of academic attention during Britain's 'retail revolution' of the 1980s. In contrast, for the ordinary consumer – the proverbial woman or man in the street – it was the emergence of ever larger superstores, at a seemingly ever more rapid pace, transforming the built environment of cities throughout the UK, together with the growing presence on roads throughout Britain of uniform fleets of delivery vehicles in the increasingly familiar colours of Tesco, Sainsbury, Safeway, Asda and Marks & Spencer, which had the most immediate impact on their lives. As *The Times* columnist Simon Jenkins wryly observed ten years later at the height of a 1997 general election which marked the end of the Conservative Party's 18-year period in power:

> Eras are rarely remembered for what most obsessed politicians at the time. Victorian Britain was convulsed over Catholic emancipation and Home Rule but we think of it as a nation of railways and Empire . . . The Tory years will be recalled not for trade union reform or privatisation . . . The cathedrals of Mrs Thatcher's Britain will be St-Tesco's-on-the-Roundabout and St-Sainsbury's-on-the-Interchange (*The Times*, 19 April 1997: 20).

In this chapter, it is to some of those organizational and technological transformations in retail distribution, not only during the 1980s and 1990s, but also in earlier decades, and in both Britain and the USA, that we now turn.

The expanding scale of retail formats – supermarkets, superstores and supercentres

As Rachel Bowlby (1997: 96) observed: 'In the nineteenth century was the department store; in the twentieth century was the supermarket.' Together, as we will see in this section and later in Chapter 11, these types of store provide the roots of contemporary large-format retailing – be it the 35–40,000 sq. ft. food superstores which had such a profound impact on Britain's built form and cultural landscape in the 1980s, the larger 70–100,000 sq. ft. continental-European hypermarkets of firms such as Carrefour, Auchan and Promodès, or the even larger 150–200,000 sq. ft. US supercentres of Wal-Mart, Kmart, Target,

Meijer and Fred Meyer which spread rapidly across the USA during the 1990s, becoming by some considerable way the fastest growing sector of US retailing in the final years of the century. But how are we to begin to understand the logic and history of the organizational transformation of store-based retailing during the last century? Here, readings of the emergence of the supermarket and its later mega-format extensions provide our route.

The emergence of the supermarket

In the USA, as we noted in Chapter 2, food store chains such as Kroger Grocery and Baking Company, the Jewel Tea Company, and the Great Atlantic and Pacific Tea Company (A&P) began to emerge in the late nineteenth century. However, by the early years of the twentieth century, such chains were still in their infancy – A&P, for example, operated only 200 stores in 1901. Rather, US 'grocery distribution was carried out largely though not exclusively by a vast network of minuscule monopolists' (Adelman, 1959: 25), each effectively protected by the distance from its nearest rivals. There was relatively little price competition between these stores, but they offered credit, delivery and other services. It was the second and third decades of the twentieth century, the years 1910 to 1930, when the fledgling US food chains began to expand at a dramatic pace. A&P alone, for example, increased its size from less than 400 stores in 1910 to a remarkable 15,700 stores in 1930, capturing 12.1 per cent of total US food store sales by 1933 in the process (Adelman, 1959: Appendix Tables 2 and 5) – a level which was never again reached in the twentieth century, even following the wave of acquisition-driven consolidation in US food retailing in the late 1990s outlined in Reading 2.4.

What powered this dramatic expansion? One observer (King, 1913), writing about the city of Philadelphia in 1913, attributed the growing success of the chains to three factors. First, their 'no credit' policies – cash sales and the consequent absence of bad-debt losses. Second, their bulk buying and physical distribution advantages – quantity purchasing at lower prices and centralized distribution – the latter being a prescient identification, as we will see below, of a critical theme of the late twentieth century. Third, their more efficient management. Other observers confirmed the significance of the way in which the retail chains began performing their own wholesaling functions, and also highlighted the increasing trend towards vertical integration by the chains into limited areas of food processing (Connor and Schiek, 1997: 322). Size of the individual stores and the principle of customer self-service were not, however, viewed as significant factors. Essentially these were large chains of very small, traditionally organized 'counter service' stores.

The rise of larger-format, self-service food stores – *supermarkets* as they became known – and the conversion of the leading chains into this format, despite its early origins in the Alpha Beta stores in California from about 1912 and the Piggly Wiggly chain in Tennessee from around 1916, dates in fact to the late 1920s and early 1930s. Early versions of these supermarkets, such as King Kullen on Long Island which opened in 1930, Big Bear in Elizabeth, New Jersey, which opened in 1932, and similar stores in Los Angeles dating from the 1920s (Longstreth, 1999) attracted – as Rachel Bowlby (1997) has reminded a cultural studies and wider social science audience – enormous amounts of attention at the time, being viewed as intensely innovative. Reading 4.1, extracted from Bowlby's much wider essay, captures some of this spirit.

Reading 4.1 – Big Bear crashes into New Jersey

On 8 December 1932, the vacant Durant automobile plant in Elizabeth, New Jersey, was the site of a much-remarked event, the result of a recent alliance between two entrepreneurs and a Hoboken wholesaler. Five years later, the occasion is enthusiastically recalled as a media event, as newspaper stories:

> flashed around the country ... described the barny structure with its crude interiors, fixtures made of rough pine lumber and its huge displays of merchandise as the center of attraction, with thousands of people who were swarming around them with baskets on their arms, content to wait on themselves (Zimmerman, 1937: v).

M.M. Zimmerman, the writer of the account ... focuses in on just one of many comparable beginnings at this time. Outfits similar to the Big Bear were appearing in many states and regions ... By 1937, Zimmerman can write:

> Contrary to the prognostications of Big Bear's early demise, it not only survived but many other Big and Little Bears have appeared in many of the large Metropolitan cities throughout the country ... Today there are more than two thousand of these Big and Little Bears operating in thirty-two states (Zimmerman 1937: vi).

There were two specific innovations [associated with the supermarket] which directly eliminated distribution costs. First, the customers travelled to the goods instead of the goods going all the way to them. The store was miles away from where the customers lived. The Model T Ford had made car-owning widespread; they would drive there themselves, parking in a special lot across the road. But the most striking way in which supermarkets turned distribution over to the customer was inside the store itself, through the 'hundreds of market baskets' piled up near the entrance. This produced the amazing spectacle of people 'content to wait on themselves'.

Self-service [was] associated not only with the saving of running costs, but also with the unprecedented idea of the food store as potentially an attractive and comfortable place for the consumer to enjoy ... For the time she is in the store, the shopper is free to move back and forth, to put into her trolley whatever she wants: the looking and touching encouraged by the open displays suggest and permit a taking that is not (yet) a commitment to purchase. In this interim, provisional space, between the entrance and the checkout, she really can have anything. This is the dreamlike face of self-service, when all is possible, possessable, until the moment of reckoning.

Extracted from: Rachel Bowlby (1997): Supermarket futures. In Falk, P. and Campbell, C. (eds) *The Shopping Experience*. London: Sage, 92–110.

Economic-oriented readings of these events, without capturing quite the same sense of the period as Bowlby conveys, similarly stress the key innovatory element of self-service selection, coupled with the economies of scale deriving from higher volume sales. They also suggest that the converted-warehouse supermarkets 'were successful initially because their lower prices appealed to consumers whose incomes were being curtailed by the Depression', and show how 'because of the success of these new stores, the major chains began converting to a smaller version of this new self-service supermarket format through the late 1930s and early 1940s' (Connor and Schiek, 1997: 323).

These themes are taken up in detail by Morris Adelman in his remarkable, but now little known, study of the corporate history of the A&P chain, published by Harvard University Press in 1959. Based on unique access to minutes of meetings held by the firm's directors and senior managers, internal letters, and detailed accounting records, Adelman is able to chart the response of what was then America's largest food store chain to the emergence of the supermarket. Reading 4.2 extracts very briefly from a book of almost 500 pages, but it serves to highlight themes which reappear frequently in the organizational transformation of store-based retailing in the twentieth century. As we will see below, those themes were being debated just as intensely 50 years later, during the building of Britain's superstores in the late 1980s, as they were in the context of the emergence of the supermarkets in the USA during the 1930s. In addition, we see in Adelman's account of A&P's initial reluctance to embrace the new format – a reluctance which he describes as astounding blindness – a clear example of the theme of corporate lock-in which, as we saw in Chapter 2, Erica Schoenberger (1994, 1997) has challenged economic geographers to understand. It provides a classic example of the way in which a dominant firm can become locked into an inappropriate corporate strategy, delaying essential reconfiguration, even though the firm may not only realize the need for transformation but understand (and indeed may have experimented with) the transformation required.

Reading 4.2 – A&P and the challenge of the supermarket

[Supermarket] development on a large scale seems to have [first] taken place in Southern California, as might be expected of an area so automobile-minded. But if large parking spaces made it possible to attract the driving public, lower costs and prices actually drew them. The Great Depression offered price competition its greatest opportunity. But supermarkets did not begin to open at a rapid pace until the latter half of 1932 ... As usually happens, the inherent advantages of this new type of store were both accentuated and disguised by some of the transitory ones. Abandoned factories on the outskirts of metropolitan areas were especially favoured as locations, partly because there was usually a second floor which could be used as a warehouse. One of the earliest stores (in Jamaica, New York) used empty boxes and cases for its fixtures. Like the chain stores, the supermarkets could buy in large amounts; self-warehoused, they could often lay down goods in their stores at a lower cost than an A&P store; self-served, they could sell at prices lower than competitors' and yet make very large profits. The proprietors of one of the best-known opened it late in 1932, in an abandoned automobile factory in Elizabeth, New Jersey. They invested $10,000; sales in the first year were over $2 million, and the net profit was $167,000, over 16 times the investment. As with the early [chain] stores, these were typical profits of innovation.

It seems impossible to draw any accurate picture of the cost-price structure of the early supermarkets . . . As the next best thing, it is possible to compare the operation of all A&P stores in 1937 with those of A&P supermarkets in the Central Western Division in 1938. This comparison [Adelman provides a detailed table] has the advantage that the method of conducting the business is the same, save only for the one critical feature of regular store versus supermarket operation. As [the table] shows, the great saving of the supermarket operation (as distinct from savings in purchasing) was in supervisory labour and in the traditional items of overhead – rent, depreciation, taxes, and the losses incurred in opening new stores . . . The table represents [therefore] the long-run [declining] cost curve [of such stores], and there is no doubt that it was so regarded by the A&P management . . . The point is simply that if we set aside, as does the [table], all the accidental advantages of opening in the depths of a depression, the supermarket could profitably sell goods at prices which would be ruinous to a conventional store.

It was this simple fact which A&P would not recognize. Their blindness is astounding, because as far back as 1927, at a divisional president's meeting, the chairman of the Southern Division reported that 'they had adopted the self-service method in some stores in Texas and it has proved successful'. And in 1928 several divisions had experimented with large self-service stores . . . [drawing the conclusion] that large economies were possible . . . and that there ought to be such experimenting in all divisions. Nothing more was heard of these early trials.

'We did not take it very seriously at first', said John Hartford [President of A&P], and this attitude was general among chain stores . . . As late as 1935, the supermarkets were still being exorcised by incantation.

Extracted from: M.A. Adelman (1959): *A&P: a study in price–cost behaviour and public policy*. Cambridge, MA: Harvard University Press, 59–64.

Gradually, however, during the late 1930s and 1940s, the major US chains responded to the challenge and began converting to the supermarket format. By May 1938, for example, A&P had opened over 500 supermarkets. These constituted only 5 per cent of the firm's stores but they accounted for 23 per cent of sales and, most significantly, almost half its profits (Adelman, 1959: 76). The lessons were absorbed, and by the end of 1941 A&P had closed over 9000 (60 per cent) of its stores, replacing them with just 1600 supermarkets – in the process regaining its profitability and share of the US market (Adelman, 1959: 91). It was a story which would be repeated time and time again as this 'unequivocally American invention . . . [and the] other features of twentieth-century consumer culture' (Bowlby, 1997: 97) which went with it, was exported first to Europe and then elsewhere.

However, the spread of larger-format food retailing and the many organizational and technological developments incorporated into the supermarket is a much more recent phenomenon than many of our readers, from an early twenty-first century perspective, might imagine. As US supermarkets in the late 1940s and 1950s began increasing in size and adding more parking, shopping carts, high-speed cash registers with automatic conveyor belts, air-conditioning, electronic-eye doors, pre-wrapped self-service meats, and so on, the food stores of a Europe recovering from the effects of the Second World War remained locked in another age. In Britain, for example, during the 1950s food retailing continued to be dominated not only, as we saw in Chapter 2, by single-outlet independent retailers but also by very small stores operating in a full-service/counter-service format reminiscent of the US food stores of the early twentieth century. Self-service food retailing developed very slowly in the UK during the 1950s, and the supermarket format even more slowly. Indeed, by the late 1950s, only 175 food stores out of more than 150,000 exceeded 2000 sq. ft. of selling space – the minimum criterion for a supermarket. Allied Suppliers, which controlled several of the leading chains of the period – and which as we saw in Table 2.3 was ultimately absorbed into the Argyll Group (Safeway) during the 1980s – was typical in operating only 12 stores of that size by 1959 (Mathias, 1967: 394). As late as the early 1960s, there were less than 600 supermarkets in the UK, accounting for under 4 per cent of food store sales (McClelland, 1962: 135) – compared to an equivalent figure of almost 60 per cent in the USA. Essentially, it was not until the 1960s and early 1970s then that this initial type of larger-format food retailing and the organizational and technological developments which accompanied it – for example changes in labour relations – began to transform the industry in the UK. Even then, UK supermarkets remained significantly smaller on average than their US equivalents.

Superstores and supercentres

Although the development of supermarkets in the UK had lagged behind the USA by 30 years, the next upward shift in format size, from conventional supermarkets to *superstores* – usually defined as single-level, self-service stores of at least 25,000 sq. ft. sales area, selling a wide range of food and related items and supported by extensive car parking – occurred rapidly and, during the 1980s, in parallel with equivalent developments in the USA. As Figure 4.1 demonstrates for the case of Tesco – one of the newly emergent chains of the 1960s which pioneered self-service supermarket retailing in the UK – by the late 1970s the average size of its stores began to accelerate upwards and the total number of stores operated began to decline sharply as the firm experimented with prototype superstores, matching earlier developments of the format in the UK by Asda.

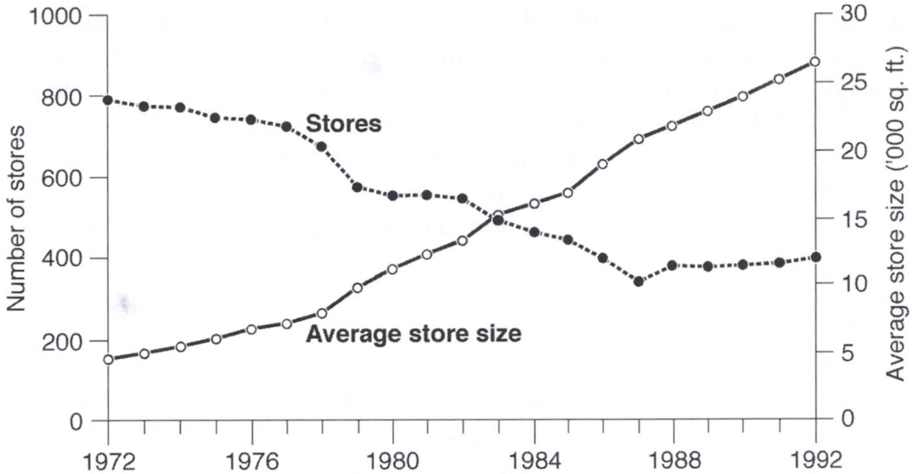

Figure 4.1: *Tesco, number of stores and average store size 1972–92*
Source: Adapted from Smith and Sparks, (1993).

During the 1980s, all the major UK food retailers became heavily involved in superstore development, increasingly on edge-of-city sites. Indeed as Figure 4.2 demonstrates, the pace of such development became ever more rapid, reaching a peak by the end of the decade. By the early 1990s (see Figure 4.1) the average size of Tesco's stores had risen to over 26,000 sq. ft. in sales area, from

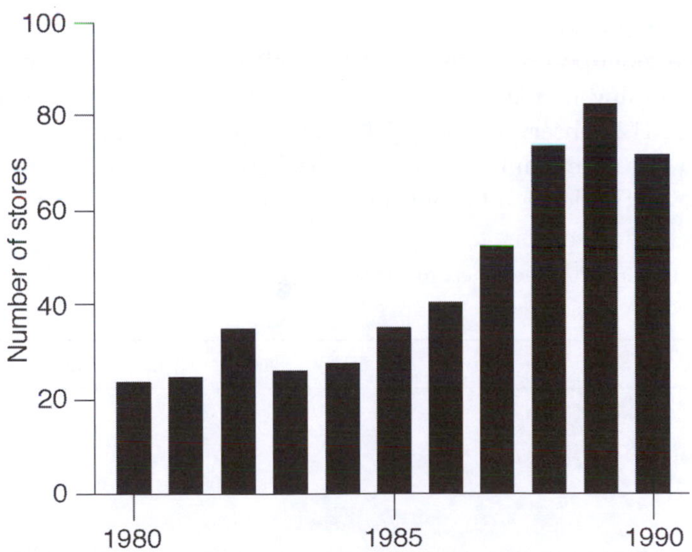

Figure 4.2: *Food superstore openings in the UK, 1980–90*
Source: Adapted from Guy (1994).

just 11,000 sq. ft. in 1980 and under 5000 sq. ft. in the early 1970s. At the same time, new Tesco stores under construction averaged 40,000 sq. ft. These larger superstores, were constructed on highly accessible sites, with significant amounts of dedicated customer parking, simple/segregated delivery vehicle approaches and, typically, had negligible catchment-area competition from rival stores of the same vintage. In practice, they became one of the most important forces powering the rapidly increasing sales and profit levels of the major UK food retailers during the late 1980s and early 1990s (see Table 2.4). In turn this created a frantic store-building boom, intense competition for development sites, and an era of *store wars*.

Just as in the case of the early US supermarkets of the 1930s, the superstores built during the 1980s in the UK were able to operate at both lower cost as a percentage of sales and with higher profit margins than the smaller supermarkets they replaced. As shown in Table 4.1 for the case of Sainsbury stores, and confirmed in a study of another of the major UK food retailers by Shaw et al. (1989), it was particularly in the area of wage costs where the economies of scale were most pronounced (replicating Adelman's findings on A&P's early US supermarket operations in the late 1930s). The higher profit margins of the superstores shown in Table 4.1 came in part from these lower relative operating costs, but also in part from the higher sales densities (sales per sq. ft.) which they achieved. In turn, these higher sales densities were a reflection of three factors. First, the ability to stock within the larger superstores a much wider product range, with more 'value-added' lines. Second, a higher average 'customer spend' within the superstores – the latter being a function of both consumer preference for such stores and the advantages of 'one-stop' shopping which the superstores offered. Third, the local spatial monopolies which were effectively conferred, by the operation of land-use planning regulation, on the superstores which were developed. As Moir (1990: 112) observed about UK superstore development at the time: 'ensconced and dominant in their own markets, the new large stores [of the 1980s] enjoy[ed] a degree of protection from all'.

Table 4.1: *Sainsbury UK supermarkets: comparative wage costs, sales intensity and operating margins by store size group, 1990*

Store size group	Wage costs (as % of sales)	Sales per sq. ft.	Operating margins
<15,000 ft^2	124	95	66
15–25,000 ft^2	102	95	96
>25,000 ft^2	91	108	114
JS average	100	100	100

Index base 100 = company average of established supermarkets.
Source: adapted from Richards and MacNeary (1991).

In the USA during the 1980s, and driven by a similar economic rationale, superstores likewise increasingly replaced conventional supermarkets – albeit that conventional US supermarkets were typically somewhat larger and had been part of the retail structure for much longer than their UK equivalents. Table 4.2 reveals, using US Bureau of the Census figures, just how dramatic that process was – with conventional supermarkets' share of sales plummeting from 73 per cent to just 28 per cent by the early 1990s. During the same period it was superstores and 'combination' superstores – larger superstores which also included a pharmacy and a more extensive health and beauty (drug-store product) section – which increased their market share most substantially, accounting for more than 50 per cent of US supermarket sales by the early 1990s.

Outside the UK and USA, similar shifts to larger-format food retailing were also taking place during the 1980s. But in continental Europe, notably in France, Germany, Belgium and Spain, the shift was to the even larger form of the *hypermarket*. These were very large superstores – usually defined as being twice as large as the smallest superstore, that is to say, with more than 50–55,000 sq. ft. (5000 sq. metres) of sales area. In addition, they offered a large selection of non-food items, typically accounting for around 35 per cent of sales, alongside food. This large-scale food/non-food format proved very successful, and the French retailer Carrefour, in particular, began to export it successfully to countries outside western Europe – initially to Brazil, Argentina and Taiwan, and subsequently to other emerging markets in south-east Asia, China, Latin America and central/eastern Europe. By the 1990s, the hypermarket format had become the primary vehicle for the entry of 'modern' western-style retailing into these emerging markets, and several of the leading European food retailers including Carrefour, Ahold, Promodès, Casino and Tesco were actively involved. However, the hypermarket format was not universally successful.

In the UK, despite extensive experiments such as Sainsbury's Savacentre hypermarkets, and a certain blurring of formats – for example, new Asda

Table 4.2: *US food retailing: share of supermarket sales by store format, 1980 and 1993*

	Share of supermarket sales	
Format	1980	1993
Conventional supermarket	73.1	28.0
Superstore/Combination superstore	21.7	52.1
Limited assortment/Warehouse[a]	5.2	16.8
Hypermarket	–	3.1

[a] Discount supermarkets with limited product variety and few services.

Source: adapted from Connor and Schiek (1997: 328), based on US Bureau of the Census data.

stores which were often developed at the threshold of the 'hypermarket' size and non-food mix criteria – development of the format was initially restricted for two reasons. First, because it proved easier to develop the food superstore rather than the hypermarket format under existing UK land-use planning regulations. Second, because early experience with the hypermarket format did not suggest either significant operating-cost advantages or enhanced consumer acceptance. Nevertheless, by the late 1990s, and despite tightened planning regulations, the market leader Tesco was actively experimenting with the format in the UK via its development of 90,000 sq. ft. Tesco 'Extra' stores. In the USA, as Table 4.2 indicates, and for rather complex reasons, related in part to the highly leveraged nature of the US food retail industry in the late 1980s (see Reading 2.4), the hypermarket format was similarly not taken up to any major extent by the leading firms in that industry. However, during the 1990s in another guise, that of the *supercentre*, and driven by a different group of retailers – the discount mass merchandisers, Wal-Mart, Kmart, etc. – larger-format, hypermarket-type retailing did ultimately begin to explode into growth in the USA.

The supercentres which began to expand in the early 1990s in the USA are essentially a combination of the discount general merchandise store, which Wal-Mart and Kmart had rolled out across the USA during the 1970s and 1980s, and a food superstore. Ranging in size from 100,000 to over 200,000 sq. ft., the most common type being approximately 190,000 sq. ft., they are stores in which the split between food and non-food products is the reverse of that typically found in a continental-European hypermarket. That is to say, around one quarter of the floor space is devoted to food, which in turn generates about 40 per cent of the store's sales, with the remaining 60 per cent being non-food sales (compared to just 30–35 per cent in the case of the hypermarket). Although pioneered in the 1980s by the food retailers Meijer in Michigan and Fred Meyer in the Pacific north-west, it was not until 1988, when Wal-Mart first began to experiment with the format, and essentially from 1991 onwards when it began to accelerate its development of such stores, that supercentres emerged as the fastest growing sector in US retailing – with growth rates triple those of the second fastest growing sector. As Table 4.3 shows, by 1999 around 1100 supercentres had opened, often through a process of conversion of existing smaller discount stores, with Wal-Mart having increased its supercentres from just 10 in 1991 to 721. Best estimates currently suggest supercentre sales increasing to $99 billion per annum from over 1500 stores during 2001, with Wal-Mart accounting for over two-thirds of these totals.

Why then did this ultra-large-format retailing suddenly explode into growth in this way during the 1990s, when the hypermarket format operated by retailers such as Carrefour had essentially failed to take off in the USA in the 1980s? The reasons lie in the slowing rates of growth experienced (and more importantly forecast) by discount mass merchandisers in the late 1980s, as

Table 4.3: *Growth of the US supercentre format, 1991–2001*

		Actual									Estimated	
		1991	1992	1993	1994	1995	1996	1997	1998	1999	2000E	2001E
Total US supercentres		225	270	316	463	601	730	811	943	1117	1318	1543
Of which:	Wal-Mart	10	34	72	147	239	344	441	564	721	886	1051
	Super Kmart	1	5	19	67	87	96	99	102	105	110	130
	Meijer	65	69	76	86	100	108	113	118	127	137	147
	Fred Meyer	70	70	77	86	95	109	130	130	133	138	144
	Target	–	–	–	–	2	8	13	14	16	31	56
Sales US supercentres ($bill)		9.3	10.3	13.5	19.0	26.9	34.6	43.4	53.3	65.9	81.3	99.0

Source: adapted from Merrill Lynch (1998a, 2000; see also Wrigley 2001a).

these giants of US and world retailing faced increasingly saturated markets for continued expansion of their traditional discount stores, together with the attractive economics of the supercentre format in comparison to those of the discount general merchandise store.

Curiously, although supercentres cost significantly more to build and operate than a traditional Wal-Mart or Kmart discount store, and although the added merchandise, primarily food, is sold at lower profit margins than the other merchandise in the store, supercentres as a whole generate better returns than the traditional discount stores which they have increasingly replaced. The keys to the success of the supercentres are, first, the increased frequency of consumer visits (relative to the traditional discount store) which food shopping generates and, second, the *cross-shopping* from food to non-food products which occurs on those visits, and which has been found to increase sales of the non-food merchandise by 25 per cent from the levels experienced in traditional Wal-Mart and Kmart discount stores. The combined result is that even if supercentres only manage to break even on the food portion of their business they can still generate better returns than traditional discount stores – a situation reminiscent of the economic advantages which Adelman in Reading 4.2 suggests the prototype supermarkets of the 1920s and 1930s held over the conventional food stores of the early twentieth century. Yet a firm such as Wal-Mart which, as we will see below, has led the US retail industry in terms of its adoption of innovations in centralized buying, logistics and IT systems, has the technological efficiency to ensure that its food sales are in fact significantly profitable in their own right – strengthening the already considerable economic advantages of the supercentre format. This then is the 'one-stop shopping' logic which drove the success of the UK superstores and the French hypermarkets of the 1980s taken to extreme levels, and given additional impetus by the particular circumstances of dominant retailers facing market maturity in their particular retail sector and seeking a new growth vehicle.

On the face of it then, the success of the giant supercentres in the USA during the 1990s appears to be just one more example of the trend towards ever larger format retailing which, as we have seen, has dominated so much of the twentieth century. However, it is vital to insert words of caution concerning attempts to extrapolate this trend. Supercentres of 190,000 sq. ft., requiring 15–20 acre development sites and extensive trade/catchment areas, face considerable difficulties in their attempts to expand into the urban and inner suburban areas of the major US metropolitan markets where suitable real estate is both scarce and expensive. In addition, in several regions of the US – notably New England – supercentres face increasingly vocal and effective community resistance, and significant difficulties in obtaining zoning approval. As a result, during the late 1990s, Wal-Mart was forced to develop a smaller, more flexible, supercentre format of 109,000 sq. ft. and, in addition, began to experiment with superstore-sized (i.e. 40,000 sq. ft.) 'small marts' or 'neighbourhood markets'.

Elsewhere, similar trends were also characteristic of the late 1990s. In several western European countries, tightened land-use planning regulation – for example, the *loi Rafferin* which in France reduced the size limit of stores requiring specific planning authorization to just 300 sq. metres (approximately 3250 sq. ft.), the revised *Planning Policy Guidance (PPG) Note 6* in the UK, and the *Ley de Ordenación del Commercio* in Spain, all introduced in 1996 (see Chapter 7) – made the development of large-format retailing much more difficult. In addition, tightened land-use planning regulation was complemented by related legislation – e.g. the *loi Galland* in France and parts of the *Ley de Ordenación del Commercio* in Spain – also introduced in 1996/7, which regulates certain aspects of the relationship between retailers and suppliers, restricts 'loss leader' selling by hypermarkets, and helps the smaller supermarket format to compete on a more equal footing. Despite their success, it is by no means certain, therefore, that the twenty-first century will be the exclusive era of the giant supercentres and hypermarkets – particularly in developed western economies. Strong counter-trends – backed by regulation in several countries – characterized the late 1990s, and smaller-format retailing has begun to re-emerge as an important growth vehicle for some of the world's most important retail corporations.

The revolution in distribution and systems

As several geographers turned business school professors – notably Leigh Sparks, John Fernie and Alan McKinnon – have reminded social scientists, the task of ensuring product availability lies at the very core of retailing. Retailers must inevitably be concerned with planning, coordinating and controlling the flows of product and information into their companies in order to get the right products in the right place at the right time. It follows, therefore, that any 'retail revolution' worthy of its name must, by default, be accompanied by a 'supporting logistics (or distribution) revolution' (Sparks, 1994: 310). Traditionally the tasks involved in that process – managing storage, inventory levels, transportation and so on – were referred to as 'physical distribution management'. However, as Sparks (1994: 313) argues, it became increasingly common during the 1980s to see that term replaced by the phrase 'logistics management', defined (after Christopher, 1986) as the process of strategically managing the supply chain, and the movement and storage of inventory from suppliers, through the firm, and on to customers. That is to say, 'management of the physical flow which begins with sources of supply and ends at the point of consumption' (Christopher, 1986: 1).

In this section we will consider in detail one example – the revolution in physical distribution, IT systems and supply-chain management which swept through UK food retailing during the late 1970s and 1980s. We will then compare the logistically refined 'demand pull' system which emerged from that revolution in the UK to the system operating in the US food retail industry during the same period. Finally, we will consider technological transformations of the late 1990s – what questions were posed by the emergence of new electronic non-store channels to market (questions we will explore in greater detail in Chapter 13), and what, in particular, does that imply in terms of retail distribution?

Seizing control of the supply chain: the logistical transformation of UK food retailing

In Chapter 2 a brief history of the rapid rise to prominence during the 1980s of a small group of UK food retail corporations was provided. Underpinning the rise of those firms, and the escalating operating profit margins which characterized the period (see Table 2.4) was a supporting revolution in distribution/logistics, IT systems and supply-chain management. Logistics and systems became a significant frontier of cost-control activity and enhanced profitability in UK food retailing, and the major firms 'progressed from simply being the innocent recipients of manufacturers' transport and storage whims, to controlling and organizing the supply chain, almost in its entirety' (Sparks, 1994: 331).

In practice, and beginning in earnest during the 1970s, that process involved taking the supply of stores out of the hands of the food manufacturers. Centralized distribution systems were developed by the retailers and, as the 1980s progressed, these increasingly became 'just-in-time' or more precisely '*quick response*' in nature (McKinnon, 1989; Smith and Sparks, 1993; Fernie, 1994; Walker, 1994). Increasing amounts, eventually over 90 per cent, of the major retailers' products were channelled via a network of purpose-built, strategically positioned regional distribution centres (RDCs), through which deliveries to stores were consolidated and optimized. Increasingly, many of these RDCs took the form of compartmentalized, multi-temperature 'composite' facilities which allowed the storage and onward distribution of products via specially designed delivery vehicles at a range of temperatures. In addition, for a number of reasons, not least to improve the return on capital employed by the retailers, operation of both the RDCs and delivery to stores increasingly became 'contracted out' to specialist logistics companies such as Tibbett and Britten and Christian Salvesen (Fernie, 1992).

The development of dedicated RDCs and retailer-led distribution systems within an increasingly concentrated food retail industry effectively eliminated the food wholesaler sector and dramatically transformed pre-existing food distribution networks. In addition, by permitting shorter and more predictable delivery lead times (frequently below 24 hours), a significant reduction in the so-called '*conversion ratio*' (the ratio of warehouse to sales space within stores) was facilitated. In turn, pressure was placed on food manufacturers to develop just-in-time, fixed-performance-specification, production and factory-to-RDC systems to match. The result was a significant and progressive reduction in retailer inventory holdings, and the amount of capital tied up in those holdings throughout the 1980s (Figure 4.3 shows the case of Tesco), as more and more of the costs (including 'uncertainty' costs) of inventory holding were passed back to manufacturers/suppliers. By the early 1990s, 'nil' inventory levels for fast-moving, short shelf-life products had become increasingly common. In essence, stock only existed in transit in the supply chain or on the shelves of the stores (Fernie, 1994). Table 4.4, from Sparks (1994: 321) after Cooper et al. (1991: 109), provides a useful summary of these developments in UK food retail logistics, stretching from the introduction of the RDCs in the 1970s to the 'quick response' systems of the late 1980s/early 1990s.

As the 1980s progressed, the centrally controlled distribution and stock-control systems of the UK food retailers provided fertile ground for the adoption of increasingly sophisticated computer-based IT systems. By the end of the decade fully integrated systems, in which EPOS (*electronic point of sale*) scanning information was used to automate reordering and the linkage between within-store stock levels, the warehouse/distribution network, and central administrative/financial functions, had become the norm. Moreover, these systems were increasingly linked back via EDI (*electronic data interchange*) into the computer systems of manufacturers/suppliers, permitting 'paperless' supply-chain control and sophisticated sales-based ordering and tracking. By the early 1990s, the EDI links of the major food retailers frequently covered more than 80 per cent of merchandise and lay at the heart of a 'demand pull' system of supply-chain management which included increasingly refined

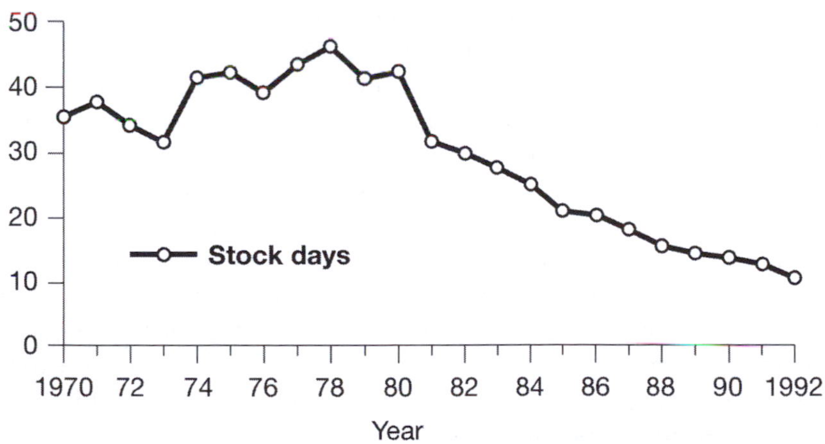

Figure 4.3: *Reductions in inventory holdings in the 1980s: the case of Tesco*
Source: Redrawn from Smith and Sparks (1993).

Table 4.4: *Developments in UK food retail logistics, 1960s to 1980s*

	Problem	Innovation	Consequences
1960s and 1970s	Disorderly delivery by suppliers to supermarkets; queues of vehicles led to both inefficiency and disruption	Introduction of regional distribution centres (RDCs) to channel goods from suppliers to supermarkets operated by retailers	Strict timing of supplier delivery to RDC imposed by retailer; retailer builds and operates RDC; retailer operates own delivery fleet between RDC and supermarkets within its catchment area
Early 1980s	Retailers becoming too committed to operating logistics services in support of retail activity	Operations of retailer-owned RDCs and vehicle fleets 'contracted out' to specialist logistics companies	Retailer can concentrate on 'core business' of retailing; retailer achieves better financial return from capital invested in supermarkets than in RDCs and vehicles
Mid-1980s	Available floorspace at retail outlets being under-used; too much floorspace used for storage	Conversion of storage floorspace at supermarkets to sales floorspace	Better sales revenue potential at retail outlets; RDCs absorb products formerly kept in store at supermarkets; just-in-time (JIT) delivery used from RDC to replenish supermarket shelves

Source: adapted from Sparks (1994: 321) after Cooper et al. (1991: 109).

waste/shrinkage control, activity-based costing, and space/category management elements. In this system, the RDCs of the 1970s and 1980s were rapidly evolving into '*cross-docking*' facilities in which there was little actual storage and 'repicking'. Instead, they were becoming mere automated marshalling or sorting facilities in which products pre-labelled by the manufacturers with store-specific stock control information entered from one side, passed through the RDC on conveyor belts, and were automatically sorted into store-specific output lanes on the other side for onward delivery.

There are several interesting and useful accounts of this logistical transformation which underpinned the retail revolution of the 1980s in the UK. Readings 4.3 and 4.4 provide two examples. The first of these is extracted from a wide-ranging essay on retail change and retail logistics by John Fernie. It provides a summary of many of the issues discussed above, characterizing the evolution of retail logistics in the UK from the 1960s to the 1980s as a three-stage process.

Reading 4.3 – Three stages of evolution in UK retail logistics

The nature of retail change ... especially the locational change, has transformed the way in which products are delivered to retail outlets. The healthy net margins enjoyed by British retailers have been partly attributed to the efficient logistical support to stores which has evolved over the last 20 years. This evolution is characterised by three main stages: a supplier-driven stage, the creation of retailer regional distribution systems in stage two, and a final shift to a retailer-controlled logistical stage ...

Until the 1970s most British retailers received products direct from the manufacturer from either their factories or one of the numerous field warehouses strategically located to supply hundreds of stores throughout the country. Store inventory levels were controlled by branch managers who bought direct from manufacturers, often with the assistance of sales representatives. Lead times were long and much stock was held in the backroom of the store. This meant that improvements in stock turn were difficult to achieve; indeed, if demand for particular lines were high an out-of-stock situation would occur because of the relative inflexibility of suppliers' deliveries and the need to redistribute stock around branches.

In order to improve stock availability and gain greater control over the supply chain, retailers began to centralise their distribution through the construction of large warehouses for the receipt of suppliers' products. Companies such as J. Sainsbury and Boots had centralised much of their stock by the early 1970s and other companies followed with the laggards, the DIY and newsagent chains, carrying out such schemes in the 1990s. Retailers negotiated volume discounts with suppliers, invariably on ex-works conditions, and consolidated stock at warehouses prior to replenishing stock at stores at greatly reduced lead times. Most of the innovation in retail logistics was carried out by the high volume operators, such as the food retailers and the mixed retail businesses. The food retailers, in particular, developed the concept of composite distribution which is unique to the UK. Composites are large, multi-compartmentalised, distribution centres which allow the storage of products at different temperature ranges under the one roof. Similarly, the trucks are also able to carry mixed product ranges when they deliver to stores. This produced a streamlined, efficient system because the superstore was the predominant trading format in the UK.

Over 80 per cent of the stock of British retailers is now centralised and in the case of the food retailers this figure is as high as 98 per cent in some companies. This development has dramatically transformed the [UK] food distribution network ... Food manufacturers' networks have been decimated with the construction of composites and other regional distribution centres, [conversely] a large new market was created in the provision of distribution services. Throughout the 1980s when major investments were made into logistical networks, British logistics providers such as NFC, Tibbett and Britten, Wincanton and Christian Salvesen won sizeable contracts to run this support function for retailers.

The final stage of logistical development is still under way as retailers continue to control the supply chain. In stage two stock was moved from the back room of stores to distribution centres; lead times were reduced but stock was just pushed further up the supply chain. Now retailers are moving towards a JIT system for the replenishment of stock. As sales-based ordering increases, products are delivered in smaller quantities from factories and are cross-docked across distribution centres.

Extracted from: John Fernie (1997): Retail change and retail logistics in the United Kingdom: Past trends and future prospects. *Service Industries Journal*, 17, 383–96.

Reading 4.4 takes up many of the themes outlined by Fernie but, in this case, examines them through the lens of a single firm. The reading is extracted from Smith and Sparks' (1993) detailed case study of the changes in Tesco's distribution system from the 1970s to the early 1990s. Smith and Sparks show how the traditional direct-to-store delivery system which the company had used during the 1960s and 1970s, in which 80 per cent of supplies were channelled direct from the manufacturers, came perilously close to collapsing during a major sales drive by the company in the late 1970s, and how, for the first time in its history, Tesco recognized that 'it was as much in the business of distribution as of retailing' (Powell, 1991: 184). They show that, as a result, in 1980 Tesco took the decision to adopt a centrally controlled distribution system. Ten years later, over 90 per cent of Tesco's supplies were passing through its RDCs, and the company was concentrating on developing 'composite' distribution based around a network of just eight multi-temperature RDCs. Reading 4.4 extracts from Smith and Sparks' case study at that point.

Reading 4.4 – Delivering quality: Tesco's logistics system by the early 1990s

The strategy of composite distribution was planned in the 1980s to take place in the 1990s . . . Each of the eight distribution centres service a region of the country and approximately 50 stores. The sites are all close to key motorway intersections or junctions which allow rapid access. Of the eight composite distribution centres only two are run by Tesco. The remainder are operated by specialist distribution companies with Glass Glover running three, Exel Logistics two and Hays Distribution one. Cross-comparisons of performance of these centres and the subcontractors enable 'league tables' to be drawn up of performance.

The composite centres are linked by computer to head office to allow the passing of data and the imposition of monitoring and control. For all products handled by the composite centres, forecasts of demand are produced and transmitted to suppliers. The aim of the system is to allow suppliers to have a basis for preparing products. This is particularly important for short-life products where the aim is to operate a just-in-time system from the factory through the composite centre to the store. To meet such targets on delivery, etc., each supplier needs information on predicted replenishment schedules.

This sharing of information is part of a wider introduction of electronic trading to Tesco. In particular, Tesco has built a Tradanet [EDI] community . . . of over 900 suppliers – the largest in European food retailing. Improvements to scanning in stores and the introduction of sales-based ordering has enabled Tesco better to understand and manage ordering and replenishment. Sales-based ordering automatically calculates store replenishment requirements based on item sales and generates orders for delivery to stores within 24 to 48 hours. This information is used via Tradanet to help suppliers plan ahead in both production and distribution.

Composite distribution provides a number of benefits. Some derive from the process of centralization of which composite is an extension. Others are more directly attributable to the nature of composite. First, the movement to daily deliveries of composite product groups to all stores in 'waves' provides an opportunity to reduce the levels of stock held at the stores and indeed to reduce or obviate the need for storage facilities at store level ... The second benefit of composite is an improvement to quality with a consequent reduction in wastage. Products reach the stores in a more desirable condition. Better forecasting systems minimize lost sales due to out of stocks ... sales-based ordering produces more accurate store orders, and more rigorous application of code control results in longer shelf life on delivery which in turn enables a reduction in wastage ... Third, the introduction of composite provides added benefit in productivity terms ... one vehicle can be used instead of the five needed in the old network. The result is reduced capital costs and less congestion at the store ... Fundamentally, the move to composite has led to the centralization of more product groups, the reduction of stock holding and its movement up the channel, information sharing via Tradanet, the reduction of order lead times, and better code control for critical products.

Extracted from: David L. G. Smith and Leigh Sparks (1993): The transformation of physical distribution in retailing: the example of Tesco plc. *International Review of Retail, Distribution and Consumer Research*, 3, 35–64.

The UK system in international perspective: contrasts with the USA

The common theme of Readings 4.3 and 4.4 and the wider literature which discusses these issues is that, in terms of integrated logistics and supply-chain management, the major UK retailers were, by the late 1980s/early 1990s, exercising an unusual degree of control over the supply chain when viewed in international terms. 'Quick response' distribution systems based on extensive EDI networks linking the major food retailers and their suppliers, and centred around purpose-built 'composite' distribution centres with rapidly evolving 'cross-docking' characteristics, were not the norm at that time in other countries (see Fernie, 1995).

In the USA, for example, because of the highly leveraged nature of the food retail industry in the late 1980s (see Reading 2.4), and the debt burdens and capital expenditure constraints which that placed on the leading firms, investment in UK-type centralized, logistically refined, 'demand pull' systems of supply-chain management was notably absent. Indeed, it was a period which even sympathetic US analysts of the industry have described as the 'dark ages of procurement' (Comeau, 1995: 12) in which forward buying opportunities and manufacturers' promotions essentially drove the system. Rather than stock being 'pulled' through the supply chain as in the UK system, with progressive reductions in retailer inventory holdings, in the USA during this period inventory was effectively 'pushed' into the retailers' warehouses by special promotional deals offered to retailers by the manufacturers and by the forward buying (buying ahead of consumer demand) and 'diverting' (passing surplus stock from one retail company to another to take advantage of the trade promotions) which this induced. Trade promotions of this type increased significantly in the USA during the late 1980s/early 1990s as cash-hungry LBO retailers, under pressure to service their debt burdens, sought to extract increased financial support of many different forms, including enhanced 'slotting fees' and 'in-house brokerage' commissions – see Marion (1995) and Hughes (1996b: 292–301, 312–16) – from the food manufacturers. The result, as Fernie (1994: 40) points out, was that these practices caused significant 'peaks and troughs in the volume of product passing through the supply chain [and] led to excessive inventories and inefficient utilization of warehouse and transport capacity' in the US food retail industry.

During the early 1990s, a number of well-known reports were commissioned and written concerned with improving the efficiency of US supply-chain management. Figure 4.4 is adapted from one of these, a report by consultants Kurt Salmon (1993) which suggested that within the dry grocery supply chain in the US, because of the fragmentation of the chain, it took an average of 104 days for products to pass from the suppliers' packing line to the consumer. The report suggested that by integrating the supply-chain, using what it termed *'efficient consumer response'* (ECR) strategies – in effect moving towards UK-type 'demand pull' systems of stock replenishment – throughtput time could be cut down to an average of 61 days. This represented a significant improvement, but it was still very inefficient in comparison to supply-chain

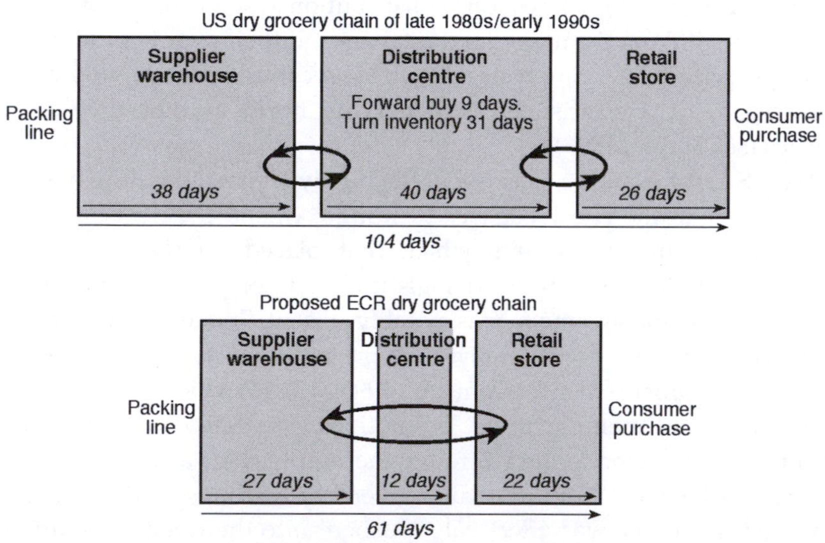

Figure 4.4: *US food retailing: comparison of unlinked (late 1980s) and integrated replenishment in the supply chain*
Source: Adapted from Fernie (1994) after Kurt Salmon (1993).

standards being achieved both in the UK and by Wal-Mart, the leading innovator in US retail logistics and systems.

During the mid-1990s, the leading multi-regional US food retailers – many of them by that time significantly less burdened by LBO debt (see Wrigley, 1999b) – began to invest heavily in improving their distribution/logistics, IT systems, and supply-chain management, and in centralizing the fragmented divisional-based operating structures which they had in the 1980s and early 1990s. Some of the attempts to remove inefficiencies from the supply chain, and to exploit synergies in administration, the buying process and logistics via the use of centralized systems were highly ambitious. American Stores' 'Delta' programme (Figure 4.5), initiated in 1994 and moving towards completion at the time its merger with Albertson's was announced in 1998, provides one of the best examples. But even the most devolved of the major multi-regional US food retailers were heavily involved by the mid- to late 1990s in many forms of centralized inventory management, coordinated buying, and systems consolidation. Integrated logistics programmes were beginning to produce significant improvements in warehouse efficiency, rationalization of distribution centre networks (including the development of specialized distribution centres devoted purely to slow-moving merchandise), and the emergence of

				Planned		
	1994	1995	1996	1997	1998	1999
Administrative consolidation						
• Centralized procurement			■	■		
• Logistics management			■	■		
• Information technology, real estate, accounting and legal	■	■				
Information infrastructure						
• Common item descriptions			■	■	■	
• Procurement systems			■	■		
• Retail price management system				■	■	■
• Centralized category management			■	■	■	
Supply chain						
• Dallas warehouse systems	■	■	■			
• Warehouse replenishment systems			■	■	■	
• Electronic data interchange (EDI)			■	■	■	
• Warehouse consolidation			■	■		
Store operations						
• Labour scheduling system			■	■	■	
• Store replenishment system				■	■	■

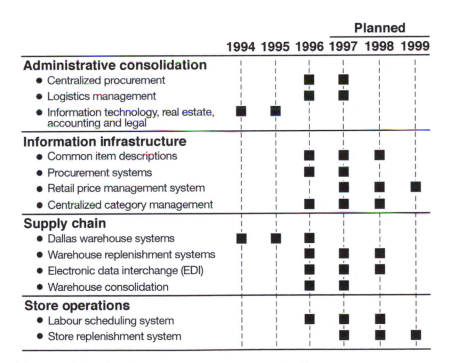

Figure 4.5: *American Stores' 'Delta' programme. Timetable and progress, 1994–99, prior to merger with Albertson's*
Source: Adapted from American Stores 1997 Fact Book.

distribution outsourcing ('contracting out') agreements with specialist logistics companies – for example, Kroger's agreement with Tibbett and Britten in 1997. US food retailers were still not as advanced in terms of the adoption of quick response systems of stock replenishment as their UK counterparts and, more significantly, lagged behind the leading innovators in US retail logistics and systems such as Wal-Mart or even Lowe's, the home-improvement retailer which had developed extensive cross-docking systems in its warehouses. Nevertheless, progressive elimination of the forward buying culture of the 1980s/early 1990s, significant reductions in supply-chain throughput times and average inventory levels, and progress towards a UK-type 'demand pull' system, had been achieved in the US food retail industry during the decade.

The emergence and logistical implications of electronic non-store retailing

Finally in this chapter we turn, briefly, to those technological transformations of the late 1990s associated with the rise of electronic non-store retailing, e-commerce, which many analysts regard as having the potential to reshape the contemporary retail landscape. In Chapter 13 we consider these new electronic channels to market at greater length – positioning our discussion within the wider context and history of home shopping and of the home as a site of contemporary consumption. Here our focus is more simply on the retail logistics and organizational implications. Will the Internet provide, as some have suggested, 'the ultimate technology for shortening many channels of distribution between the supplier and the consumer' (Jones and Biasiotto, 1999: 77)? Will traditional store-based retailers, particularly those with large property portfolios and the potential sunk costs which those portfolios might represent, be sufficiently flexible to take advantage of the opportunities which the new electronic channels to market represent? And, if electronic non-store retailing takes off in the twenty-first century, what will be the nature of the retail logistics changes necessary to support this development – will 'a new logistical network based on [something similar to] the current mail-order system become the norm' (Fernie, 1997: 395)?

Estimates of the total volume of e-commerce retail sales on the Internet place them at around $1 billion in 1996 and between $7 and $8 billion in 1998: that is to say, a rather insignificant share of total retail sales, which in the USA alone were $2.7 trillion in 1998, and were dwarfed even by the retail sales of single sectors, for example the $441 billion of total US food store sales in 1998. Projections (Jupiter Communications, in Merrill Lynch, 1999a) suggest that Internet retail sales might grow to between $40 and $45 billion by 2002 – a fivefold increase in just four years – although less conservative estimates place that figure at a much higher level. The bulk of those Internet retail sales will be generated in the US market where household personal computer ownership levels in the late 1990s were significantly higher than in other countries (40 per cent of households compared to 22 per cent in the UK and 12 per cent in France) and where the number of Internet users is predicted to grow from 29 million in 1998 to almost 50 million in 2002. Best estimates (Forester Research in Merrill Lynch, 1999a) suggest that US Internet retail sales accounted for a 79 per cent market share of the world's e-commerce in 1998, followed by western Europe with 9 per cent and Japan and Canada at 4 per cent each.

Traditional store-based ('bricks and mortar') retailers were initially rather slow to move into e-commerce, with Jones and Biasiotto (1999: 77) reporting that in 1997/8 'most of the leading retail websites are associated with suppliers/manufacturers (e.g. Dell, Compaq, Microsoft) or non-store, electronic retailers (e.g. Amazon Books) and/or catalogue merchants (e.g. Lands' End)'. By the end of the 1990s, however, store-based retailers were either gearing up to exploit, or moving rapidly into, these new electronic channels to market, having learned from the mistakes of the first movers into the market and how to avoid 'the very heavy losses that the first movers incurred to create awareness of e-commerce activity' (Merrill Lynch, 1999a: 124). What then are the implications of this movement for those organizational and technological transformations of retailing which we have discussed in the chapter? Quite simply, what is the future of the store-based retailing and retail logistics systems we have discussed above in a world in which a greater proportion of products will be purchased 'on line' in the twenty-first century?

A key issue in answering this question relates to the fact that retail distribution currently, and for any foreseeable future, is in essence concerned with providing 'real' (rather than 'virtual') products to consumers – in particular, providing the right products in the right place at the right time. At the heart of e-commerce, therefore, is the logistical problem of what is termed *'fulfilment'*. Products ordered via the Internet must be sent to individual homes – a very expensive process, involving specialized distribution systems and the creation of a network of fulfilment centres, the numbers and locations of which depend on the size and density of the market and the nature/perishability of the products being distributed. As analysts at Merrill Lynch (1999a: 125) have noted:

> Fulfilment is very expensive. So expensive in fact that it is one of the major reasons why the catalogue industry, whose genesis goes back to the late 1800s with the Sears and Montgomery Ward catalogues, only generates about $55 billion today or 9 per cent of total US general merchandise retail sales. Fulfilment costs keep catalogue prices from being significantly different than prices at bricks and mortar retailers. Catalogue companies also do not give you instant gratification of being able to take the item home when you purchase it. Nor do they enable you to touch, feel or wear the item before you buy it. These limitations of the catalogue industry also seem to apply to most e-commerce operations.

In this context, as Fernie (1997: 394) has observed, any suggestion of 'the demise of the shop is premature'. Rather most store-based retailers view e-commerce operations as complementary to their existing businesses, and are likely to develop them as part of a dual strategy in which home shopping channels are used both to extend (or at least protect) their customer franchise and markets and also to bring people into their stores. Indeed, extensive store networks offer major advantages in terms of the fulfilment problems of e-commerce, with the capacity to be used in a fulfilment-centre/distribution role but, more importantly, being available to consumers as sites for 'pick-up' and/or return and exchange of products.

As we have seen earlier in this chapter, major store-based retailers in many countries have increasingly seen themselves as being as much in the business of distribution and inventory management as retail selling. As such, they potentially have considerable competitive advantage over the purely non-store electronic retailers who, during their rapid emergence in the late 1990s, struggled to achieve profitability, not least because of the costs of fulfilment and the challenges and scale requirements of the buying process and inventory management. And it is these issues which lie at the heart of any assessment of the future of the store-based retailing and retail logistics systems described above, during a period in the early twenty-first century when a greater proportion of products can be expected to be purchased 'on line'.

Some geographers/business school professors take the view that, whilst

> like many technological innovations, it may take a generation for Internet shopping to attain mainstream status . . . finally, every innovation creates new industry leaders. The lead firms that emerge are committed to the new technology in a way that firms pre-dating the technology can never be. Unless established, store-based retail organizations invest significant resources and create completely fresh, separate and Internet-dedicated organizations, their ability to take advantage of the market opportunities that the Internet presents will be limited (Jones and Biasiotto, 1999: 78).

Others take a more positive view of the potential of store-based retailers in an era of expanding Internet shopping, arguing that the first-mover competitive advantages of the non-store electronic retailers are confined to those products where distribution costs are low as a percentage of the total cost of the product, which are standardized (i.e. potential quality variability does not have to be assessed by the consumer by inspection and touch), and where the 'sociality' of the purchasing/consumption process (see Chapter 13) is relatively unimportant. These commentators see traditional 'bricks and mortar' retailers as being capable of absorbing the new electronic channels to market as a complementary part of their retail offer. In the process those retailers are likely to exploit the significant competitive advantages offered by the 'fulfilment' potential of their existing store networks and by their distribution, inventory management, buying and branding expertise and, as in the case of the UK food retailers' move into the retailing of financial services during the late 1990s (Alexander and Pollard, 2000), the extensive customer databases which their EPOS systems and loyalty card initiatives have produced. In this view of the future, 'dual distribution systems' (Fernie, 1997: 394) are likely to be created by the traditional retailers with fulfilment centres and fulfilment-centre-to-home delivery systems being grafted onto their existing logistics networks. Indeed, home delivery systems of this type (based on consumer store visits or ordering by telephone/fax) were a rapidly expanding feature of retailing in the late 1990s, independently of the growth of e-commerce. (In the UK, for example, Iceland the frozen food retailer, was able to reinvigorate its corporate strategy in 1998 on the basis of offering such a service.) The extent to which the distribution support function of such fulfilment centre/home delivery systems is likely to be contracted out to specialized logistical providers is, at this stage, unclear. Nevertheless, what is certainly the case is that, however insignificant in terms of total retail sales e-commerce currently is, and however 'defensive' in motivation is the response of the store-based retailers to the growth of electronic channels to market, the challenges posed by that growth ensure, as Fernie (1997) notes, that retail logistics entered a significant new phase at the end of the twentieth century.

5

Changing retail employment relations

Retailing is quintessentially a labour-intensive industry. Indeed, after the cost of buying the merchandise which the retailer sells, the cost of employing staff to run the business and to serve the customers almost invariably represents the retailer's next largest expense. Not surprisingly, the retail sector provides a major source of employment in the advanced capitalist economies on which we focus in this book. In the USA, for example, by the beginning of the 1990s retail employment as measured by the US Department of Labor had outstripped employment in the entire manufacturing sector and accounted for over 20 million workers or approximately 17.5 per cent of total US employment (Marshall and Wood, 1995: 11), whilst in the leading European economies it accounted for around 10–12.5 per cent depending on the precise definitions of the sector used (Marshall and Wood, 1995: 12; Freathy and Sparks, 1996). Despite this, as Susan Christopherson (1996) argued, the 'production of consumption' via retail work and in 'consumption workplaces' remained a relatively neglected topic in academic writing across the social sciences (although there were important exceptions including Benson, 1986). However, in the late 1980s and 1990s, as retailing began to be perceived by increasing numbers of social scientists as 'spearheading the progressive penetration of the market into all areas of . . . life, encouraging and facilitating the spread of consumer culture' (du Gay, 1996: 116), and as having been transformed 'from dull distributive cipher' to 'leading edge sector' (du Gay, 1996: 97), so writers from several disciplinary perspectives began to take a fresh look at retail work and workplaces. They sought to understand not only how employment practices were being 'engineered' by the leading retailers as a source of competitive advantage but, more broadly, how retailers were 'simultaneously making up new ways for people to be, both *inside* and *outside* of the workplace' (du Gay, 1996: 119). It is to these readings of the organizational transformations and 're-imaginings' of retail work, retail employment relations and consumption workplaces that this chapter now turns.

The consequences of labour intensity

Before we can consider those broader readings of retail work and consumption workplaces, however, it is first necessary to explore the consequences and imperatives of the labour intensity of retailing, together with the characteristics of retail employment which emerged in the late twentieth century as a result of those imperatives. We can begin this task by considering Figures 5.1 and 5.2.

Figure 5.1 displays across a range of industrial sectors in the UK during the late 1990s the labour costs of firms in those sectors as a percentage of their operating profits. It can be seen that retailing is characterized by very high labour costs in relation to profits – indeed food retail is second only to support services in this respect. What this implies for the retailer is demonstrated in Figure 5.2 which considers the likely impact of a 1 per cent increase in labour costs on the operating profits of firms in those sectors. In the case of food retailers a 1 per cent increase in UK labour costs is seen to translate into a decline in operating profit of 1.5 per cent – whereas across all sectors the average decline is merely 0.6 per cent. What this implies, of course, is that the retail sector (and food retail in particular), in the UK and in all the advanced economies considered in this book, is amongst the most highly *geared* industries in relation to labour costs. Indeed there is a strong case for suggesting that Figure 5.2 might significantly underestimate the true position retailers face, as they have less capacity to respond to increasing domestic labour costs than manufacturers who can rapidly shift their production to lower cost economies – for a retail business it is the location of its sales that determines its labour costs.

For these reasons – essentially because of the differentially high gearing which the retail sector has consistently been exposed to in relation to its labour costs when compared to other industrial sectors – the containment and reduction of labour costs and the enhancement of labour productivity has provided a perennial theme in retailing over the last century. The introduction of self-service large-format retailing in the USA during the 1920s and 1930s (see Chapter 4; also Longstreth, 1999), and its subsequent export to Europe and elsewhere in the second half of the century, in particular, had a dramatic impact on retail labour costs and the organization of retail work. Indeed, as Adelman (1959) was able to show in his account of the response of America's largest food retail chain of the era (A&P) to the emergence of the self-service supermarket in the 1920s and 1930s, one of the key reasons 'the supermarket could profitably sell goods at prices which would be ruinous to the conventional store' (Adelman, 1959: 60) was because of the considerable savings the format offered in labour costs – particularly in the cost of skilled supervisory/managerial labour as a percentage of a store's sales.

As Ducatel and Blomley subsequently reminded geographers in their pioneering attempt to reposition the study of retail capital and its transformation into the mainstream of economic geography, the increasing centralization of retail provision associated with large-format self-service retailing – first supermarkets, then superstores, then hypermarkets/

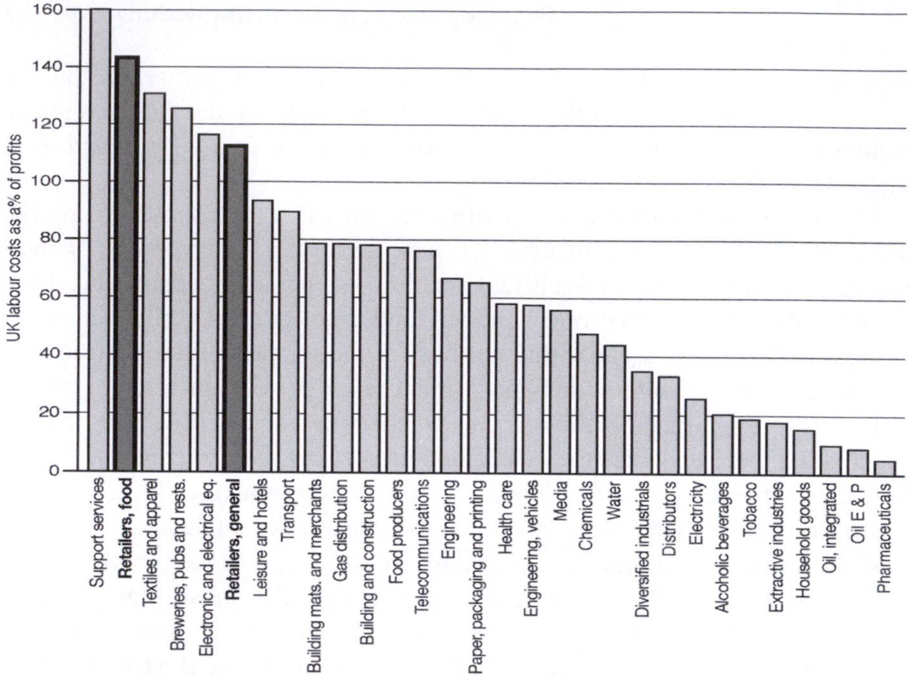

Figure 5.1: *UK industrial sector labour costs as a percentage of UK operating profits*
Source: Adapted from Credit Suisse First Boston (1999).

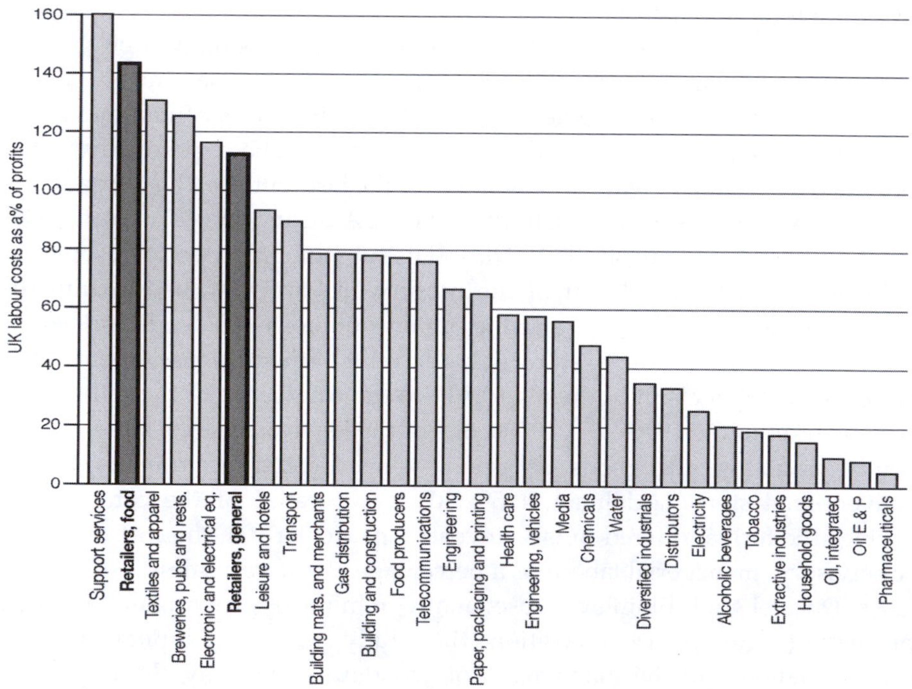

Figure 5.2: *Impact of 1 per cent increase in UK labour costs on UK operating profits*
Source: Adapted from Credit Suisse First Boston (1999).

supercentres (see Chapter 4) – required 'a renegotiation of the work of consumption between retailers and consumers'. Consumers were obliged to undertake more of 'the work of exchange in return for the lower costs and wide range of goods which the cost structures and sheer size of larger stores can accommodate' (Ducatel and Blomley, 1990: 223). As consumers took on more of the work historically performed (in the days of 'counter service' retailing) by retail labour, so retailers were able progressively to transform the organization of retail work – deskilling many of the personal service functions of retail workers into more repetitive and routine functions (shelf filling, check-out/till operations, etc.). They could then 'recruit labour for occupations demanding little or no specialized knowledge and easily acquired skills' (du Gay, 1996: 105), employing cheaper sources of labour – particularly women, part-time workers and the young – with significant implications for the composition of the retail workforce. Indeed, in Ducatel and Blomley's (1990: 222) view, the 'steady dismantling of a relatively skilled, male-dominated labour force, and its subsequent replacement by a low-skilled, lowly paid, feminized workforce, coupled with the increasing use of part-time and juvenile labour' that this implied, provides one of the key features in the attempts by retail capital throughout the twentieth century to reduce its 'circulation costs' and to speed up the turnover time of capital. It is to these characteristics of retail employment which emerged most strongly in the second half of the twentieth century as a result of the imperatives of the labour intensity of retailing, and to their theoretical conceptualization, that we now turn.

Characteristics of retail employment – feminization and part-time working

Table 5.1, from Freathy and Sparks (2000), illustrates the highly interrelated nature of the feminization and part-time working which characterized the UK

Table 5.1: *Composition of the UK retail labour force, 1997*

	Numbers employed ('000s)		
	Male	Female	Total (%)
Full-time	483.2	528.0	1011.2 (43.9)
Part-time[a]	250.7	1042.9	1293.6 (56.1)
Total (%)	733.9 (31.8)	1570.9 (68.2)	2304.8 (100.0)

[a] People normally working for not more than 30 hours per week.

Source: adapted from Freathy and Sparks (2000: 91); original data from *Labour Market Trends*, November 1997.

retail sector by the late 1990s. By that time, 68 per cent of the UK retail workforce was female and 56 per cent of employees worked less than 30 hours per week. However, of these part-time workers the vast majority (80 per cent) were female. During the 1980s and 1990s, as Table 5.2 demonstrates, part-time employment in UK retailing had increased progressively, with certain sectors such as food retail leading the way, in an attempt both to reduce labour costs and to match labour in stores to temporal fluctuations in consumer demand in a flexible and more precise way.

What this implied at the level of the individual retail firm can be seen (Table 5.3) in the case of Sainsbury, the UK's second largest food retailer. During the period 1987–98, despite variable commercial success and the loss of its UK market leadership, Sainsbury increased its workforce by more than 100,000 – 20,000 of these being in its USA subsidiary Shaw's (see Chapter 8; also Wrigley, 1997a, 1997b, 2000a) and 17,000 in its Homebase DIY subsidiary. In the process, it continued to push up its proportion of part-time workers from the already high level of 60 per cent to 69 per cent by the late 1990s – or 70 per cent if its UK food retail operations only are considered. And within those UK food retail operations, 65 per cent of its employees by the late 1990s were female.

Increasing use of part-time (predominantly female) labour – what Christopherson (1996) refers to as 'contingent workers' – in this way during the 1980s and 1990s, both to reduce labour costs and to increase the flexibility of retail managers to match employee input to fluctuating daily and weekly customer demand and to extended/variable trading hours, was a trend

Table 5.2: *Part-time employment share (%) in UK retailing, 1981–1995*

	1981	1985	1989	1993	1995
All retailing	40.4	42.6	44.4	48.2	54.1
Food retailing	–	50.0	–	–	66.0

Source: adapted from Freathy and Sparks (2000: 145), original data from *Employment Gazette* (various issues) and *Labour Market Trends* (1995).

Table 5.3: *Composition of workforce, J. Sainsbury plc, 1987–1998*

	1987		1993		1998	
	No.	%	No.	%	No.	%
Full-time	26,967	39.9	42,776	35.6	54,308	31.3
Part-time	40,653	60.1	77,343	64.4	119,467	68.7
Total	67,620		120,119		173,775	

Source: J. Sainsbury plc, Annual Report and Accounts, various years.

common to many of the advanced capitalist economies on which we focus in this book. Baret, Lehndorff and Sparks (2000), for example, compare retail working time practices in France, Germany, Japan and the UK, and document the rise of retail part-time employment as what they term the 'standard employment relationship'. Nevertheless, substantial national variations in retail part-time employment rates and in working time structures continued to exist by the late 1990s. There was also evidence of a slowing (even a reversal) in the shift towards part-time work in the retail industry in some countries as some employees began to 'shift away from numerical flexibility and towards functional flexibility, using full-time workers to do multiple tasks' (Christopherson, 1996: 173).

Table 5.4: *Distribution of hours normally worked per week by female part-time retail employees – France, Germany and the UK, 1994*

	France	Germany % of total female retail part-timers	UK
Fewer than 10 hours	10.2	15.0	22.6
11–20 hours	40.9	46.3	42.4
21–24 hours	9.3	8.9	12.8
25–30 hours	23.0	24.2	16.4
More than 31 hours	16.6	5.7	5.8

Source: adapted from Gadrey and Lehndorff (2000: 161); original data from Eurostat 1996.

In Reading 5.1 by Paul Freathy and Leigh Sparks (both geographers turned business school professors), some of this uncertainty about the future trends in part-time UK retail employment by the mid- to late 1990s is captured, together with the rather unusual regulatory environment which had encouraged the progressive rise of part-time working in UK retailing to very high levels during the previous 20 years. It is important to preface this reading with an inspection of Table 5.4. This shows that it was not only part-time female employment which characterized UK retailing through the 1980s and 1990s but also very *short hours* of work. As Gadrey and Lehndorff (2000: 160) explain, and Reading 5.1 expands upon, 'in the UK, certain legislative and fiscal arrangements have long discriminated in favour of part-time work and have helped to make this employment form, in particular part-time working with very short hours, preferable for both employers and employees'.

Reading 5.1 – Regulation and employment in 1990s UK retailing

In the UK, the trend of national policy (until the change of government in 1997) has been to facilitate competition on the basis of low wages and limited employment protection. Retailers have been well aware of the cost of labour and are always keen to reduce their costs. For a long time this has been combined with a positive encouragement to employ part-time labour because of its cost advantages through the non-obligation to provide either pro-rata pay or any of the normal benefits of employment. Combined with a steady erosion and then abandonment of collective bargaining and sector-wide agreement apparatus in the 1990s, employers have been free to treat employees pretty much as they wished. In retailing, with pay such an issue due to cost considerations and the need for various degrees of flexibility to meet operational and customer demands, the dynamic has been to employ part-time labour on short, frequent shifts. The Sunday trading deregulation in the 1990s added to this tendency.

Traditionally, in retailing as in other sectors, part-time employees have had access to less discretionary benefits and, indeed, have often been denied pro-rata 'entitlements'. With only basic or statutory obligations covered, employers have gained cost advantages. Until relatively recently, part-timers working less than 16 hours, for example, required five years' continuous service with the same employer before they were entitled to employment protection. The national insurance system in the UK also encourages employees to use short hours working, as contributions are paid only above a certain rate of pay. In short, it has been both easy and attractive for UK employers to use part-time contracts which are cheaper and offer lower levels of protection...

[But by the mid-1990s] the 'straws were in the wind' and the Conservative Government was being forced to implement European legislation and to reduce qualifying periods for benefits. Many of the leading retailers, aware of these trends and the likelihood of an incoming Labour government with its employment protection agenda, had positioned themselves to be amenable to adopting more standard and equitable practices ... The background for the sector was therefore one of relative freedom to pursue their own employment agendas ... tempered by the knowledge that change was on the way.

Extracted from: Paul Freathy and Leigh Sparks (2000): The organisation of working time in large UK food retail firms. Chapter 6 in Christophe Baret, Steffen Lehndorff and Leigh Sparks (eds) *Flexible Working in Food Retailing*. London: Routledge, 83–113.

Conceptualizing retail employment

The emergence of part-time working as the standard employment relationship in many sectors of retailing vividly illustrates, as many academic commentators have observed (see Noyelle, 1987; du Gay, 1996; Freathy and Sparks, 1996), an industry characterized by a marked *dualism* in its methods of labour utilization. Driven, as we have seen, by the imperatives of labour intensity and the increasing need, as du Gay (1996: 108) puts it, for retailers to purchase 'labour, much as they purchase goods from manufacturers, on a "just in time" basis', long-term employment relations and the 'internal labour market' (Doeringer and Piore, 1971) within the industry have declined. The result is an industry characterized by a growing duality of employment – that is to say an industry which has developed a highly segmented labour market in which 'a large majority of those employed in retail find themselves in jobs demanding limited skills, offering few opportunities for on-the-job training, and extremely limited opportunities for upward mobility' (du Gay, 1996: 108).

A primary/secondary labour market?

These themes of duality/polarization and deskilling in retail employment, set within a wider context of attempts by retailers to reduce their 'circulation costs', were highlighted by Ducatel and Blomley (1990: 222) in their pioneering attempt to explore issues of retail restructuring within a reconfigured retail geography. As a result, the theoretical conceptualizations of retail employment most commonly used by geographers as a basis of debate since the early 1990s have been ones rooted essentially in Piore's (1975) view of the labour market as *hierarchical* and *segmented* – divided broadly into *primary/core* and *secondary* segments. As du Gay (1996: 118) observed with respect to British retailing, the contrast drawn has essentially been between

> a 'core' comprising a small group of mostly white men who exert a large degree of control over access to well-paid managerial and professional occupations and, on the other, a quite finely differentiated 'secondary' group which comprises both full-time and, increasingly, part-time and temporary, unskilled, low-paid (on the whole) non-unionized, mainly female workers (but also some groups of men, especially minorities and young men).

The 'core' has been seen as consisting of central and line (store) managers and head office professionals administering such functions as finance, personnel, purchasing, marketing, estates, research, etc., who enjoy extensive remuneration packages, varying degrees of autonomy and responsibility over their work practices, and classical 'internal labour market' benefits such as career development opportunities and progression and relative security of employment. The 'secondary' group, in contrast, has been seen as composed of the shop-floor workers – dominantly female (and/or young), part-time, lowly paid, exhibiting as a group rapid labour turnover and limited opportunity for career development/advancement.

Table 5.5 shows Freathy and Sparks' (1996) summary of some of the remuneration/reward implications of this contrast between the primary and secondary labour market for the case of a British food retailer in the early 1990s. Updating the contrast to the mid-1990s, Freathy and Sparks (2000: 110) report very similar results when comparing a typical food superstore manager earning between £30,000 and £60,000 per annum and enjoying a package of benefits including: company car, company pension scheme, an exercisable share option equivalent to basic salary, private healthcare for the whole family, employee profit-sharing, staff discount on goods purchased, PEPs scheme, subsidized staff restaurant, travel agent discounts, paid holidays, and substantial relocation expenses when moving from store to store, with a shop-floor assistant on pay rates varying from £2.88 to £3.82 per hour and a very limited range of benefits.

Table 5.5: *Primary/secondary segment rewards – examples from a UK food retailer, 1992*

General manager	
Category 8 Manager	(Band from £39,000–£49,000, performance related)
Salary	£45,000 per annum consolidated (no premiums)
Overtime	Double time for Sundays Fixed payment of £100 for bank holidays
Benefits	Employee profit-sharing – average 6 per cent per annum[a] Executive share options Annual bonus of up to 30 per cent of annual salary[a] Private medical insurance Annual medical (Harley Street) Company car (ranges from BMW 520-Audi Coupe, Volvo 960-Senator, includes all costs including petrol) Staff discount card – 10 per cent off over £3.00 29 days paid holiday per annum 26 weeks company sick pay PEPs schemes Company pension scheme (including life insurance) Subsidized staff restaurant 5 per cent discount at Hogg Robinson Travel Agents
Shop-floor worker	
General assistant at age 18+	£7,374.54 per annum
Cashier at age 18+	£7,743.21 per annum
Benefits	10 per cent staff discount Full-time and part-time pension scheme Save As You Earn scheme Profit-sharing Subsidized staff restaurant 5 per cent discount at Hogg Robinson Travel Agents Four weeks and two days paid holiday as soon as joining company Full uniform provided Staff social club

[a] Dependent on company profit performance.
Source: adapted from Freathy and Sparks (1996: 191).

To what extent, however, is this conceptualization of the retail labour market as being highly segmented and dualistic an adequate one? What are some of the important nuances that are revealed by more detailed examination? And is the primary/secondary thesis an overstatic view set within the context of an industry experiencing dynamic change in its competitive conditions and corporate structures, and subject to the types of organizational and technological transformations described in Chapter 4?

A multiply segmented and shifting retail labour market

As the 1990s progressed it became increasingly apparent to many geographers that the simple dualistic primary/secondary conception was becoming an increasingly inadequate means of capturing the emerging subtleties of a multiply segmented retail workforce. In that workforce there were not only clear and fundamental divisions within the 'secondary' group between voluntary 'career' part-time workers (often women with children at school) and involuntary part-time workers (see Christopherson, 1989; Freathy and Sparks, 2000), but also important shifts in the way part-time workers were expected to perform their duties in an industry in which customer service and customer care were becoming increasingly perceived as the frontiers of competitive advantage (Sparks, 1992; Lowe and Crewe, 1996; Shackleton, 1998b). Furthermore, the impact of, and contradictions involved in, the deployment of the extensive computer-based management information systems in retailing discussed in Chapter 4, provided additionally important challenges to the simplistic core/periphery conceptions of retail employment relations. It is to some of these more subtle and nuanced views of the retail workforce that we now turn.

Technological shifts in particular had an increasingly significant impact on retail employment relations. The impact of ever more sophisticated IT systems built on EPOS scanning information and EDI-linked supply chains were not only felt by those traditionally peripherized in the retail labour market – the part-time shop-floor workers – but also impacted on other groups including retail managers. EPOS technology allowed on the one hand, as Smith (1988), du Gay (1996), Freathy and Sparks (1996) and others have noted, greater management surveillance and control of the secondary workforce – facilitating staff monitoring, scheduling and control of pilferage. On the other hand, it also resulted in what Smith (1988: 154) described as 'extreme deskilling' among retail warehouse workers and led, in particular, to a major shift in the nature of retail management at the store level. Centralization of buying, advertising, pricing and merchandising decisions, facilitated by the development of EPOS data-driven management information systems had the effect of dividing the 'core' workforce into two segments – a head office/managerial group who largely managed to maintain the autonomy of their work practices in functions such as finance, strategic planning, personnel, property development and marketing, and a store management group who were increasingly focused only on managing the store's labour force – what Smith (1988: 149) terms issues of 'punctuality and presentation' with much reduced autonomy concerning selection and ordering of product lines, layout of stock, and other traditional areas of decision-making responsibility.

Reading 5.2, taken from the work of Steve Smith and initially presented at a conference on the 'organization and control of the labour process' held in the mid-1980s, captures the lack of control typically emerging during that period amongst store managers over their labour process. Baret, Lehndorff and Sparks (2000: 10), reviewing this position in the late 1990s, confirm that the removal of areas of decision-making responsibility from store managers has meant a requirement for 'less entrepreneurialism, but more practical management'. However, they stress that it is important not to overemphasize a deskilling thesis for this segment of the 'core' workforce for 'while centralisation has circumscribed aspects of the store manager's job flexibility, the level of flexibility remains higher than for other store staff in terms of both content and of the hours worked'. In addition, they suggested that the functional complexity of the ever larger retail stores of the late 1990s had increased the requirement for better-educated and better-trained store managers.

Reading 5.2 – Taylorism for retailers?

The majority of retailers which I visited tend to cluster around the Taylorian ideal type. They practise an intense division of labour, a pronounced centralisation of control and a lack of discretion among not only shopfloor personnel, but also among store managers too ... most of the main features of store performance were centrally determined from locally supplied sales data.

Typically I found that store managers had little or no say in any of the following: number of lines carried, selection of lines carried, store layout of lines carried, promotions, price, window display, staff budgets, marketing, store design and decoration and delivery dates. This left them with responsibility for staff punctuality and presentation – usually devolved to the senior store supervisor and some control over day-to-day staff deployment. The store manager (like the bank manager) will have some effect on the 'tone' and morale of the branch, and will make sure that tidiness is maintained. In one leading chain the manager hands out detailed work schedules to the individual staff which he (sic) receives from the headquarters mainframe. These will indicate what tasks each worker should be doing for each hour of each working day. The manager has little part to play at all and has already been automated out to a considerable extent.

The store manager will devolve training to the training supervisor who may in any case be a periodic visitor from regional head office. The content of the training will again be set centrally in order to ensure consistency throughout a national chain. Finally, the store manager will be expected to control 'shrinkage' (theft) and to ensure that the necessary trading information is communicated to the centre for stock control and auditing purposes. With the gradual diffusion of EPOS systems even the latter functions are eroded within the Taylorian ideal type, as data may be transmitted without intervention of store personnel.

Extracted from: Steve Smith (1988): How much change at the store? The impact of new technologies and labour processes on managers and staff in retail distribution. Chapter 7 in David Knights and Hugh Willmott (eds) *New technology and the labour process*. Basingstoke: Macmillan, 143–62.

In assessing Reading 5.2 it is important to note, however, that Smith acknowledged that not all retail managers during the mid-1980s were experiencing this same trajectory of deskilling. Indeed, for some of the retailers Smith studied, EPOS and other technological changes of that period were leading to an extension of what he termed 'the art of retailing'. More specifically, these retailers – which Smith referred to as 'craft retailers' – continued to sell on the basis of ambience, style and special look, and it was their view that they 'would undermine this should they deskill and regiment the retail functions' (Smith, 1988: 157). Buyers' decisions were seen by such retailers as 'acts of faith not safe computer bets' (Smith, 1988: 157).

It is in relation to these issues regarding the 'art of retailing', the ambience and style of stores, and attempts by retailers to commit their employees emotionally as part of 'a far more profound managerial control strategy' (Smith, 1988: 160) that the 'peripheral' retail workforce experienced dramatic alterations to its everyday working experience during the 1990s. As Shackleton (1998b: 229) notes in relation to the UK food retail industry, the 1990s saw 'a radical departure in the way that part-time employees [were] expected to approach and perform their duties. The main developments [were] the encouragement of team working with employees being made to feel that they actually have a voice in the running of their department or store.' The principal aim of this strategy was 'to engender a customer service ethos throughout the store' (Shackleton, 1998b: 229) but in the process, as Smith (1988: 160) had observed in the 1980s, it had the effect of creating employees who were 'much less able to leave work behind' – 'shop-workers at all times'. It is to these strategies, and their implications for work and consumption, that we now turn in the final section of this chapter.

Consumption work and consumption workplaces

Although customer care and customer service strategies emerged forcefully in the 1990s, and although their wholesale and vigorous application had become a hallmark of retailing by the end of that decade, it is important to remind our readers that the department stores, retail pioneers at the turn of the nineteenth/twentieth centuries, had adopted exactly these kinds of strategies as part and parcel of their pioneering approach to selling (see Benson, 1986; Leach, 1993; Domosh, 1996a; and Wood, 2001a for discussion of the US department store industry at the time). Reading 5.3, taken from Susan Porter Benson's *Counter cultures* – a study of saleswomen, managers and customers in American department stores in the late nineteenth/early twentieth centuries, emphasizes the unique role of the salesperson who had a 'formidable influence over the success of their stores' (Benson, 1986: 124). In direct contrast to the self-service stores emerging in other sectors of American retail at the same time (see Chapter 4), 'the unique element in department store labour policy was the encouragement of *skilled selling*' (Benson, 1986: 125), because 'the interaction between customer and salesperson could be neither subdivided nor standardised' (Benson, 1986: 127). In particular Benson, a feminist cultural historian, highlights the specific role of women in department store selling and the difficult interaction between the gender and class of the 'shopgirl' and the customer in the department store.

Reading 5.3 – The shopgirl: class, gender and selling

The selling staff of the typical department store by the 1890s was overwhelmingly working class and overwhelmingly female, and these facts powerfully shaped labour policy. Class and gender in selling interacted in extremely complex ways, sometimes contradictory. Executives set out to change the class-based characteristics of their salespeople and to co-opt their gender-based characteristics . . .

Qualities which had for a century been encouraged in women – adeptness at manipulating people, sympathetic ways of responding to the needs of others and familiarity with things domestic – fit nicely into a new view of selling. Managers urged saleswomen to transfer skills from their domestic to their work lives; during the early 1920s, Filene's tested aspiring coat, suit and dress saleswomen on their knowledge of style and fabrics and their ability to choose 'the correct style' for different types of customer. Making the store more and more like a home, executives encouraged saleswomen to act more and more like hostesses, to treat their customers as guests. Empathy and responsiveness constituted the irreducible core of selling skill. A writer in 1911 urged, 'shop with the customer, not at her'; Macy's training director affirmed in 1940 that 'interest in the customer's problems' was the key to selling success. Twentieth-century selling centred on the salesperson as a lay psychoanalyst of the counter, the evangelist of the therapeutic ethic of the culture of consumption . . .

In department stores' formative years women's cheapness and cultural characteristics dovetailed nicely. But as executives pursued their goal of skilled selling more energetically, a contradiction between women's position in the labour market and their role as skilled saleswomen emerged. Saleswomen constantly heard their supervisors emphasize the critical importance of skilled selling, and understood from their daily experience their ability to make or break a sale, but as women workers they remained low-paid and low-valued in the labor-market hierarchy.

Extracted from: Susan Porter Benson (1986): *Counter cultures: saleswomen, managers and customers in American department stores, 1890–1940*. Urbana and Chicago: University of Illinois Press, 128–31.

Many of the themes captured in Benson's book have important parallels with the near universal adoption of customer care and service strategies during the 1990s – in particular issues relating to the presentation of the sales staff and the selling techniques employed. Reading 5.4 by Michelle Lowe and Louise Crewe illustrates this for the case of US clothing retailer, Gap Inc. The reading outlines Gap's utilization of customer service as a weapon in the front line of retail competition and in doing so echoes many of the issues highlighted by Benson ranging from 'suggestive selling' to the use of 'secret shoppers'. Benson had noted how department store managers in the early twentieth century had devoted much of their attention to 'remodelling' the people whom they hired, and this was certainly the case in the 1990s and remains so today. Interestingly, at the Gap the technological transformations of retailing discussed in the previous section, rather than simply deskilling, can be seen to have facilitated the scope for sales staff to concentrate on developing their customer service skills.

Reading 5.4 – The Gap, Inc.: a case study in customer care

Customer service is considered to be one of the Gap's strongest competitive advantages. Knowledgeable, helpful 'sales associates' are one of the company's signature attributes. The company recognizes that 'it is our dedicated dynamic staff that makes real our most ambitious plans' (Gap, Inc., 1991). It stresses that:

> *As a company we are measured by sales and profit and earnings per share but these measurements are simply extrapolations from how good we are at satisfying the customer. Every day in our 1200+ stores our sales staff make hundreds of thousands of impressions. Each one of these represents the future of the company. Friendly, helpful service has become one of the quality features which markedly distinguishes our company from other retailers. We spend a great deal of time and effort ensuring that all our store employees understand the critical role they play. Our customers want to be treated exactly as we do when we are shopping. To always meet and hopefully to exceed their expectations is our constant daily plan (Gap, Inc., 1991).*

Interestingly at the Gap the utilization of EPOS is viewed as allowing retail 'sales assistants' (staffers) the opportunity to concentrate on serving customers effectively. 'Through bar codes and optical scanners sales information is entered into the company's database and used to manage inventory levels at each location. These systems free up sales staff to focus on serving the customer' (Gap Inc., 1991). Significantly then, technological change (a critical focus of retail restructuring in the 1980s) which has taken away decisions regarding merchandise range, pricing, space allocation from the level of the store has allowed 'customer care' at a local level to become the critical frontier of competitive advantage . . . The so-called Gap Lexicon (Kahn, 1992) is used to instruct 'staffers' how to operate while in the store.

The Gap Lexicon

Gappers	People who work at the Gap
Gap Act	How a Gapper is supposed to deal with a customer. Each letter stands for an action: **G**reet the customer in the first 30 seconds in the store. **A**pproach and ask if there's anything you can find. **P**roduct information. Know the merchandise. **A**dd-ons. Suggest a great top to complete the outfit. **C**lose the sale. If something looks terrible, don't lie. **T**hank the customer.

Super sales Sales of more than $200 to one customer gets your name in the company paper, *The Gap Rap*.

L.Y. Stands for last year. Refers to the same-store sales on the same day last year. The aim is to top the L.Y.

Goal The total sales amount the store expects you to make on a given day.

Secret Shoppers and Mystery Callers Anonymous testers from headquarters.

Five in, two out Gappers should encourage the customer to take five items into the dressing room and come out with two to purchase.

Color blocking Laying out the tables so that the colors are pleasing to the eye.

U.P.T.s Units per transaction, or the number of items the customer buys. The Gapper with the most U.P.T.s in a day per store can win a Gap T-shirt.

Extracted from: Michelle Lowe and Louise Crewe (1996): Shopwork: image, customer care and the restructuring of retail employment. Chapter 10 in Neil Wrigley and Michelle Lowe (eds) *Retailing, consumption and capital: towards the new retail geography*. Harlow: Addison Wesley Longman, 196–207.

Perhaps some of the most important writing on consumption work and consumption workplaces is that of Paul du Gay, a British sociologist whose book *Consumption and identity at work* includes within it a specific plea for the importance of retail and consumption research in the social sciences. For du Gay (1996: 98), 'the marginalization of retailing from the research agenda' – 'only those industries that *really make* something are important' – is inhibiting. More specifically, du Gay is interested in 'exploring the ways the economic *folds into* the cultural in the practice of retailing'. His study concentrates on retailing as cultural practice and focuses on various logistical/technological developments taking place in the industry in the late 1980s and early 1990s, allowing retailers to stay closer to the customer than ever before. Most importantly, though, he concentrates on 'the subjectivizing aspects of contemporary retail change, both for customers and, increasingly, for retail employees', arguing that 'attempts by retailers to make up the consumer have consequences for the way in which the social relations of employment are imagined within the retail sector' (du Gay, 1996: 98). Reading 5.5, taken from du Gay's book, centres on this final point. Here du Gay uses the example of the development of Harrods' customer care strategy in the late 1980s. A key component of the strategy's development was a visit by Harrods' personnel director to the Disney Corporation in the USA in order to internalize the Disney corporate ethos.

Reading 5.5 – The culture of the customer in retailing

At Harrods, a stated 'commitment to excellence' in customer service, as part of a 'wider strategic change programme', led in the late 1980s to their personnel director visiting the Disney Corporation in the USA in order to explore the possibility of transposing certain elements of their 'service through people' theme to the Knightsbridge store. On his return, senior management were informed that the 'language' of Disney was absolutely right for Harrods. 'There are big similarities between Orlando and Knightsbridge', it was argued; 'like Disney, Harrods is really theatre'. It was important, the personnel director felt, for staff to realise that they were a part of an 'amazing' and 'spectacular' show, a living piece of British history, and that they were engaged in producing that show for their 'guests', the customers, every minute of every trading day. It was therefore essential for the company to recruit the right sort of staff, people who would 'internalize' the 'Harrods' culture, who could 'believe in it and become part of its history and prestige'; people that would learn to feel 'ownership'. By following the Disney corporate ethos, Harrods and its 'family' of employees would therefore become more cohesive, productive, efficient and effective.

Extracted from: Paul du Gay (1996): *Consumption and identity at work*. London: Sage, 120.

Of course, Harrods' choice of the Disney Corporation as a destination for its personnel director to learn about customer care was by no means accidental. Indeed retailers (and many other service sector companies) have drawn repeatedly on the more *performative* aspects of Disney's corporate culture as part and parcel of their inculcation of a specific set of employee–customer relations (see Sorkin, 1992; Ritzer, 1993; Leidner, 1993; Hochschild, 1983). Du Gay (1996: 121), for example, finds that UK food retailer Asda utilized the perceived excellence of Disney as a provider of 'service through people' in an attempt to implement a new culture-of-service ethos in the firm during the late 1980s. We have little space here to expand in detail on these issues. Suffice it to say, though, as du Gay's work indicates, the consumption workplace increasingly acts as a 'stage' upon which the everyday drama of consumption takes place, and this has important implications for the way consumption workplaces are studied. In this respect Philip Crang's (1994) research on the workplace geographies of display is exemplary. In particular, his sensitive treatment of spaces ('front and back' regions) in consumption workplaces (see also Gregson and Crewe's (2001) use of similar concepts in relation to what they term 'interiorities' of charity-shop retail) has considerable, but as yet relatively unexplored, potential. We examine these geographies of the retail workplace further in Chapters 9, 11 and 12.

6
Retail regulation and governance

We began our discussion of the changing character of retailer–supplier relations in Chapter 3 by considering an Office of Fair Trading (OFT) announcement in 1998 of an investigation into the increasing power and market dominance of the leading UK food retailers. By March 1999, after an intensive nine-month investigation, the director general of the OFT, although unable to reach a conclusive decision as to whether the four major food retailers were making excess/supranormal profits to the detriment of the British consumer, had decided that there were indeed sufficient grounds for concern to refer the whole UK food retail sector to the Competition Commission (the successor to the Monopolies and Mergers Commission) for a more formal appraisal. In the process (although in the event this did not occur – see Wrigley, 2001b) it opened the possibility that action of varying degrees of severity would be taken against those food retailers whose practices and conduct were considered to be an abuse of their dominant market position. Likewise, in Chapter 4, our assessment of the future development of large-format retailing in continental Europe highlighted the significance in France and Spain of legislation introduced in the mid-1990s – the *loi Galland* in France and parts of the *Ley de Ordenación del Commercio* in Spain. That legislation regulates certain aspects of the relationship between retailers and suppliers, restricts 'loss leader' selling by hypermarkets, and helps the smaller supermarket format to compete on a more equal footing.

The above are merely two examples of many which could be quoted of the way retail industries worldwide are subject to regulation at a variety of spatial scales ranging from the 'local' (including regional/state levels) to the national, supranational and international. Yet traditionally, as we argued in *Retailing, consumption and capital* (Wrigley and Lowe, 1996: 13),

> orthodox retail geography was remarkably silent about regulation and the complex and contradictory relations of retail capital with the regulatory state. With the exception of a rather one-dimensional discussion concerning the constraining influence of land-use planning regulation and some limited debate about shop opening hours regulation, the transformation of retail capital appeared to take place in a world devoid of a macro-regulatory environment shaping competition between firms, the governance of investment, the use of labour, and the overall extraction of profits from what Appadurai (1986) calls the 'situation of exchange'. Neither was there any attempt in orthodox retail geography to conceptualize the nature of the contemporary

regulatory state, and to use such a conceptualization to inform analysis of the challenging structures and geographies of channel relations within production–consumption chains.

Fortunately, as a reconstructed sub-discipline of retail geography began to emerge during the 1990s, and began to engage with and be engaged by the wider debates on contemporary consumer culture and consumption (see Miller (2001) for an assessment of the contributions to those debates of the 'new retail geography'), this lacuna was progressively closed. Indeed, it was closed to such an extent that, in reaction, a number of geographers began to turn their attention instead to the non-regulated or loosely regulated fringes of the consumption spaces of retail capital (Gregson et al., 1997; Crewe and Gregson, 1998).

In this chapter we attempt to guide our readers through some of the debates and literature concerning retail regulation. We leave, however, for Chapter 7 issues concerning land-use planning regulation, and for Chapter 9 debates concerning the regulation and contestation of retail spaces that are privately owned and policed. Here we focus initially on three aspects of retailing and regulation. First, attempts by geographers to address the importance of contested regulatory practices operating at a variety of spatial scales, and the way such practices are critical in shaping the economic landscape. Second, debates about private-interest versus public-interest regulation and the privileging of corporate retail capital. Third, attempts to locate debate about retailing and regulation within broader theorizations of the nature of the contemporary regulatory state. Finally, we consider issues relating to the *internal* regulation of the retail firm, returning to the matters of 'corporate governance' which we first raised in Chapter 2 (Reading 2.7), and to links between the internal and external regulatory environments of retail firms.

Spatial scales of retail regulation and market consequences

Of the many spatial scales at which retail regulation occurs, it is the local and the national which have attracted the greatest amount of attention from geographers. Here we provide examples of that work which focus on the regulation of retail trading hours, food safety regulation in retailing, and the regulation of competition between retailers and their suppliers and between retailers themselves. These examples, in their turn, raise much wider theoretical issues concerning an appropriate conceptualization of the nature of regulation and the practices of the regulatory state.

Retail regulation at the local scale – contradictions and tensions

One of the pioneers of the 'new retail geography' of the 1990s, Nicholas Blomley, was first drawn into his attempts to retheorize the subject, exploring in the process 'the subtlety and importance of retail capital, consumption and space' (Blomley, 1996: 256), via a Ph.D. study (Blomley, 1986a) in the mid-1980s concerned with the attempts of the Thatcher Conservative government in the UK to repeal the legislation which had governed shop opening hours and conditions of retail employment in England and Wales for several decades. That legislation – the Shops Act (1950) – which had been subject to no less than 19 previous attempts at repeal or reform was, by the mid-1980s, held in disrespect by a large and increasing number of retailers who blatantly flouted its restrictions on Sunday trading. Local government, charged with enforcing the provisions of the Act within its districts, was increasingly reluctant to enforce legislation that was viewed as anachronistic and unpopular, but was often forced to take action after formal complaints. As a result, enforcement of the Act varied enormously at local level across England and Wales (Blomley, 1985, provides a map of prosecutions under the Act at the county level).

Blomley's objective, in a paper published in *Society and Space* (1986), was to explore the contradictions in retailer–regulatory state relations which emerged from enforcement of the Act. But, in doing that, he was forced to confront two wider issues of considerable theoretical importance. The first of these concerned what he termed the 'legitimatory crisis' faced by the state. In simple terms, the fact that the widespread flouting of the law by retailers and the unwillingness of many local authorities to enforce the Act raised important issues concerning the 'rule of law' and the legitimacy of state intervention in the market – a topic of considerable theoretical importance to which we will return later in the chapter. The second concerned the tensions which differential enforcement of the Act created between the central and local state, raising issues about the degree and nature of autonomy of the local state from the central state. Clearly the local state does not operate merely as the central state writ small (Cockburn, 1977). In considering these issues, Blomley provided what is arguably the first consistent attempt to 'read' retail regulation through the lens of the wider theoretical debates shaping contemporary human geography.

Reading 6.1 extracts from Blomley's essay at the point at which he considers a case study of the contradictions surrounding enforcement of the Shops Act in the mid-1980s in one local authority – disguised in the extract under the name 'West' but actually the City of Bristol. His more general theoretical perspectives on the nature of regulatory state action will be returned to later in the chapter.

Reading 6.1 – Contradictions of enforcing the Shops Act at the local level

The crisis over the legitimacy of a law which is widely ignored both by the enforcers and by the subjects of that legislation is at its most pronounced at the local level. 'West's' officers and councillors often find themselves in opposition to the very principle of the

Act, yet feel themselves obliged to enforce it 'because it is law'; a process of enforcement for enforcement's sake ... This has led to a sterile enforcement policy, an almost ritualized going through the motions. When questioned as to the objectives behind enforcement, councillors and officers seem unclear. One is told of issues such as the 'unfair competition' suffered by law-abiding traders, for example, but the main concern seems to be that the Act is still on the statute book and, as such, must be enforced ... The local state, although reluctant to enforce the Act vigorously because of the moral ambivalence surrounding the Act, coupled with its evident dissonance with retail structural change, has a desire to 'be seen' to enforce the law. When called into question, the legitimacy of the law and the state must be safeguarded. If unprovoked, a response is not likely ...

A second related constraint within which enforcement policy operates is derived from the relationship between the local state and the demands of certain sectors of retail capital. [Indeed] there is an extensive literature describing the manner in which enforcement agencies are 'captured' by those they seek to control ... In the case of the enforcement policy of 'West', it is apparent that certain well-placed sectors of retail capital have proved partly successful in influencing enforcement policy to suit their own ends ... The more traditional stores, by virtue of their 'respectability', historical function as wealth generators, and powerful national backing enjoy good relations with the City Council. This powerful trader lobby has thus been successful in arguing for the prosecution of a number of the larger offenders under the Shops Act, most notably the out-of-town retail warehouses and the Sunday markets ... [and] have even gone so far as to threaten manamus proceedings against the local authority, thus legally obliging them to enforce the Act. Faced with such pressures, 'West' has had little choice but to prosecute regularly a number of the larger offenders ... The net result of an enforcement policy based upon specific complaints is, of course, that the Act is enforced against one section of the retail trade at the behest of another. In other words, the Act is being used as a 'commercial weapon' to constrain that fraction of retail capital which, by virtue of its lower labour costs per unit sales is able to open out of hours and thus cream off the profits of the more traditional retailer ... Certain retail interests are active at national level in securing the maintenance of the Act. Similar interests at local level, benefiting from the constraints of the local state, are using the Act against competitors.

Extracted from: Nicholas K. Blomley (1986b): Regulatory legislation and the legitimation crisis of the state: the enforcement of the Shops Act (1950). *Environment and Planning D: Society and Space*, 4, 183–200.

Both the theoretical themes and the substantive focus of Blomley's mid-1980s work have attracted attention during the emergence of the 'new retail geography' of the 1990s. In a substantive context, the deregulation of retail trading hours considered in the UK by Blomley has provided a focus for research by Australian geographers (see Baker, 1994, 1995). In a theoretical context, it has been the nature of contested regulatory practice at the local and national scales, the conceptualization of local modes of regulation, and the insight that provides into central–local state relations which has provided the focus.

Two brief examples will serve to illustrate the latter literature. The first is taken from the work of one of the authors (Wrigley, 1997c) and extends at a local level themes first discussed in a comparative study of regulation and corporate restructuring in the food retail industries of Britain and the USA which we will consider in much greater detail below. Here we concentrate simply on one aspect of the local–national regulatory interface in US food retailing in the mid-1990s (for an update to the end of the 1990s/early 2000 see Wrigley, 2001a). In Reading 6.2 we consider how a deregulatory interpretation of competition (antitrust) regulation which emerged at the national level in the USA during the 1980s was eventually challenged at the local level, and how, by the mid-1990s, the local state (in this case Connecticut) sought to deal with regulation of merger-induced structural change in food retail competition. Structural change, in this case, resulted from an announcement by the Dutch retailer, Royal Ahold (by that time already the eighth largest food retailer in the USA and with an existing market presence in Connecticut), of its agreement to acquire Stop & Shop, the leading regional food retail chain in New England.

Reading 6.2 – Implementing competition regulation at the local scale in US retailing

The Celler–Kefauver Act of 1950, which amended and strengthened Section 7 of the earlier Clayton Act, was designed to forestall anti-competitive mergers in their incipiency, in particular to prevent 'merger-induced structural changes falling far short of actual monopoly' (Mueller and Paterson, 1986: 373). From the late 1950s to the 1970s the statute had a very considerable impact on the structure of both US industry in general and the US food retail industry in particular. During the 1960s, for example, the FTC initiated a series of actions designed to halt geographical market extension mergers in food retailing, and a 1966 Supreme Court decision (United States v. Vons Grocery Co.) made it clear that even mergers involving small combined market shares were prohibited by Section 7 when they involved leading sellers in a market experiencing a trend towards concentration (see Wrigley, 1992).

During the early 1980s, the regulatory climate changed. The first Reagan administration came to power heavily influenced by the intellectual challenge to US antitrust legislation mounted by the 'Chicago School' (Posner, 1979) and committed to slackening the regulatory constraints under which business operated. Enforcement of the antitrust laws fell dramatically, providing the necessary conditions ['deals could be done which in the past would have been challenged and probably stopped on antitrust grounds' (Magowan, 1989: 12)] and, in part, stimulating the changes in capital market conditions which then propelled US food retailing into the intense period of hostile takeover attempts and LBOs between 1985 and 1988. By the late 1980s, however, growing public and Congressional criticism of the FTC's relaxed merger-enforcement policy began to have effect. In particular, the California Attorney General launched a vigorous challenge to American Stores' merger with Lucky Stores Inc. (at the time one of the leading US chains with more than 350 stores in California). The challenge was carried successfully to the Supreme Court in 1990 and American was forced to divest its entire 145-store Alpha Beta chain in the state. The case provided a precedent for similar action by other state Attorneys General. The pattern which has subsequently emerged during the 1990s has two components. The FTC has essentially adopted a fix-it-first policy, agreeing not to oppose those mergers where the acquiring retail firm commits in advance (under the spirit of Section 7/Celler–Kefauver) to divest itself of all clear horizontal market overlaps. The relevant state Attorney General may, however, seek to impose far more rigorous antitrust enforcement – as in the case of the Massachusetts Attorney General and the 1995 Stop & Shop acquisition of Purity Supreme – with much stronger impacts on the nature of retail competition at the local level.

And so it was in the case of Connecticut. Initial announcement of the Ahold agreement to acquire Stop & Shop brought an immediate statement (*The Hartford Courant*, 29 March 1996) by the Connecticut Attorney General, Richard Blumethal, that the magnitude of the merged company within the state (92 stores and a 65 per cent market share) demanded a rigorous review. Expressing serious concerns about possible coordination of pricing across the two previously competing chains, Blumethal stated his intention to press for divestments of Ahold stores in the state to preserve competition and protect the consumer interest. His concerns were echoed by consumer group representatives – 'the Attorney General and FTC are all that stands between the merger's few winners and its main losers' (*The Hartford Courant*, 27 June 1996: A19).

In the event, the Attorneys General of the three states (Connecticut, Rhode Island, Massachusetts) in which Stop & Shop had its core markets combined to press the case for divestment with the FTC. With the help of an expert adviser to assist their antitrust negotiations, a deal was finally agreed in closed sessions in late May at the FTC in Washington DC. The competitive position in each local market was considered in turn, and a trade off established between the small number of divestments considered necessary by the FTC and the much larger number sought by the Attorneys General to mitigate the monopolistic potential of the merger. The deal finally struck involved the divestment of thirty-one Ahold stores and sites ... At a stroke the retail landscape of the state was transformed.

Extracted from: Neil Wrigley (1997c): Foreign retail capital on the battlefields of Connecticut: competition regulation at the local scale and its implications. *Environment and Planning A*, 29, 1141–52.

Our second example of contested regulatory practice at the local/national interface is taken from a wide-ranging study of the regulation of food quality and provision in the UK by Terry Marsden, Andrew Flynn and Michelle Harrison (see Marsden et al., 1997, 1998; Harrison et al., 1997; Flynn et al., 1998). We will return to this study below when we consider debates concerning the emergence of 'private-interest' styles of regulation in retailing. Here we extract briefly from a paper published in *Transactions of the Institute of British Geographers* in 1997 in which the authors seek 'to gain a "window" on the interrelations between the state at central and local levels, and retailers operating on similar spatial dimensions' (Harrison et al., 1997: 474) by focusing on an analysis of the construction and implementation of regulatory practice concerning food quality in one local authority in inner London.

Reading 6.3 – Contested regulatory domains: the local/national interface

At the macro-level of national regulation, at least four regulatory domains in which British retailers have managed to maintain and shape their competitive space and flexibility have been identified: competition and pricing policies, planning and environment, food law, and food quality . . . it is the latter two that are the focus here. During the 1990s, these domains have been far from stable or predetermined. Indeed, they are very much contested and constantly evolving as a result of their reliance upon the relative strengths of the main actors and agencies involved, and in reaction to food safety 'scares', the onset of a regulatory 'onslaught' from Brussels, and a 'deregulatory' initiative within UK government. [Our] paper encapsulates this dynamism as it explores the regulatory 'battle' of the early 1990s in the food policy arena, and depicts the dialectical response between (national) policy formulation and (local) implementation . . .

Local research was based upon two lines of enquiry that fitted within a broader analysis of retailing and regulation. First, what was the nature of food retailing in the [inner London local authority] and how, and to what extent, were its separate 'tiers' able to (privately) regulate their quality control systems? Secondly, what was the nature of local authority (public interest) food regulation in the locality, and how do enforcement officers implement national legislation? . . .

Our study of the interactions between food policy-making and its implementation over the early 1990s illustrates the process by which the nature of local-level regulation has been redefined to accommodate macro changes in the nature of retailing and the government's deregulatory stance . . . As a result of corporate ascension in the food retail sector and a period of contestation between policy construction and implementation, a revised system of food regulation has emerged reflecting the relationships between corporate and non-corporate retailers, and the national and local state. Local food law enforcement officials have been de facto assigned two principal roles. For the privately regulated 'superleague' retailer, they are the largely external insurers of internal quality control. For smaller independent food businesses, however, they remain central to the provision of safe food and the protection of both the honest trader and the consumer interest. To do this, they vary their enforcement strategy from persuasive to insistent according to circumstances . . .

The local forms of regulatory practice do not simply function as an arena in which rules are implemented, enforced and reinterpreted. They also act as an important source for the reformulation of rules and regulation in their own right. They begin to focus upon how the contests for regulatory authority (between corporate retailers, central government, local regulators) are played out and territorialized. The shape and diversity of food provision in space – in this case an inner London borough – is thus an outcome of these dynamic and contested regulatory practices . . . The nature of the 'economic landscape' is crucially tied to the spatiality of combinations of national and locally derived regulatory policies and practices . . .

Extracted from: Michelle Harrison, Andrew Flynn and Terry Marsden (1997): Contested regulatory practice and the implementation of food policy: exploring the local and national interface. *Transactions of the Institute of British Geographers*, NS22, 473–87.

Retail regulation at the national scale – market rules and spatial outcomes

Harrison, Flynn and Marsden's view that the character of the economic landscape is tied crucially to what they refer to as the 'spatiality of combinations of national and locally derived regulatory policies and practices' is one that has been echoed by many geographers. At the scale of national regulation, Susan Christopherson (1993: 274), for example, has argued that the regulatory 'rules' governing investment within and competition between firms 'constitute *environments for capital accumulation*' and produce 'quite different patterns of economic behaviour within and across national boundaries'. As a result, she argues that comparative cross-national studies of market rules and their spatial outcomes are essential.

There are an increasing number of examples in the geographical literature, drawn from the retail industry, of how contrasting national regulatory environments can produce very different corporate and spatial structures (see Wrigley, 1992; Shaw et al., 2000). One case studied in detail in a paper by one of the authors of this book was published in *Environment and Planning A* in 1992 and involved a comparison of the food retail industries of Britain and the USA. What that paper demonstrates is that, by the mid-1980s, a number of rather unexpected disparities had emerged between the UK and US food retail industries. The UK industry, as we saw in Chapter 2, was experiencing a very rapid increase in concentration and the emergence of a small number of powerful food retail corporations. In turn, those firms were, as discussed in Chapter 3, gaining the ability to shape and control supply relations as the balance of power shifted progressively away from the food manufacturers and towards the major retailers. In contrast, the US food retail industry had remained stubbornly unconsolidated throughout the post-war period and was characterized by far less shift in power from manufacturers to retailers. The argument presented in the paper is that in essence,

> the somewhat unexpected disparities which emerged between British and US food retailing in the 1980s – in corporate concentration, power relations, profitability, productivity, *and also* in geographical structures – owe a considerable amount to the *differential* nature of the regulatory environments in which the industries operated (Wrigley, 1992: 746).

It is shown in the paper that for a period of almost 50 years in the USA – from the 1930s to the early 1980s – competition ('antitrust') regulation was hostile to the development of 'big' retail capital and to market share being concentrated into the hands of a small number of major chains operating multi-regionally and enjoying considerable purchasing leverage. The effect was to privilege the dominant position of the food manufacturers. Via price discrimination legislation (the Robinson–Patman Act, 1936) aimed at protecting the smaller trader, via criminal and civil indictment of the leading

US food retailers of the time (A&P was taken to criminal trial in 1945/6 by the US government and was convicted and fined), and via long periods of stringent anti-market-extension merger regulation (the Celler–Kefauver Act, 1950), a food retail industry had emerged by the 1980s that was far less consolidated than might have been expected, and in which there had been far less shift in power from manufacturers to retailers than in Britain. In addition, it was an industry that had largely become structured, in spatial terms, as a set of regionally dominant chains.

In contrast, in the UK, the post-war regulatory environment was conducive to the concentration of retail capital, the emergence of national chains, and the increasing retailer dominance of the supply chain. In particular, two major OFT and Monopolies and Mergers Commission (MMC) reviews of competitive conditions in the retail industry during the late 1970s and early 1980s, *Discounts to retailers* (MMC, 1981) and *Competition and retailing* (OFT, 1985), somewhat controversially concluded that the apparently anti-competitive buying practices of the increasingly powerful food retail corporations were not harmful to the public interest. As a result, they provided regulatory conditions which many academics have regarded as being supportive of the emergence of that period of rapidly escalating profitability, increasing concentration, and frantic new store development in the food retail industry which characterized the late 1980s and early 1990s in the UK. As the chairman and chief executive of a by then much reduced A&P perceptively observed in the late 1980s, when contrasting the food retail industries of the USA and UK,

> in the post-war years ... the US market place, because of Robinson–Patman, moved to a regional structure and the old large chains lost out ... [but] the UK without this disadvantage moved to the consolidation route with the advantages of purchasing leverage driving the success of a few national chains (Wood, 1989: 15).

In choosing to compare in this fashion the differential *impact* of regulatory regimes on corporate and market structures in retailing, the paper makes it clear, however, as Harrison, Flynn and Marsden also stress in Reading 6.3, that the regulatory environment should not be viewed analytically as somehow prior and exogenous:

> Although the process of corporate restructuring in retailing is clearly contingent upon the legislation which governs competition in the industry, corporate restructuring and its spatial expression, in turn, transform that regulatory environment. There is no simpler example of this reflexive relationship than the impact of increased concentration and geographical market extensions of the leading firms in the 1930s in the USA on the initiation and passage of the Robinson–Patman legislation, and the subsequent impact of Robinson–Patman actions on restraining the cumulative growth of power of the largest food retailers, ossifying the balance of power between manufacturers/suppliers and retailers, and creating a regionally structured US food market (Wrigley, 1992: 748).

In addition, it should be borne in mind when considering this paper that it interprets the conundrum of the stubbornly unconsolidated US food retail industry only up to a point in the mid-1980s. As we saw in Chapter 2, during that period the regulatory constraints under which the US food retail industry operated began to be relaxed and conditions became more conducive to consolidation. In practice, however, because of the pressure imposed by financial re-engineering in the industry (see Reading 2.4), that consolidation took more than a decade to occur, and it was not until the late 1990s that the US food retail industry began to move much closer to that of the UK in terms of its corporate and market structures (see Wrigley, 1999b, 2001a).

Private-interest versus public-interest regulation in retailing

As we saw in Reading 6.1 by Nicholas Blomley, there has long been considerable documentation in studies of regulation of the manner in which enforcement agencies can become 'captured' by those they seek to control. However, as the studies of retailing and food policy regulation in the UK by Terry Marsden and his co-authors have demonstrated, a far more subtle version of this 'capturing' of regulatory agencies can also be observed – based on what Marsden et al. (1997: 212) describe as 'a growing *regulatory embeddedness* on the part of the retailers'. What Marsden and his co-authors identify is the development of regulatory policies and structures in which private interests (in our case the corporate retailers) have become more generally empowered in the formulation and implementation of state regulation. As such, those interests are not required to 'capture', in any crude sense, the regulatory agencies. Rather, they are increasingly delegated by the state many of the key regulatory responsibilities which previously accrued to those agencies.

This form of regulation, whose emergence Marsden and Wrigley (1995, 1996) position within the period of intense deregulatory/re-regulation pressures during the 1980s, has been referred to as 'private-interest' regulation and contrasted, as shown in Table 6.1, with more traditional models of 'public-interest' regulation. Rather than conceiving of regulation, therefore, as simply a state function carried out in the public interest through policies which intervene in the market, an interpretation of regulation as encompassing both public and private activities, in which the state will often empower particular private sectional interests to act on its behalf, would appear to offer richer possibilities. Markets, as Susan Christopherson (1993: 275) has reminded geographers, will always be regulated; the question is merely 'by whom, in whose interests, and at what scale?'

What then are the implications of these rather abstract arguments about the emergence of private-interest forms of regulation for the practice of retail regulation? Here it is useful to return to the study of food quality regulation in

Table 6.1: *Comparing public-interest and private-interest regulation*

	Public-interest regulation	Private-interest regulation
Type of regulation	Legislation; guidance	Voluntary
Agent of regulation	Central or local government	Private or third party
Who pays?	Taxpayer	Manufacturer or retailer
Core characteristics	(i) baseline standards (ii) obligations enforced on behalf of public	(i) individual consumer choice (ii) informed customers and quality definitions

Source: adapted from Marsden et al. (1997).

the UK considered in Reading 6.3. What Harrison, Flynn and Marsden suggest in that study in general terms is that:

> Coinciding with the rise of a 'super league' of food retailers [in the UK in the 1980s/early 1990s] has been the emergence of a private-interest style of regulation in the food system. This contrasts with the traditional regulatory style based upon notions of the public interest. In the latter case, it is central government that sets standards through legislation that is enforced locally on behalf of the public by officials (e.g. Environmental Health and Trading Standards Officers and Public Analysts). In so doing, similar baseline standards for all consumers are ensured. In its more traditional form, food regulation has sought to ensure a combination of security of supply, accessibility, affordability and safety. Conversely [in private-interest style regulation] the food retail corporations voluntarily regulate their own systems at their own expense, promoting individual choice based on their own hierarchy of quality definitions. So, as the major retailers have become principal actors in the food market, they have negotiated key responsibilities in the management and policing of that system. Like all models, this public/private dichotomy is an abstraction, but it does highlight the changing nature of regulation and the role of the state in relation to food (Harrison et al., 1997: 476).

However, the authors also suggest that the maintenance of this private-interest relationship between the major food retailers and the national state in the UK has, during various periods in the 1990s, became extremely problematic and riven with very considerable tensions. In particular, they note that, in response to successive food safety scares in the UK, the reaction of government has frequently been to re-emphasize the role of 'public-interest' forms of regulation at the expense of the self-regulation which the major food retailers had come to expect as an entitlement – particularly under the private regulatory terms of the Food Safety Act of 1990. That legislation which charged the retailers with demonstrating the exercise of 'due diligence' in the manufacture, transportation, storage and preparation of foodstuffs had, in practice, recognized the ability of the major retailers to self-regulate their own supply chains and thus guarantee food standards without the burden of unnecessary regulatory control.

Marsden, Harrison and Flynn, in a paper published in *Environment and Planning A* in 1998 explore some of these tensions in the private/public regulatory relationships which emerged between government and food retailers in the UK during the 1990s, and ask how those tensions lie at the heart of what they term the shaping of competitive spaces in retailing. Reading 6.4 extracts very briefly from that paper. It provides merely a flavour of a rich and multi-layered assessment of the evolving regulatory cultures in UK food provision in the 1990s. We encourage our readers to explore these arguments in more detail in their own reading of the original.

Reading 6.4 – Private-interest regulation and the shaping of competitive space

The intense competitive environment in which corporate retailers are located in Britain means that they are forced to participate in and attempt to shape a whole series of regulatory domains: land-use planning, environmental regulation, food law, crime and social order, competition and employment policy, food-safety and hygiene issues. Corporate retailers cannot afford [not] to become involved in the shaping of competitive spaces which 'prise open' new territories and terrains of exchange and profit making ...

[However] the growing corporate power of retailers in the provision of food, together with the decreasing public confidence in food products, provides a major conundrum for government. How can the state best regulate food provision given its priorities for continuing to 'deregulate' the economy but encouraging the 'health of the nation'? ... One important consequence is that the maintenance of corporate retailers' market power is increasingly dependent upon their social and political actions both towards state agencies and towards consumers. It is through these actions that market 'spaces' can be kept open and legitimated ...

The growth in the role of corporate retailers inside as well as outside government has been one expression of the developing private-interest model of regulation, whereby significant former government functions are bestowed for what is seen as more effective management, to private sector interests. New forms of 'private-interest government' have been developed in the British case both as a cause and as a consequence of the growth and maintenance of retail power. It is important to realise that such models do not deny the continuing role of government in food regulation. Rather, they change its shape ...

Because of the potential lack of public confidence generated by self-regulation, retailers and the state have evolved working relationships which maintain public legitimacy and market power. This is achieved less through the establishment of new institutional structures as it is through the mutual agreement and construction of what constitutes 'the consumer interest' in the provision of 'quality' goods and services ... The state has not simply been 'hollowed out'. Rather, the political space has been reoccupied by new microcorporatist relations which stress the delivery of public policy through hybrid, corporate and state relationships and procedures.

Extracted from: Terry Marsden, Michelle Harrison and Andrew Flynn (1998): Creating competitive space: exploring the social and political maintenance of retail power. *Environment and Planning A*, 30, 481–98.

'Real' regulation, retailing and the nature of the regulatory state

As the work of Terry Marsden and his co-authors demonstrates, the retailer–regulatory state relations which emerged in the UK food system in the 1990s were highly complex, riven through with tensions, internal contradictions, and the potential for crisis. On the one hand, faced with pressure to justify their market power and self-regulatory entitlements, the major retailers were forced to become more adroit at projecting the benefits of their role in the regulation of food provision, at sustaining in social and political terms the lucrative markets which they had created in the 1980s, and at developing mechanisms, arguments and ideologies which would both protect and enhance their competitive space. On the other hand, the regulatory state which had, in many senses, become dependent on the economic dominance of the retailers – the 'new masters of the food system' (Flynn and Marsden, 1992) – for the delivery of public policy, was forced to seek periodically to control the private recipients of its delegated power. It faced, therefore, the difficult task of balancing the dual pressures involved in both sustaining corporate retail capital accumulation whilst maintaining a divergent public interest.

Inevitably, such considerations of the tensions and contradictions in retailer–regulatory state relations have resulted in attempts by geographers to locate discussion of retailing and regulation within broader theorizations of the nature of the contemporary regulatory state. In attempting to do this, Marsden and Wrigley (1996) draw attention to the inadequacy of many traditional conceptualizations of the state which view it is a separate and functional entity capable of acting consistently in pursuit of particular class interests. They quote Marden (1992: 757) who believes that:

> the state cannot be viewed simply as representing some sort of functional integrative entity regulating the regime of accumulation but, rather, is itself the object of struggle, and therefore cannot resolve the contradictions of capital because by its very nature it reproduces them in a political form.

As a result, they attempt to locate their various analyses of retailing and regulation (e.g. Wrigley, 1993b; Marsden et al., 1998) within what Gordon Clark (1992) has referred to as a 'real regulation' stance which places emphasis on the administrative manner, style and logic by which the state regulates the economic landscape. In this way, the socio-political *practices* involved in regulation become the prime focus for research, and attention is shifted to the highly contested nature of regulatory space, and to the powerful actors who dominate that space at any given time (Hancher and Moran, 1989).

However, by adopting such a focus on the social and administrative practices of the state, and seeking to understand regulation as a dynamic process actively being interpreted and reinterpreted by participants in the regulatory structure,

Marsden and Wrigley (1996) stress that they are not implying that broader theoretical questions about the role of the state should be neglected. Indeed their discussions about the problems faced by the regulatory state in the UK when confronted with growing public unease about apparent anti-competitive practices in food retailing, highlights the structural limitations imposed on the state by its relation to capital, and the dilemma this poses for regulation. Rather, they are suggesting that an attempt is made to redress what Clark (1992) has perceived as the 'privileging of economic meta-imperatives over institutional interests' in discussions of regulation in geography – replacing it with sensitivity to the institutional cultures of regulation (see Marsden et al., 1998: 491), to the web of complex contingencies involved in regulation, and to the spatial differentiation of regulation; not least between nation states. Marsden and Wrigley's analysis of retailer–regulatory state relations in the UK suggests that corporate reorganization – particularly in a sector such as retailing in which the boundaries of competition and accumulation are constantly being tested but in which the regulatory state has become, in critical areas, dependent upon the major firms – requires what they refer to as 'contingent state practices'. The regulatory state, as Wrigley (1993b) notes in his assessment of the proposals for legislative harmonization of UK competition policy regulation with that of the EU, will often seek to retain a certain amount of adaptability, creativity and capacity for action within its administrative practices. As a result, Marsden and Wrigley (1996) suggest that what is required in the study of retail regulation is a broader theoretical perspective which stresses a variable and contingent interpretation of the nature of the state.

Corporate governance – the internal regulation of the retail firm

We have seen then that the neglect of retail regulation and the complex and contradictory relations of retail capital with the regulatory state has been to a large extent overcome during the development of the 'new retail geography' of the 1990s. However, issues concerning the *internal* regulation of the retail firm, and the links between the internal and external regulatory environment of such firms, remain topics which are, as yet, largely unexplored. This is surprising given the very considerable literature relating to matters of 'corporate governance' of the firm which exists in the social sciences, particularly at the interface between corporate law and financial economics. (See Schleifer and Vishny (1997) for a well-known, wide-ranging survey of the field; Monks and Minow (1995) for a text which has an activist-investor orientation and provides case studies of corporate governance failure in the USA; and Prentice and Holland (1993) for discussion of corporate governance issues from a UK perspective.)

In a narrow sense, corporate governance, as we saw in Chapter 2, essentially deals with the separation of ownership and control – the fact that ownership rests with shareholders of the firm, and control with its managers. As Berle and Means (1932) who originally defined the problem observed, what justification is there for assuming that those in control of a firm (the managers) will choose to operate it in the interests of the owners? Or, alternatively, as Schleifer and Vishny (1997) prefer to express it, how do suppliers of finance to corporations – the people who sink capital in the firm – control managers? Systems of corporate governance in this narrow sense, therefore, aim to secure the alignment of the interests of owners and managers. As a result, they involve matters such as the appropriate structure and composition of boards of directors, methods of monitoring managerial performance, the design of optimal managerial incentive packages, ensuring the adequacy of financial disclosure, and the ability of capital markets to impose 'discipline' on managers via the market for corporate control (i.e. via takeovers).

However, there are also broader 'stakeholder' perspectives on corporate governance (Karmel, 1993; Blair, 1995) which place emphasis on other non-shareholder constituencies – employees, suppliers, customers and the communities in which the firm is located – that are crucial to a firm's market success: constituencies which have increasingly been recognized as having claims on the firm's corporate assets and prospects. These broader stakeholder perspectives place emphasis, therefore, on ensuring that the governance mechanisms of the firm can foster all of the reciprocal relationships through which the future success of the firm will be realized, and that the structure, composition and processes of the board of directors can represent multiple stakeholder interests. In addition, these perspectives recognize that the corporate activities of the firm may produce *unintended* social, individual and environmental effects outside the company, and that the relationship between a firm and the community is built upon an implicit contract or licence to operate. As a result, they place emphasis on the ways such unintended results of the pursuit of the firm's economic goals and objectives can be handled *within* the governance structure of the firm.

On the basis of these narrow and broader perspectives on the corporate governance issue, Figure 6.1 suggests a diagrammatic representation of systems of corporate governance, highlighting some of the relationships which exist between boards of directors, the suppliers of finance to the firm, the strategic business units (constituent parts/companies within the firm), the wider stakeholders, and the external regulatory environment. As yet, there has been little consideration within the debates of the 'new retail geography' of the 1990s of the implications of such systems of corporate governance and their relations to the regulation of retailing (though, more generally, an attempt by Sparks (1996c) to consider the role investment analysts play in the relations between the suppliers of finance and the retail firm is worthy of note, and we refer our readers once again, and in particular, to the paper from which Reading 2.7 is drawn). However, amongst all industrial firms, store-based

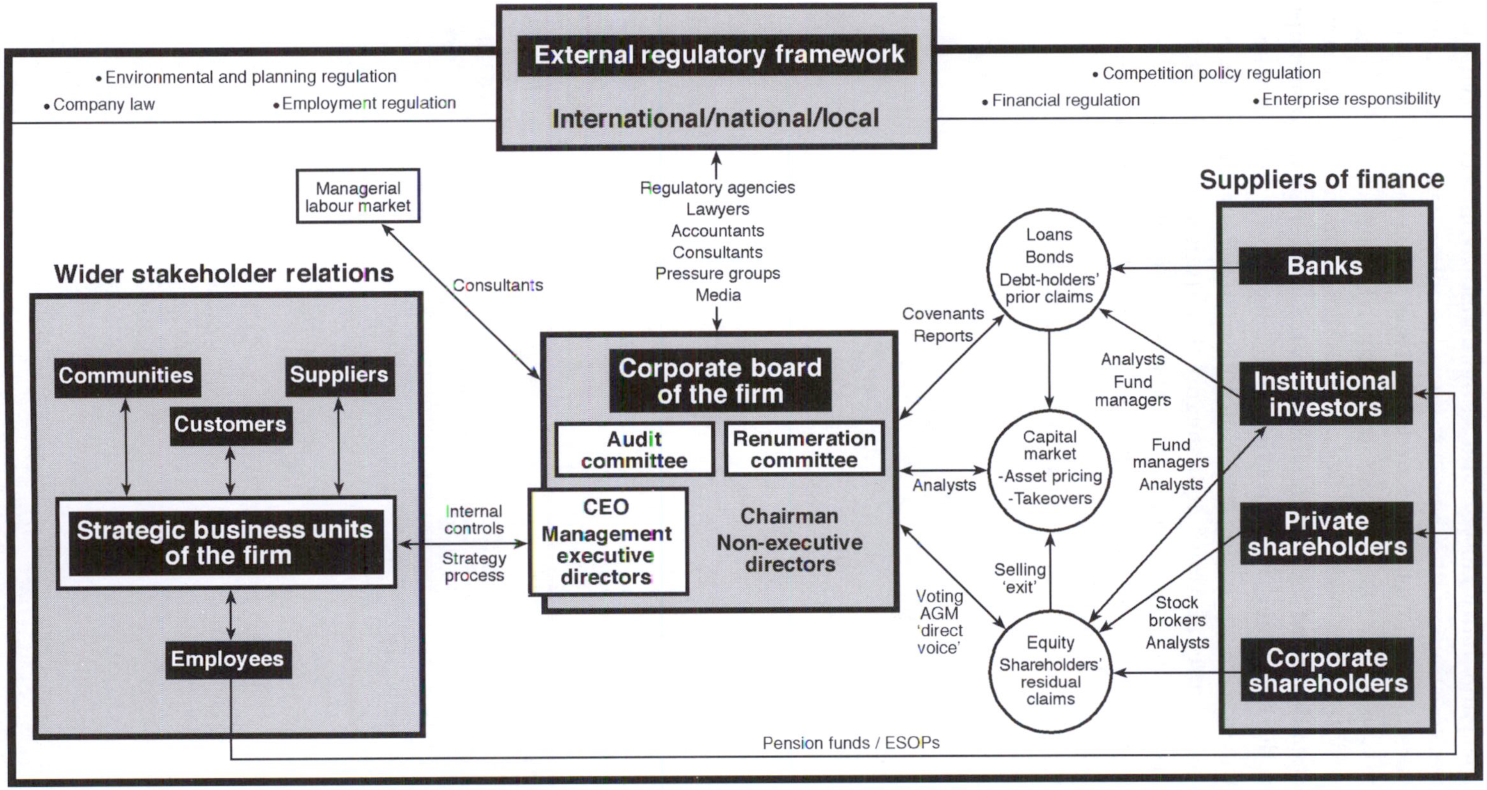

Figure 6.1: *Systems of corporate governance*
Source: Author, adapted from material discussed at University of Cambridge, Corporate Governance Workshop, July 1996.

retailers are particularly closely, indeed symbiotically, involved with wider stakeholder constituencies – their customers, suppliers, employees, and the communities in which their stores are located and which provide the essential patronage base which sustain those stores. Indeed, Marsden et al. (1998: 481) stress how vital it has been for UK food retailers in attempting to maintain their politically embedded custodial role in the food system to attempt to 'represent the consumer' by conveying and constructing 'what are seen by government officials to be legitimate conceptions of the consumer interest'. As a result, Marsden et al. (1998: 483) suggest that studies of retail regulation: 'need to incorporate notions concerning the *corporate governance* of the firm, [alongside] the interrelated sets of linkages between firms, and between firms and state agencies' (Marsden et al., 1998: 483). Figure 6.1, we suggest, offers a framework within which our readers might begin to explore that task.

PART 3

Making and re-making the geographies of retail capital

7
The inconstant geography and spatial switching of retail capital I: urban issues

Our purpose in the previous section of the book has been to provide an appreciation of the dynamic nature of the transformation of retail capital and the processes of retail restructuring which power it. By considering the rise of corporate retail power, the reconfiguration of corporate structures in retailing, the character and shifting power balance of retail–supplier relations, organizational and technological transformations of the industry, changing employment relations, and issues of retail regulation and governance, our intention has been to lay the essential groundwork for 'a more sophisticated understanding of the *reciprocal* nature of the relations between space and corporate retail activity' (Clarke, 1996: 292). It is that understanding which we now pursue much more explicitly in this section of the book.

Here then we focus, in the terminology of Storper and Walker (1989), on what we term the 'inconstant geography' and spatial switching of retail capital, together with issues relating to the configuration, manipulation and contestation of retail space. We show how corporate retail capital 'actively explores and penetrates specific spaces at a number of scales' (Ducatel and Blomley, 1990: 225), the way it manipulates spatial layout and design to induce consumption, and the way it often remakes an inherited spatial configuration via a process of capital switching, but sometimes gets locked into (finds it difficult to abandon) existing geographies. In contrast to 'orthodox' retail geography, we show how space in the new economic geographies of retailing is interpreted in a far more dynamic way – being conceived as the product of social/political activity, in which the active creation and *re*creation of markets, the 'grounding' of capital (see Reading 2.5), the relations between spatial configuration, spatial discipline and control, and so on, are viewed as central issues in both the capital imperatives facing corporate retailers and in the contested retailer–consumer relation.

We begin in Chapter 7 with one of the recurrent themes in contemporary economic geography – the 'creative destruction' of the built environment by the spatial switching of capital – in our case, retail capital. Our focus is on the shifting of retail investment over the past 50 years from the downtown and inner city to the suburban periphery and back again. This, we acknowledge, is one of the most familiar stories of economic geography. But there are constantly new readings of the story and new themes to explore, and we offer here some of the possibilities. In Chapter 8, we then turn our attention to the internationalization and globalization of retail capital. This was a topic which barely rated a mention in even the most comprehensive of the treatments of the geography of retailing in the late 1980s/early 1990s (Jones and Simmons, 1990). Yet, a decade on, we find emerging markets throughout the global economy being transformed by the reach of rapidly emerging retail multinational/transnational corporations, and geographical writing on retailing and consumption attempting to engage with the key theoretical debates in the vast and ever expanding globalization literature. In addition, we also consider in Chapter 8 the relatively neglected regional dimension of the inconstant geography of retail capital. Finally, in Chapter 9 we focus on the configuration and manipulation of retail space at the level of both the shopping centre/mall and store showing how, in Blomley's (1996: 239) terms, retail spaces rather than being viewed as passive surfaces 'are increasingly being cast as actively produced, represented and contested'.

The creatively destructive nature of the spatial switching of retail capital

As Sharon Zukin (1991: 19) suggested in *Landscapes of power*:

> the task falls to economic geography to provide a sense of landscape's 'structured coherence'. For a radical economic geographer, landscape is the tabula rasa of capital accumulation. It reflects the 'spatiality' of the capitalist mode of production in each of its historical phases. From this perspective, the underlying cause of repetition and singularity in the landscape is the profit motive, shifting capital between investment in industry and in property, cycling it into new construction or reconstruction, shuttling it between the downtown and the suburban periphery.

What Zukin is drawing attention to here is what David Harvey (1985) in *The urbanization of capital* and many other human geographers have referred to, building on Schumpeter (1942), as the creatively destructive nature of the spatial switching of capital in the built environment. That is to say, in Harvey's (1985) terms, the 'perpetual struggle in which capital builds a physical landscape appropriate to its condition at one particular moment in time only to have to destroy it at a subsequent point in time'.

Indeed the retail landscape provides one of the clearest examples of the strength of Schumpeter's 'perennial gale' of recurrent capitalist innovation, with retail investment continually on the move, creating, then abandoning, then rediscovering, spaces of profit extraction – in the process relocating downtowns, shifting retail investment to the suburban frontier, devalorizing and polarizing retail environments of the inner city, then revisiting, rehabilitating and revitalizing some of the most degraded of these areas, perennially reversing the pre-existing logic and location of profit extraction and the sources and flows of creativity in the retail landscape.

Although our focus in the following sections of this chapter will be on the post-war period and the shifting of retail investment linked first to suburbanization and subsequently to the gentrification/rehabilitation of central cities, it is important to preface that discussion by stressing an appreciation of the continuity of the process. Two brief examples will suffice.

The first relates to the emergence of the department store as a major focus of retail investment in cities in the late nineteenth century (see Chapters 9 and 11 for further details). As Domosh (1996a) has stressed, department stores not only depended on location and being near their clientele but also on their *association* with the fashionable classes. 'It was essential therefore that they created new landscapes associated with the domestic enclaves of those classes – both locationally within the city and also via the internal configuration of their retail space' (Wood, 2001a: 52). In New York City this implied, as Domosh (1996a) has discussed, a spatial shifting of the emerging department stores (see Figure 7.1) up Broadway during the 1870s and 1880s, onto 6th Avenue by the 1890s, and by the turn of the century to a new retail focus centred on 5th Avenue between Union and Madison Squares and extending east to 6th Avenue and west to Broadway. As Domosh (1996a) notes, the shifting location of A.T. Stewart's department stores illustrates these trends. His first department store, conspicuous for its white marble façade, opened at Chambers Street and Broadway in 1846. By 1862 Stewart, anticipating the uptown movement in the focus of retailing, had abandoned his 'marble palace' for a six-storey cast-iron and glass palazzo located between 9th and 10th Streets. And, as Figure 7.1 illustrates, the spatial shifting of department stores in New York City continued during the early twentieth century, to leave its focus in the post-war period firmly in the mid-town area.

The second example relates to the transformation of commercial space in the Los Angeles metropolitan area during the 1920s and 1930s. In this case, Richard Longstreth (1999) provides a lavishly illustrated account of the emergence of two radically new forms of retail space – the supermarket and the drive-in shopping centre ('stores the road passes through') which profoundly transformed the twentieth-century American city. Nurtured in the Los Angeles area during this period, these two innovations remained rather separate during the pre-war period. By the 1950s, however, they had gradually come together, producing in turn the shopping centres, and regional and supra-regional malls, which transformed the landscape of post-war America and which we discuss in the following section. Longstreth brilliantly illustrates this evolution, concluding that:

> By this circuitous route, the two spatial orders pioneered by the drive-in and the Los Angeles super during the inter-war decades were fully integrated . . . The collective results

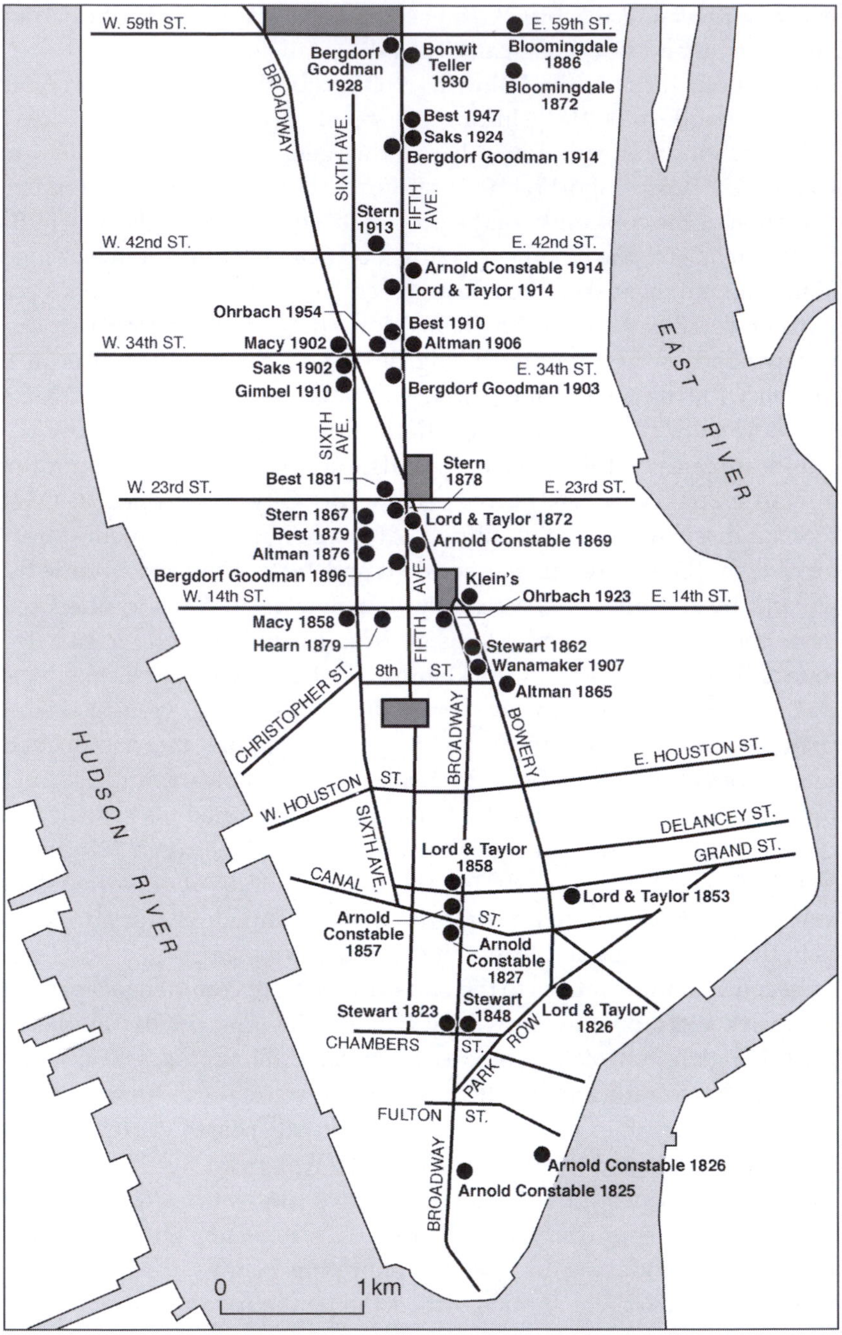

Figure 7.1: *Uptown relocation of New York City department and speciality stores, 1825–1960*
Source: Redrawn from Ferry (1960).

> were revolutionary. The location of retail facilities, the scale and arrangement of their spaces, the underlying merchandising strategies, and the shopping routines that were induced all had fundamentally changed. From metropolis to small town, Main Street would never be the same (Longstreth, 1999: 179–80).

Throughout Longstreth's account, the sense of the perennial and creatively destructive nature of shifting retail investment is constantly present. No sooner had the drive-in market revolutionized commercial space in the rapidly expanding automobile-orientated Los Angeles of the 1920s – a construction rate of about a dozen a month in the mid-1920s – than it was eclipsed by the emergence of the large-format supermarket (see Chapter 4) and the rethinking of both retail location and the internal configuration of retail space that this necessitated. Reading 7.1 by Longstreth, an American architectural and cultural historian at George Washington University, takes up the theme.

Reading 7.1 – Stores the road passes through: the drive-in market and the Los Angeles supermarket in the early 1930s

Four years after the drive-in market began its rapid ascent, [it] suffered an equally swift eclipse. The number of drive-in markets planned for the Los Angeles area in 1931 amounted to about one-third that undertaken the previous year . . . [and] only three drive-in markets are known to have been built in 1932. So pronounced a change might easily be related to the concurrent economic downturn . . . but it was not the downturn that brought the demise of the drive-in market. Dollars spent on constructing supermarkets during the early 1930s may well have equalled, if not exceeded those spent on food stores of all kinds in the years immediately previous . . . Once established, the tendency to build supermarkets in Los Angeles progressed, in the words of one observer, with 'almost cyclonic momentum': between 1930 and 1932 hardly a week passed without a new unit opening. Four years later, it was estimated that over 180 supermarkets were operating in the area and that they commanded 35 to 40 percent of the retail food business . . .

The drive-in market helped pave the way for these developments . . . Southern Californians quickly became accustomed to the convenience of one-stop shopping. Equally important, consumers had gotten in the habit of not just driving to the store, but of choosing one store among many based on the strength of its location, goods and pricing. However, the drive-in was structured to fit the traditional routine of shopping on a more or less daily basis and purchasing a modest quantity of goods on any given trip. Higher-volume sales necessitated a larger selling area, a larger on-site storage facility and, since the car was a central agent in the process, a larger parking lot as well. Size was not the only issue. The drive-in's configuration came to be seen as fundamentally unsuited to the shifting demands of the field. Flexible space was now considered of primary importance, so that the quantity of goods of any given type could be adjusted according to available supplies and market demand. The linear arrangement of the drive-in was far less conducive to the unobstructed, rectangular space . . . that came to typify the supermarket . . .

The drive-in did not face immediate extinction with the supermarket's rise; while many closed during the 1930s, many others continued to operate as sound businesses.

Nevertheless, the supermarket became the unquestioned outlet of choice among retailers, whose lead the real estate interests now followed.

Extracted from: Richard Longstreth (1999): *The drive-in, the supermarket, and the transformation of commercial space in Los Angeles, 1914–41.* Cambridge, MA: MIT Press, 79–81.

Decentralization: learning from Chicago

> The downtown of department stores, retail districts, and popular entertainment survived through the end of World War II, when a new set of circumstances in the United States began to tear it apart (Frieden and Sagalyn, 1989: 11).

We acknowledge that the post-war suburbanization of retail capital in the United States and the decline of downtown retailing is one of the best-known stories in human geography, and that its systematic documentation represents one of the lasting achievements of spatial-analysis-oriented 'orthodox' retail geography. The classical work is that of Brian Berry and his associates working at the University of Chicago in the 1960s and 1970s, who captured in countless publications 'the rolling tide of residential segregation' (Berry et al., 1976: 57) and decentralization which swept the focus of retailing in that city out to the newly constructed suburban regional shopping centres.

These centres first appeared in the 1950s, and towards the end of the decade (Figure 7.2) there were just four operating – 'open malls with plenty of parking . . . each centre more than 20 miles from any of the other three and at least 13 miles from the Loop [Chicago's CBD]' (Berry et al., 1976: 44). The response in the retail map of Chicago (Figure 7.3), as detailed by Berry in *Commercial structure and commercial blight* (1963), was, however, quite dramatic. Retail capital, as Figure 7.3 illustrates, had begun rapidly to flow out of the traditional centres of the inner city, led by the department store chains which had begun to realize the competitive advantage of shifting their corporate focus to the expanding middle-class suburbs. And as the expressways which opened up the suburban frontier were gradually completed, a ring of regional and smaller shopping centres – typically enclosed malls, following the pattern of Victor Gruen's pioneering Southdale centre in Minneapolis (Lowe, 2000), rather than open malls as in the 1950s – was built around the city by an all-powerful economic alliance which linked together developers, department store chains and the lenders of finance. As Frieden and Sagalyn (1989: 80) observed:

> Being able to move fast was a high priority. Once developers convinced department store executives to go to the suburbs, those retailers lost little time gearing up for expansion. Life insurance companies wanted to write mortgages for what they considered the Cadillac of investments, secured not just by real estate but by long-term leases from national retail

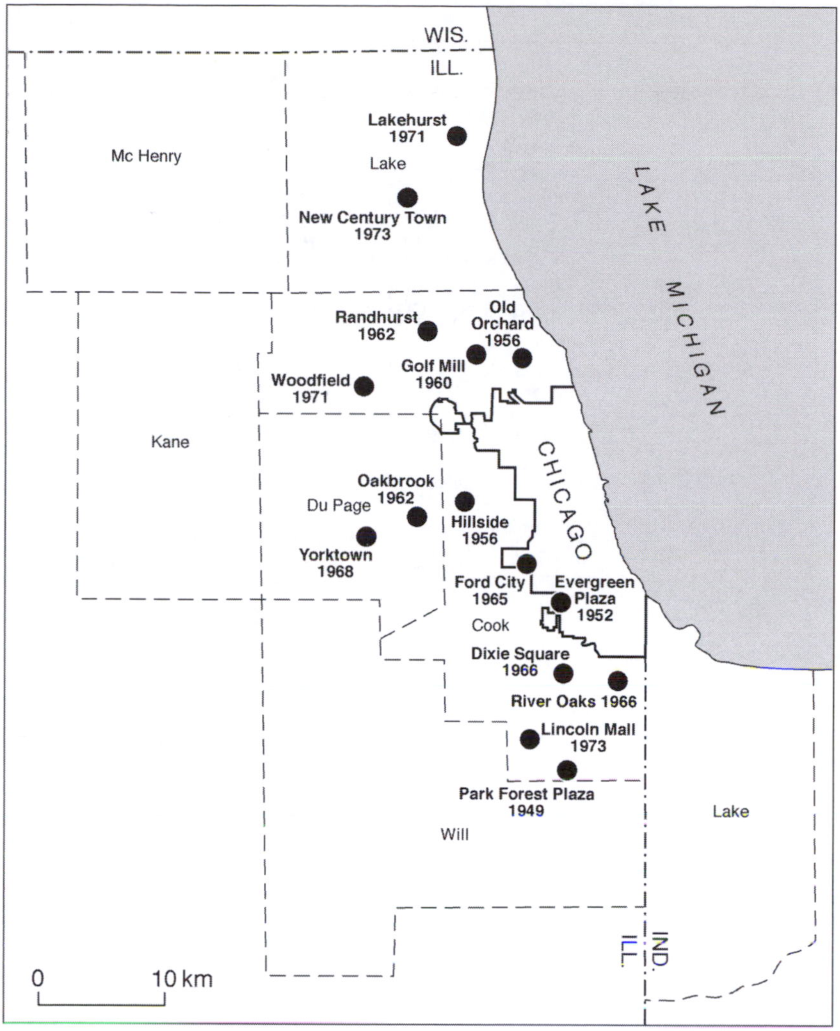

Figure 7.2: *The development of Chicago's suburban regional shopping centres, 1949–1974*
Source: Redrawn from Berry et al. (1976: 45).

> chains. Local regulators and citizens were in step with growth, and zoning conflicts were rare. In that environment, most malls went up fast . . . Time was money and competition was intense. 'Every good developer had a backlog of 15 projects . . . The problem was to get them built . . . to get there before someone else did.'

By the end of the 1960s, 11 suburban regional shopping centres were open for business with combined retail sales rivalling Chicago's CBD. By 1974, following an explosion outwards to the beginnings of a second ring (Figure 7.2), there were 15 regional centres. Those centres contained over 40 branches

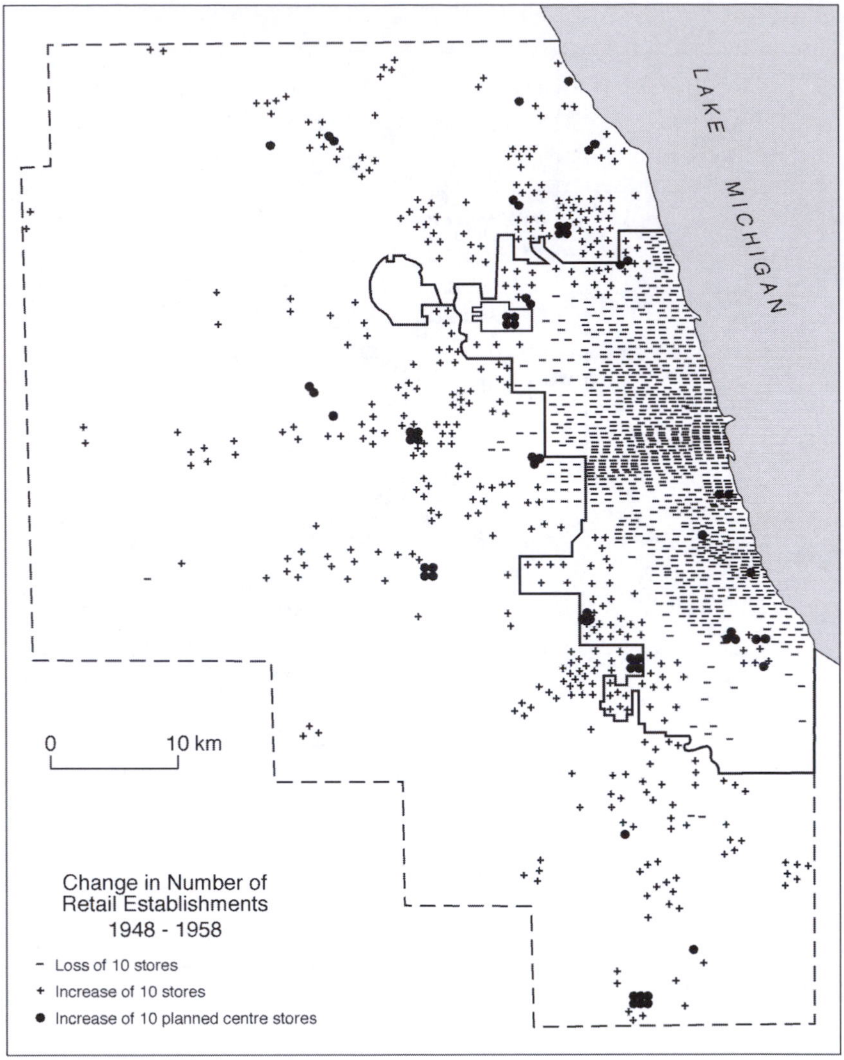

Figure 7.3: *The shifting retail map of Chicago, 1948–1958*
Source: Redrawn from Berry (1963).

of the seven department stores (Marshall Fields; Carson, Pirie, Scott; Goldblatt's; Wieboldt's; Sears Roebuck; Montgomery Ward; J.C. Penney) that had once dominated the downtown CBD but whose corporate profits were now tied to sales in the suburban centres.

Left behind by this cycling of investment into suburban construction and the 'shuttling' of retail capital to the suburban periphery were the traditional retail centres of the inner city – particularly those caught up in the 'rolling tide of residential segregation' as neighbourhoods experienced dramatic shifts in ethnicity and income. Reading 7.2 from Canadian geographers Ken Jones and Jim Simmons' book *The retail environment* (1990) but based, in turn, on a report of the City of Chicago Planning Department and the earlier work of Berry and his associates, outlines the case of one of those inner city shopping areas known as 63rd and Halsted. Prior to the Second World War and in the immediate post-war period, 63rd and Halsted was the largest retail centre in Chicago outside the CBD, but by the late 1970s/early 1980s, it had been reduced to a shadow of its former importance.

Reading 7.2 – Devalorization of inner city retail: the case of 63rd and Halsted

Forty years ago, 63rd and Halsted was the largest retail centre in Chicago outside the Loop. It was located about 8 miles from downtown, at a point where the elevated transit line connected with a network of bus and trolley routes, in the centre of mile after mile of blue-collar and middle-class neighborhoods. A block-long Sears store was the main attraction, but there was also Weiboldt's and Goldblatt department stores, a movie theatre, and hundreds of smaller stores. In 1958 retail sales amounted to $250 million (in 1982 dollars), more than any of the new regional plazas that were springing up in the suburbs.

However, in the 1960s retail activity began to decline and in the 1970s it collapsed. Blacks replaced Whites in the immediate neighborhood in the 1960s and household incomes declined. A recent market analysis suggests that while 45,000 households still live in the trade area, the average household income is less than $10,000. Forty per cent of the households earn less than $7500 and 45 per cent of the population is 25 years old or less.

It is difficult to maintain a retail structure when there is so little money to spend. Sears moved out in the mid-1970s; Goldblatt's went broke. Of the 100 stores that remain, the only chains are Walgreen's, Jewel Foods, and Payless Shoes. People who have cars shop at Evergreen Plaza, a successful Black mall 6 miles away. The 63rd and Halsted center survives in an empty shell of abandoned stores and vacant lots. The pedestrian concourse that came with an urban renewal scheme two decades ago looks shabby and inappropriate. Declining income leads to fewer stores and fewer customers – a reversal of the growth cycle that sparked North Michigan Avenue.

Even so, relative to other retail areas on the south side of Chicago, 63rd and Halsted is a kind of success. It still has 100 stores, many of them locally owned, and sells $50 million dollars' worth of merchandise. It has infrastructure: all it needs is customers with money.

Extracted from: Ken Jones and Jim Simmons (1990): *The Retail Environment*. London: Routledge, 273. Based on original reports by the City of Chicago Planning Department (1987) and Berry and Tenant (1963).

Of course, as Wilbur Kowinski (1985) graphically described in *The malling of America*, what was true of Chicago was true of much of the United States in the 1960s and 1970s – 'the go-go years of shopping centre growth' in which 'cities were barely visible on map[s] contoured by swirls of suburban activity linked together by highways' (Frieden and Sagalyn, 1989: 79). The regional shopping malls, 'signature structures of the age' – those 'meticulously planned and brightly enclosed structures, those *ideas* conveniently located just off the great American highway' (Kowinski, 1985: 22) – spread their 'suburban alchemy', their 'middle American magic' (Kowinski, 1985: 22) far beyond their original boundaries. Downtowns throughout the USA, with their traditional hit or miss retailing mix and their chaotic and often threatening public spaces, wilted against the competition of the suburban centres and awaited a new vision, in some cases sustained by business coalitions which retained hopes of some form of downtown retail revival. Other inner city retail centres were more frequently left, however, as in the case of 63rd and Halsted, to a process of gradual or sometimes extremely rapid devalorization. And outside the USA, particularly in Europe, central and city governments and urban planners observed the developments with trepidation – often vowing to use regulation to prevent similar processes, at least in their most extreme form, from decimating the vitality and viability of their city centre retailing.

Reading 7.3 from the work of Cliff Guy and Dennis Lord, a city and regional planner and geographer from the UK and USA respectively, explores some of these issues in the context of a comparison between two cities with very similar population and retail market sizes, Charlotte in the USA and Cardiff in the UK, at the beginning of the 1990s. Although both cities had experienced similar pressures towards suburbanization of retail investment during the 1970s and 1980s, the retail structures of their central areas are shown to have evolved in radically different ways during these years. Whereas Charlotte's city centre retailing had shrunk to almost complete insignificance during the period, in Cardiff UK land-use planning regulation had been used to restrain the shifting of retail investment to the suburban periphery and the city centre remained the focus of retail investment.

Reading 7.3 – City centre retail transformation: a US–UK comparison

In the early 1970s, both city centres had substantial retail areas: about 2.6 million sq. ft. of gross retail floor area in Cardiff and 1.6 million sq. ft. in Charlotte . . . [Since then] Charlotte has experienced massive disinvestment . . . The three large department stores, which had accounted for more than half of the total central-area retail space in the 1960s and 1970s, had closed by late 1990. Sears was the first to abandon downtown operations when it closed its store in 1977, following the opening of a new store in Eastland Mall, a suburban regional shopping centre six miles east of the city centre. Belk followed in 1988 with the closure of its 400,000 sq. ft. store, now the site of the city's tallest office tower . . . Iveys finally closed its doors in late 1990 . . . Between 1970 and 1990, 74 planned shopping centres were built in the Charlotte area, most having suburban locations. These contained almost 10 million sq. ft. of retail floorspace, more than the total in Cardiff. Each of the three former department store companies in the central area is now represented by stores in the city's three large suburban malls . . .

In Cardiff the city council and county council have consistently opposed major retailing development outside the city centre. They have used their powers under town and country planning legislation to refuse permission for such development on many occasions. While some of these decisions have been overruled on appeal to the Secretary of State, there is no doubt that the council's attitudes have encouraged retail in the city centre . . . The city council has also been involved directly in city-centre retail development [using] its powers under town and country planning legislation to purchase land required for large-scale redevelopment schemes . . . The largest of these areas, a site of some 8 acres to the south of [the principal shopping street], was bought in 1977 by the council for £4.5 million. The council then entered into an agreement with the Heron Corporation (the developer) and Coal Industry Nominees (the financier) to build the St David's Centre on this site. The council is thus ground landlord of the centre, and also has a share in the centre's operating profits. The council has also been involved indirectly in other recent schemes through detailed negotiation with the developers . . .

[In Charlotte, in contrast,] officials, organizations and local planning authorities have not shown the will, nor do they have the legal means, to stop the conversion of central-area land from retail to other uses. Neither have they made any effort to stop suburban retail development, a position that would not be politically feasible because of the strong pro-development sentiment in the business and political communities.

Extracted from: Cliff Guy and J. Dennis Lord (1993): Transformation and the city centre. Chapter 5 in Rosemary Bromley and Colin Thomas (eds) *Retail change: contemporary issues*. London: UCL Press, 88–108.

As Guy and Lord discuss in much greater detail in their paper, in both cities the enclosed shopping mall – the format which had so successfully spearheaded the exodus of retailing to the suburban periphery – had during the 1980s been brought into the downtown to revitalize its retail offer. In the UK and continental Europe, where retail decentralization was frequently controlled by special laws governing new development or via the implementation of the land-use planning system (Guy, 1998e), the result was often highly successful. In the US, as exemplified by the failure of Charlotte's attempt to develop 'Cityfair', a downtown 'festival'-style retail centre, in the late 1980s, the results were far more mixed and the route to downtown retail revival frequently long and tortuous. It is to this story of downtown retail rehabilitation during the 1980s and 1990s, and the reversal of the shifting of retail investment to the suburbs in American cities, that we now turn.

Rehabilitation: would the mall play downtown?

During the 1970s the development climate which had powered the suburban shopping centre boom of the previous 20 years in the major US cities gradually and subtly began to change. The pipeline of development opportunities began to shrink – in some areas mall construction had jumped well ahead of population growth – and the build-up of suburban growth regulations began to stretch out development times and costs.

> Developers groped for new strategies to adapt to the dwindling number of suburban sites and keep their big organizations active. Many built smaller malls in secondary retail markets that had been passed over before . . . Others made profitable improvements to older, existing malls by enclosing open air centres, adding new department stores and renovating the common areas. A handful began to think about urban sites (Frieden and Sagalyn, 1989: 93).

At the same time urban planners, city officials and downtown business coalitions, after two decades of brutalizing, clearance-orientated urban renewal and highway programmes, began to reach 'a new consensus about what was appropriate for downtown, and the shopping mall was an important part of their vision' (Frieden and Sagalyn, 1989: 94). The idea of reviving downtown retailing was in the air – not least as a means of humanizing the sanitized spaces of the central city redevelopment projects of the 1950s and 1960s. Early experiments in cities such as New Haven, Buffalo, Sacramento, Worcester, Hartford and San Bernardino produced rather mixed results, but the opening of Faneuil Hall Marketplace in Boston in 1976, the historic three-building complex that included Quincy Market, was a landmark. Developed by a partnership of the previously successful builder of suburban malls, James Rouse, and Cambridge architect Ben Thompson, it was the first nationally recognized popular success in downtown retail development. In turn, it spawned countless imitations and the genre of the so-called 'festival marketplace'. Reading 7.4 from US geographer Jon Goss provides a brief summary of these developments and centres, extracted from a much wider critique of a retail form which spread rapidly across America during the next decade.

Reading 7.4 – The Faneuilization of America

As early festival marketplaces drew unforeseen crowds, liberal commentaries celebrated the return of public life to the city . . . and the 'triumph of 'natural urbanism'. They also were a surprising commercial success, generating record sales per square foot and catalyzing further real estate development in decaying downtown districts, in turn producing enhanced tax revenues and a positive civic image for urban government. Ironically, for he had made his fortune developing suburban shopping centres, James Rouse was hailed as an 'urban visionary' and the 'saviour of downtown'. During the 1980s, the 'festival setting' became a keystone component of urban redevelopment and cities throughout the US sought to lure this 'urban Midas' and 'wizard' into a public–private partnership to construct a festival marketplace in their city. There are now at least 25 cities in the US that have jumped on the 'festival bandwagon', and many other retail developments incorporate some of its hallmark features, such that observers refer to the 'Rousification' or 'Faneuilization' of America.

> As festival marketplaces have been widely reproduced according to the more or less clichéd formulas developed by Rouse and Thompson, and as they have failed financially in some contexts, some cities have signalled a concern that public subsidies have been too generous and that promises of revitalized public space are false. They argue that the festival marketplace contrives to be everything that it is not – it is a 'carefully orchestrated corporate spectacle' masquerading as a genuine public space ... and better represents a 'triumph of commercialism' over natural urban life ...
>
> Neither celebration nor condemnation of festival marketplaces is appropriate, however, partly because so little is known about how they work for the majority of patrons – rather than for the critics who form a semiotically privileged and disproportionately vocal minority – but also because they are profoundly contradictory spaces that articulate both desire for genuine urban life and rational strategies of social control.
>
> **Extracted from: Jon Goss (1996): Disquiet on the waterfront: reflections on nostalgia and utopia in the urban archetypes of festival marketplaces. *Urban Geography*, 17, 221–47 (citations removed from extract, see original pages 222–3 for details).**

We will take up many of the contradictions which Goss draws attention to in Chapters 9 and 12. Here it is sufficient to note that the festival marketplaces – retail centres which lack as Goss (1996: 221–2) notes: 'a conventional anchor tenant and rely on a sense of historic public life, represented in architecture, cultural exhibits, concert programs and ethnic festivals, to attract customers to the speciality produce markets, restaurants and speciality shops' – were not the only prototypes for downtown retail revitalization. Others, beginning with Water Tower Place which opened in Chicago in 1976, urbanized a model set in 1970 by the suburban Galleria centre in Houston and mixed together retail (including a department store anchor), hotel, office, recreation and car parking space into a single high-density development. Indeed downtown retail revitalization increasingly went hand in hand with office development, which began to boom in the late 1970s/early 1980s. It also went hand in hand with the planning and building of convention centres, with the development of downtown hotels to meet the needs of office and convention trade, and with the revival of in-town living by young professionals who began to gentrify long-neglected inner city neighbourhoods in the mid- to late 1970s. As MIT urban planners Bernard Frieden and Lynn Sagalyn (1989: 284, 287) noted in their well-known book *Downtown Inc: how America rebuilds cities*:

> The downtown shopping malls were well suited to a time of transition in the recovery of downtown. They made a start at bringing people back to the city and helped attract tourists, conventioneers and business visitors. They made in-town living more attractive. By changing the development signals, they stimulated at least some other construction and possibly a great deal ... Even a public that clearly preferred to live in suburbs or small towns recognized that cities had advantages as well as problems.

However, as Goss indicates in Reading 7.4, and as we previously noted in relation to the failure of Charlotte's attempt to develop a downtown 'festival' centre in the late 1980s, the route to downtown retail revival in US cities in the late 1970s and 1980s was frequently long and tortuous, and sometimes never accomplished. It involved high-cost development, working with rigid building codes, difficult land assembly problems, and significant city subsidies frequently involving 'off budget' financing. Moreover, it invariably involved complex public–private partnerships in which cities had to serve both as partner and as regulator at the same time, and in which the influence of local politics loomed large. Operating costs of the downtown malls were also higher than the suburban malls and the margin of error between commercial success and failure was frequently wafer thin. Gradually, however, both cities and developers learned to manage the process. By the late 1980s the playing field, that had so favoured suburban retail from the 1950s to the mid-1970s, was more level again with 'lenders prepared to lend, speciality stores prepared to pay high rents, and department stores that wanted to expand willing to go to the city' (Frieden and Sagalyn, 1989: 314). Nevertheless, there was an inevitable shake-out of some of the more marginal downtown marketplaces, and the politics and public–private financing of downtown retail development in the US remained an intensely complex process throughout the 1990s.

Reading 7.5 by Marilyn Lavin, a US business studies professor from the University of Wisconsin, illustrates and expands this latter theme in the context of attempts to rehabilitate retailing in the Harlem district of New York City. By the early 1990s, as Lavin points out, the retail development agenda in US central cities had broadened out from the downtown malls and festival marketplaces of the 1980s to much wider attempts to regenerate the retail facilities of inner city low-income districts which had been stripped of even the most basic retail facilities over the past 40 years. On the face of it the attempt by Pathmark, the leading food retail chain in the metro New York/New Jersey area, to develop a moderately large-format (53,000 sq. ft.) store on 125th Street in the heart of Hispanic East Harlem provided the opportunity for 'a perfect partnership between the public sector, the private sector and a non-profit group' (*New York Times*, 28 April 1995: A16). Harlem, with its huge low-income population (no less than 45,000 households in a one-mile radius from the site to be developed) had been so neglected by the major retailers that it had not a single large supermarket – the largest was barely 10,000 sq ft. Moreover, the site proposed for development had been vacant for more than 20 years and was owned by the City of New York. However, despite considerable development cost support (including $1.1 million of state and federal funds) and the sale of the site on a mere 1 per cent mortgage by the City of New York, the project became embroiled in a protracted and highly charged process of local politics involving intense ethnic divisions in which the mayor of New

York (Giuliani) had to play a major role in establishing a compromise. As a result, a proposal first announced by Pathmark at the end of the 1980s did not gain approval until 1995 and was not completed until 1997. Clearly, despite the lessons which Frieden and Sagalyn suggested had been learned in the 1980s in terms of public–private partnership development of downtown malls, the wider revival of US inner city retail during the 1990s continued to be beset with many of the same problems which had led the suburban mall developers of the 1960s to regard central city retail as a development quagmire. Indeed, these facets of US inner city retailing and the size and characteristics of the unserved inner city retail markets were to become a central focus of Michael Porter's well-known 'Initiative for a competitive inner city' during the late 1990s.

Reading 7.5 – Pathmark comes to Harlem

In recent years, large format retailers have become increasingly involved in establishing operations in the 'inner cities' of the United States. Retailers' efforts in the 'inner cities' are not altruistic . . . in such area(s), competition is low, consumer need is evident, and urban governments, in recent years, have become more receptive to large-scale stores . . . Although economic and competitive factors have led mainstream supermarket operators to reconsider city locations, the return of such food retailers to urban sites – even those in degraded shopping areas – may not, however, be without controversy. Zukin (1995) has argued that corporate revitalization may prompt disagreement and resistance with current inhabitants regarding 'who owns, who occupies, and who controls the city's public spaces'. [She] further contends that the upgrading of buildings and facilities can be seen as an intrusion of international market culture, and a middle class appropriation of areas that have traditionally serviced low-income people . . . The experience of Pathmark Inc., as that retailer attempted to establish a store in . . . Harlem presents a unique opportunity to examine these issues . . . Harlem with its vast low-income population, had no full-service supermarket, and the site selected for the store had been vacant for more than 20 years. Nonetheless, the supermarket chain encountered intense opposition . . .

Pathmark put forward its Harlem project at a time when, in theory, the governmental climate was highly favourable to such an undertaking;[a] nonetheless, the proposed supermarket, in reality, became entangled in a quagmire of ethnic/racial politics that delayed its approval for six years . . . The owners of small bodegas and mid-sized grocery stores in the area surrounding the 125th Street site were vociferous in their opposition to Pathmark. They argued that they had invested in East Harlem when that area had been neglected by major supermarkets. Their position, as a group likely to lose if the development project proceeded, was forcefully articulated by their lobbyist who contended that 'it's unfair to subsidize one of the richest food corporations when you have minority entrepreneurs who have invested their livelihoods to service this neighbourhood when big supermarket chains wouldn't touch it'. Moreover, the local grocers were able to enlist the assistance of local politicians and to utilize long-standing ethnic tensions in their effort to block the Pathmark store[b] . . .

The Retail Initiative [an equity fund to assist community groups in their efforts to obtain necessary financing for urban retail developments and to encourage major food retailers to participate in such projects, and which had invested $1.5 million to trigger the Pathmark project] expected that the Harlem project would serve as a model for its programme ... helping overcome retailers' reticence about the viability of inner city undertakings ... [Its president] commented: 'The Harlem project has taught us all a valuable lesson ... the level of resistance caught us off guard'.

Extracted from: Marilyn Lavin (2000): Problems and opportunities of retailing in the US inner city. *Journal of Retailing and Consumer Services*, 7, 47–57.

Notes:

[a] The leakage of retail sales to large-format stores in the suburbs had prompted the Giuliani administration to reconsider during this period the city's policy towards retailers and to suggest a raft of initiatives including: zoning modifications, liberalization of regulations on retail development of stores up to 200,000 sq. ft. on wide streets, elimination of the higher parking/square footage ratio for supermarkets, elimination of licence requirements no longer deemed in the public interest, and reductions of many of the city's business-related taxes.

[b] These tensions were essentially between the Hispanic community of East Harlem and the Black African–American community of Central Harlem. A Hispanic development partner could not be found for the Pathmark project and it was offered instead to the development arm of the African–American Baptist Church of Central Harlem. That resulted in ongoing antagonism with all nine members of the Hispanic caucus of the New York City Council opposing the establishment of the store and putting pressure on Mayor Giuliani, who owed his election in part to the Hispanic vote. After a tense and evenly divided vote on the sale of the site by the city to the development group, Guiliani offered a compromise to assuage his Hispanic supporters and the political obstacles to the project were finally but rather uneasily overcome.

In the UK and continental Europe, despite the regulation which had largely protected central city retailing from the most extreme forms of US-style retail decentralization (Guy, 1998e), withdrawal of retail capital from low-income inner city areas had similarly, though to a much lesser extent than the US, become a pressing social issue by the 1990s. In the UK, as Guy's (1996) study of the consequences of building a ring of edge-of-city large-format superstores around one British city illustrates, and as Thomas and Bromley (1993), Wrigley (1998c) and others debate, food retailing in particular had been unevenly stripped out of parts of British cities or repositioned downwards in quality terms in those areas. The result – though not to the extreme levels in Harlem prior to the opening of the Pathmark store or found elsewhere in the US by the 'Initiative for a Competitive Inner City' – was the emergence of underserved and socially excluded urban markets which in UK public-policy debate terms became known as *food deserts*. Indeed, the public health problems of those areas saw an important linkage of the retail access, social exclusion and health inequality debates in the UK by the late 1990s (for a summary of these debates see Wrigley (2002a) and for broader commentary Clarke (2000)).

And could 'the street' play in the suburbs?

At the heart of revitalization of US downtown retailing from the late 1970s to the 1990s was the subjection of 'a key part of downtown to the discipline of suburban malls ... For middle-class people accustomed to suburban shopping, the downtown marketplaces were different enough to be exciting but familiar enough to make them feel comfortable' (Frieden and Sagalyn, 1989: 236, 240). And an important feature of this process was the heavy reliance of the downtown malls and marketplaces on national chainstores as their tenants – not least because these were the dependable tenants that the developers of downtown centres had come to rely on most heavily elsewhere. In the process, however, the innovation and energy of the small independent urban retailers were often lost to the downtown centres. Yet, ironically, it was that very promise of urban authenticity, not simply in terms of the architecture and ambience of historic public life but also in terms of the excitement and creativity of the retail mix, which was being offered by the downtown centres. As John Goss notes in the paper from which we extracted Reading 7.4, the result was the creation of retail spaces that were profoundly contradictory. Moreover, embedded within them were the seeds of yet further reversals of what in the introduction to this chapter we referred to as 'the pre-existing logic and location of profit extraction and the sources and flows of creativity in the retail landscape'.

If US downtown retailing had been revived by transplanting the suburban mall and its disciplines and clothing it in a pastiche and promise of revitalized public space, could not suburban retail centres reverse the process? That is to say, could a suitably disciplined version of the urban 'street' be transplanted to the suburbs and the flow of retail innovation that had favoured the downtown during the late 1970s and 1980s be reversed? During the late 1990s several commentators in the USA began to suggest that this was exactly what was beginning to occur – a recycling of retail innovation and inevitably investment, to the retail centres of the inner-ring suburbs that were in the process of remaking themselves to feel more quirkily urban. This process involved taking many of the architectural features of the urban 'street', rebuilding parts of their retail centres to look like the Greenwich Village neighbourhoods of New York captured by Jane Jacobs in *The death and life of great American cities* (1961) and, perhaps most importantly, catering to a disenchantment amongst particular groups of affluent consumers with the creeping chainstore 'suburbanization' of downtown retailing. To service these essentially anti-mall and anti-chainstore-orientated groups, a new form of suburban *anti-mall mall*, masquerading as urban 'street' and tenanted by *anti-chainstore chainstores* that look and feel independently owned, but are often not, has developed.

Academic writing on retail and consumption, particularly by geographers, has yet to catch up fully with these trends – although there is some initial consideration of the selective rediscovery of the 'street' by corporate US retail during the 1990s in the work of Sallie Marston and her students (Marston and Modarres, 2002; Hankins, 2002). However, US journalists have begun to produce insightful commentary on the issue. In Reading 7.6 we provide an extract from one of these articles, produced as part of a much larger feature on 'The suburban nation' by the *New York Times* (9 April 2000).

Reading 7.6 – Turning suburbia into SoHo

Urban exiles disapprove of the suburbs in principle but find themselves living there in practice ... It turns out there are so many urban exiles that an entire retail sector has grown up to meet the demands of antisuburban suburbanites. And these stores, congregated in affluent inner-ring suburbs and spreading their ethos outward, are changing the face of suburbia. Bethesda [Maryland] has quickly become inundated with anti-chain chain stores. These are little boutiques that cater to people who consider themselves too refined and individualistic to shop at the mall or the mass-market big-box stores. The boutiques look and feel independently owned, but they are not ...

The nonconformist instincts that used to drive people from the suburban subdivisions into the bohemian city now impel them to remake the suburbs so they feel more quirkily urban. The result is that even big-box stores ... are trying to divide themselves into little boutiquey areas. And inner-ring suburbs, the hot cauldrons of this trend, are rebuilding themselves to look like little urban streetscapes, with wider sidewalks, small stores and small restaurants ... Towns like Bethesda have become nice places to hang out in. And another strange thing has happened. The need to design places for antisuburban suburbanites has unleashed a lot of creativity ... The urban-exile suburbs are replacing the cities as the hubs of retail innovation. Now if you want to go to a weird store that sells Tibetan dining-room stuff, you go to Bethesda. But if you want to go to a huge chain store, like a Gap or a Disney or a Hard Rock Café, you go downtown to Washington, or in the New York area, to Times Square or 57th Street.

The whole point of living in the city used to be that you were at the cutting edge. Life might be stressful and unpleasant, but at least you were where the action was. But with the chains invading the cities and rents rising, it is cheaper to experiment at a suburban location. The urban exiles, unlike their suburban predecessors, are proud to be early adapters; they like new things and have money to spend. Stores like Anthropologie open first in the suburbs and then trickle down to locations like SoHo and Fifth Avenue.

Extracted from: David Brooks (2000): Exiles on main street. *New York Times*, 9 April 2000: Section 6, 64–8.

In assessing Reading 7.6, it is important to note that the tradition of crossover from journalism to academic writing in this area is a long one. The most notable example in the 1990s is the well-known book by Joel Garreau of the

Washington Post on *Edge city: life on the new frontier* (1991). Reading 7.6 follows in that tradition, and although journalistic in orientation, captures in our view an important essence of the 'inconstant geography' of retail capital in the late 1990s, as the shuttling of retail innovation and investment in US cities moved into an interesting new phase.

8

The inconstant geography and spatial switching of retail capital 2: regional issues and internationalization

If the spatial switching of retail capital between the downtown and suburban periphery was, at least in part, one of the most familiar stories of economic geography, the two issues that we turn our attention to in this chapter were certainly not. The internationalization of retail capital barely rated a mention in the discipline until the 1990s, and geographical writing on retail and consumption has been slow to engage with the rapid transformation of markets by an emerging group of retail transnational corporations. Similarly, the regional dimension of the inconstant geography of retail capital, despite some pioneering work in the 1980s, remained for a long period relatively untouched. Fortunately in both cases, as we will see below, geographical readings of retail have recently broadened to incorporate these issues. It is to these themes that we now turn.

A regulatory take on issues of regional expansion and consolidation

During the 1980s, Swedish geographer Risto Laulajainen (1987) in his book *Spatial strategies in retailing* considered the range of regional expansion strategies undertaken by leading US retail chains during the twentieth century as they grew from local to multi-regional, and in some cases to proto-national operations. Although couched very much within the orthodox spatial-science-oriented tradition of the time in retail geography, Laulajainen did begin to explore in that book, and in two subsequent journal papers (Laulajainen 1988, 1990), some of the regulatory context of regional expansion by retail chains. In the process, his work provided a link to the themes of retail regulation that we examined in Chapter 6 and which, as we saw, have represented a significant area of theoretical debate in the 'new retail geography' of the 1990s. In considering regional expansion of US retail chains by a process of merger and acquisition, Laulajainen (1987: 170) in particular noted that: 'It is not just the willingness of the target to get acquired or the aggressor's capacity to place a hostile bid. It is as much a question of the FTC's [Federal Trade Commission's] attitude to the deal which, in turn, is dependent on the political climate in general.'

During the 1990s that regulatory constrained nature of regional consolidation processes in US retailing and the making of competitive space in different sectors of the industry has provided a focus for research within a 'new retail geography' framework. In particular, Neil Wrigley and Steve Wood have considered these processes within the food retail and department store sectors respectively.

In the case of the US food retail industry (Wrigley, 1999b, 2001a), we discussed in Chapter 2 and again in Chapter 6 the conundrum of an industry which remained stubbornly unconsolidated until the late 1990s and the role of regulation within that process. By the late 1990s, as Reading 2.4 outlines, a wave of regional consolidation on the east and west coast of the USA had occurred. This was followed rapidly by mega-mergers between the leading multi-regional firms creating, as Figure 8.1 illustrates for the case of

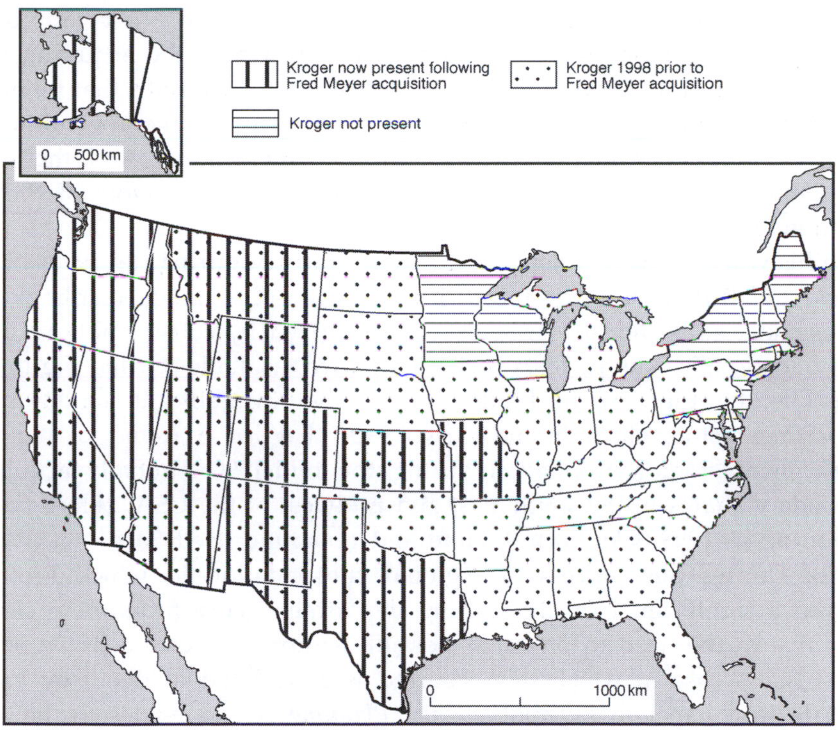

2200 supermarkets and 820 convenience stores

Major trading names:

Kroger	**Fred Meyer**
Dillon Food Stores	**Smiths/Smitty's**
Fry's	**Ralphs/Food-4-Less**
King Soopers	**Quality Food Centers/Hughes**

Figure 8.1: *Kroger following merger with Fred Meyer*
Source: Author, based on information supplied by Kroger and Fred Meyer.

the merger of Kroger and Fred Meyer, embryonic national chains. During this process, the geography of acquisition clearly involved, as Laulajainen had suggested, an important regulatory dimension. Acquisition targets for the major firms were not only required to offer strategic fit in terms of potential synergies, market position and scope for improved financial and operating performance, but also regional fit in respect of offering limited risk of rigorous antitrust enforcement by the FTC. Potential acquisition targets, in their turn, had to be conscious that the pre-existing geographies of the multi-regional consolidators and the regulatory risks posed by market overlap, implied a strictly limited number of potential suitors – knowledge that they had to factor into the timing and price of their decision to sell or merge.

During the consolidation wave that swept through the US food retail industry during the late 1990s, therefore, merger and acquisition deals were structured by the leading firms in anticipation of the two-component regulatory framework outlined in Reading 6.2: that is to say, a framework in which the FTC essentially operated a so-called 'fix-it-first' approach, agreeing not to block mergers and acquisitions where the acquiring firm committed itself in advance to divest itself of market overlaps that might be deemed to be uncompetitive, whilst the state Attorneys General would often seek to impose far more rigorous antitrust enforcement. And the same regulatory framework essentially also applied to other sectors of the US retail industry.

In his work on the US department store industry during the 1990s, Steve Wood (2001a) examined these issues. Like US food retailing, the department store industry was also characterized by considerable regional consolidation. This is perhaps best exemplified, as Table 8.1 and Figure 8.2 illustrate, by the case of Proffitt's Inc. – an insignificant southern US department store chain in the early 1990s that led a round of regional department store consolidation during the mid-1990s, initially acquiring secondary chains in the south, east and midwest of the USA, before finally acquiring in 1998 Saks Fifth Avenue and renaming itself Saks Inc. (Wood, 2002). During this process of regional consolidation in which Proffitt's played a significant role, the leading department store firms were clearly sensitive to the need to pre-empt possible antitrust enforcement action by the FTC. A case in which this occurred is considered in detail by Wood (2001b) in a paper in *Environment and Planning A* and relates to the 1998 $2.9 billion acquisition by Dillard, the third largest conventional department store firm in the US (see Table 2.2), of Mercantile Stores, the seventh largest operator in the industry. The deal was announced by Dillard in May and was followed in July and August 1998 by a series of store swaps and divestitures to competitors, in part designed to pre-empt FTC regulatory action.

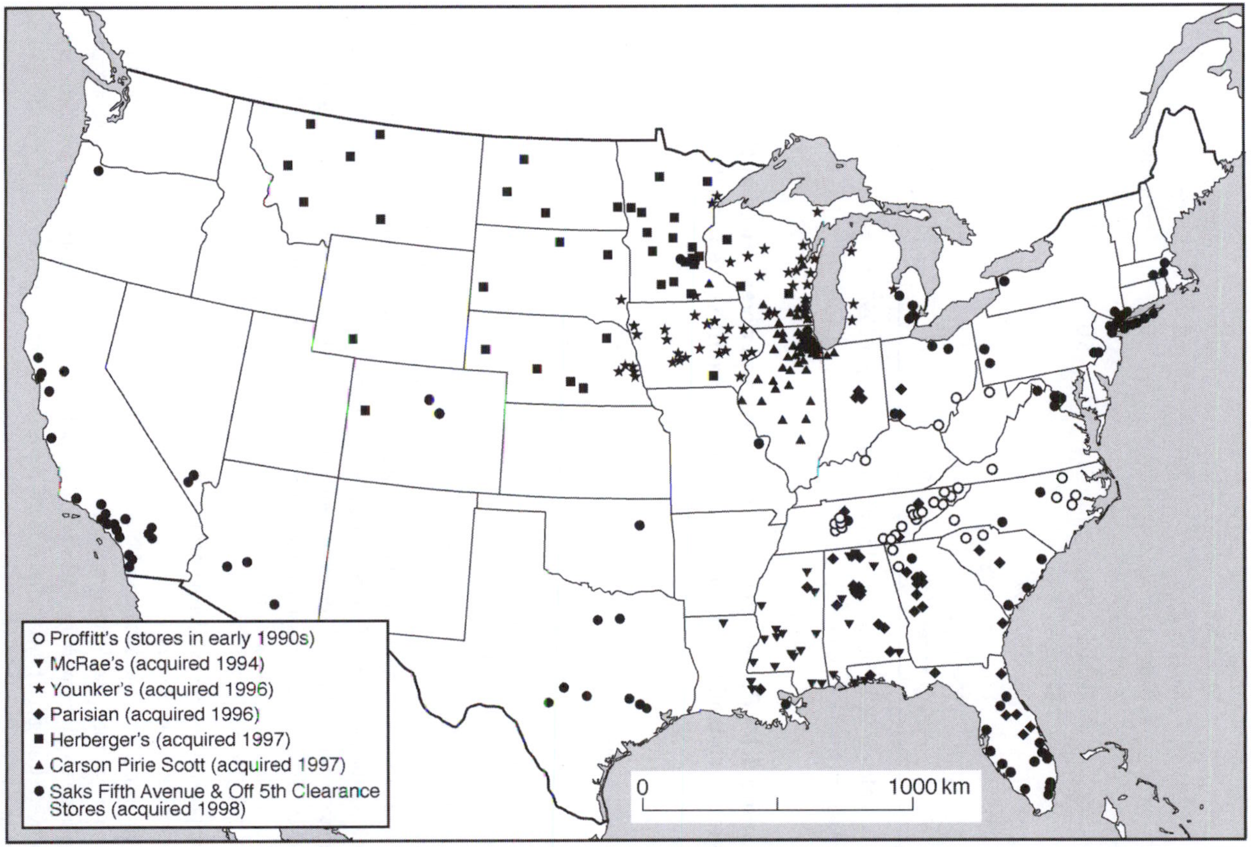

Figure 8.2: *The department store acquisitions and regional expansion of Proffitt's Inc. during the mid-1990s*
Source: Adapted from Wood (2002).

Table 8.1: *US department store chain acquisitions by Proffitt's Inc.*

Chain acquired [a]	Date	Stores acquired	Gross sq. ft (millions)	Locations
Lovemans	May 1988	5	0.3	Tennessee
Hess	July 1993	18	1.2	South-east
McRae's	March 1994	28	2.8	South-east
Parks-Belk	April 1995	3	0.2	Tennessee
Younkers	February 1996	51	5.0	Midwest
Parisian	October 1996	38	4.1	South-east and midwest
Herberger's	January 1997	40	2.8	Midwest and Great Plains
Carson Pirie Scott	February 1998	55	8.2	Midwest
Brody's	March 1998	6	0.3	North Carolina
Saks Holdings	October 1998	96	6.7	National

[a] Plus 15 ex-Mercantile stores acquired from Dillard's, August 1998.
Source: adapted from Wood (2002).

Reading 8.1 – Pre-emptive store swaps and divestitures in the regional consolidation of the US department store industry

Retailers learned in the 1990s to operate within the remit of the FTC's 'fix-it-first' policy . . . Two events, in particular, demonstrate the willingness of Dillard's to take action ex-ante of regulatory enforcement . . .

On July 19th 1998 Belk, the largest privately owned US department store chain with coverage principally in the Southeast US, agreed to exchange with Dillard's seven Mercantile stores located in Florida and South Carolina for nine Belk stores [principally] in Virginia. For Dillards, the transaction represented an agreement that would avoid duplication of stores in certain geographical areas, improving the strategic fit of the Mercantile acquisition for the corporation . . . [whilst at the same time] appeasing the FTC. In August 1998, Dillard followed the coup of the Belk store swaps with the announcement that it was divesting [26] of the acquired Mercantile locations to Proffitt's Inc. and May Co., again to pre-empt any FTC regulatory enforcement and to improve the geography of the acquisition . . . Critically, the stores sold were located primarily in markets in which Dillard had a strong presence prior to the Mercantile acquisition, especially around Kansas City, Nashville and Orlando. Such action prevented Dillard, therefore, from enjoying a localised monopoly in conventional department store retailing in those markets but equally ensured no cannibalisation of sales . . .

Although Dillard's acquisition of Mercantile was of itself strategically questionable, the execution of the consolidation provides an interesting case study of a firm tailoring an acquisition to conform to the FTC's 'fix-it-first' policy whilst at the same time producing an improved strategic fit beyond that evident in the pre-merger geography . . . it was this dual motivation that drove the divestitures and store swaps.

Extracted from: Steve Wood (2001b): Regulatory constrained portfolio restructuring: the US department store industry in the 1990s. *Environment and Planning A*, 33, 1297–1304.

In assessing Reading 8.1 it is important to note, however, that regulation and the inconstant geography of restructuring which it generates is subject to continuous re-interpretation. At the very end of the 1990s, for example, evidence began to mount that the FTC had begun to rethink important features of its prevailing 'fix-it-first' approach to horizontal merger and acquisition in the US retail industry (Cotterill, 1999; Wrigley, 2001a; Balto, 2001). Suddenly, and particularly in the food retail industry, regional consolidations which might have progressed earlier in the 1990s with 'consent orders' requiring some degree of divestment were effectively blocked by FTC opposition. Wrigley (2001a) and Balto (2001) discuss several examples of these 'landmark' cases, including an injunction obtained by the FTC to block Kroger's proposed acquisition of Winn-Dixie stores in Texas and Oklahoma and the failure of Ahold's attempt to acquire Pathmark in 1999. In the latter case, as Figure 8.3

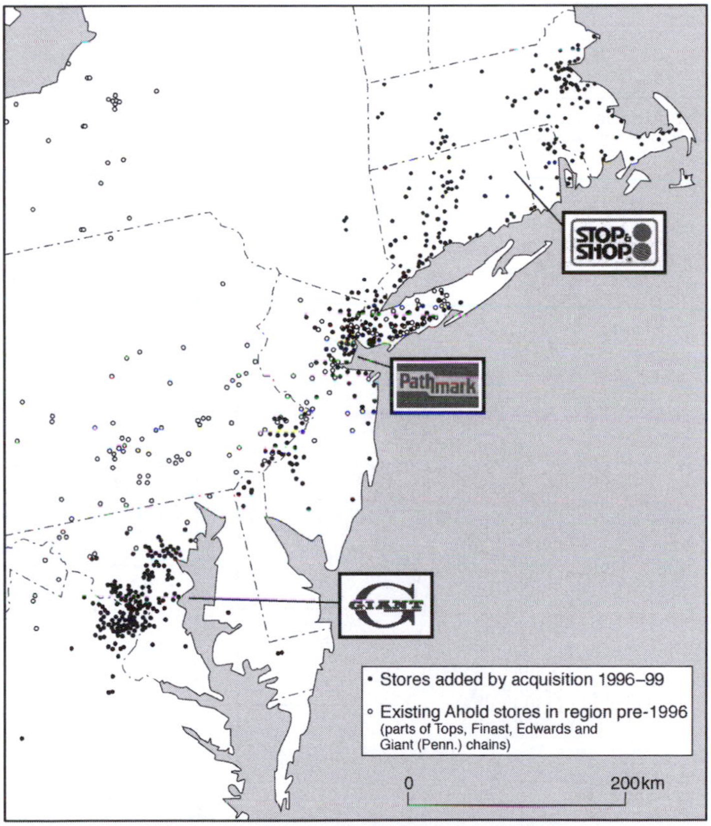

Figure 8.3: *Royal Ahold operations in north-east USA 1999, showing chains acquired (and proposed for acquisition) 1996–99*
Note: Ahold withdrew from acquisition of Pathmark, Dec. 1999, following opposition to acquisition from Federal Trade Commission.
Source: Author, from data supplied by Royal Ahold, Giant Food, and Pathmark.

illustrates, Ahold, the Dutch global retailer which had grown to become the fourth largest US food retailer by that point, would have significantly strengthened its core Boston–Washington DC area of market dominance by adding Pathmark and the leadership of that chain in the New York/New Jersey metro area. However, despite Ahold's willingness to divest a considerable number of stores in the area to eliminate competitive overlap, the acquisition collapsed in the face of FTC concerns about the effectiveness of the proposed divestments. The blocking of the Pathmark acquisition was quickly followed in mid-2000 by Delhaize America (the Belgian operator of US chains Food Lion and Kash n' Karry) being required, in effect, to divest the entire south-east division of Hannaford Brothers (see Figure 8.4) in order to obtain FTC approval of its plans to acquire Hannaford and build a multi-regional chain along the US east coast. However, no sooner had these landmark decisions in 1999/2000 signalled the shifting of the bar towards regulatory tightening by the FTC, than the Democratic Clinton administration was replaced by the Republican Bush

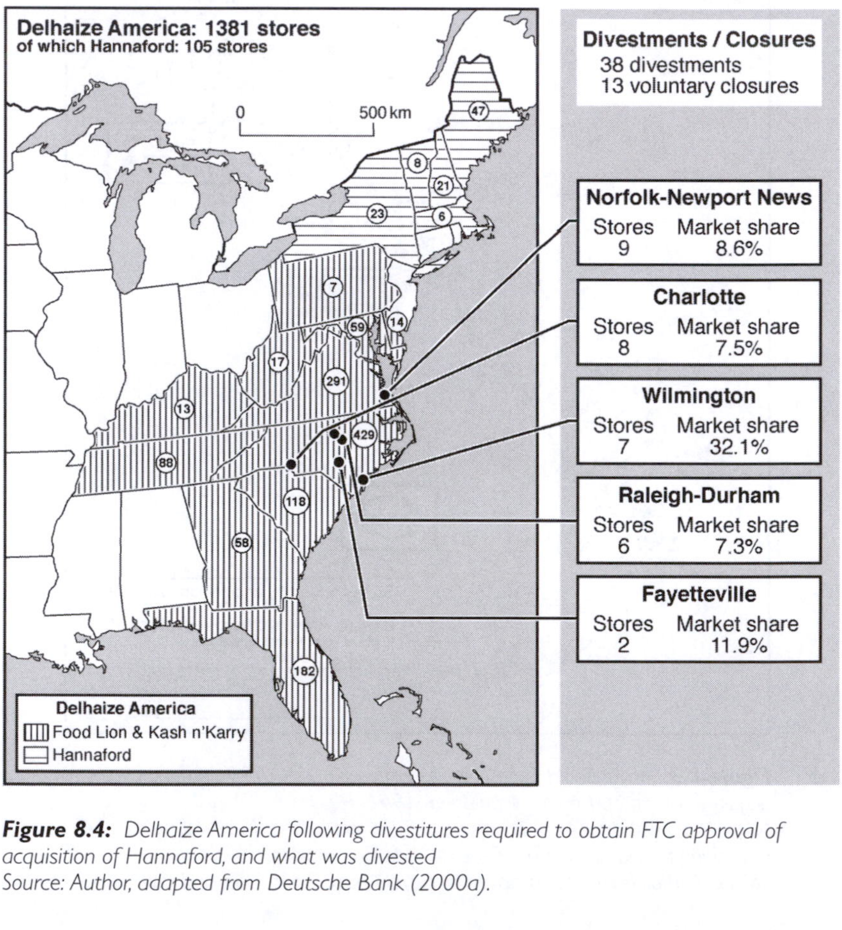

Figure 8.4: *Delhaize America following divestitures required to obtain FTC approval of acquisition of Hannaford, and what was divested*
Source: Author, adapted from Deutsche Bank (2000a).

administration. New political appointees were installed to head the FTC and the antitrust division of the US Department of Justice and corporate America began to anticipate a loosening of antitrust enforcement.

Competition and regional monopoly debates

Turning to the UK, we find work on the regional dimension of the inconstant geography of retail capital focused to a large extent during the 1990s on the emergence and consequences of an uneven geography of retail provision – in particular as it involved the leading food retailers. The major contributions in this area come from a group of geographers at the University of Leeds led by Graham Clarke, and some of that work (Langston et al., 1997; 1998) is most appropriately considered in the context of our discussions in Chapter 2 relating to the rise of the major food retail corporations in the UK and their expansion during and after the era of the so-called 'store wars'. By the late 1990s, however, the uneven geography of corporate retail provision in the UK and the extent to which it might possibly involve 'monopoly' or 'duopoly' situations in regional and local market provision had become an important policy issue. Indeed, it provided a significant focus of the Competition Commission's (2000) extensive enquiry into competitive conditions in the UK food retail industry and involved parallels with similar concerns exercising the FTC in the USA (see Wrigley, 2001b).

When the Competition Commission investigated regional shares of grocery sales in stores of 15,000 sq. ft. or more in the UK during the late 1990s, it found (Table 8.2) five firms accounting for almost 90 per cent of a market worth £40 billion in 1998/99. Each of these firms had market shares exceeding 25 per cent in at least one standard region of the UK and the national market leader, Tesco, had regional shares exceeding 40 per cent in East Anglia, Wales and Northern Ireland. Across southern England the regional shares of the top two food retailers (Tesco and Sainsbury) alone consistently lay in the 60–70 per cent range. Not surprisingly, the Commission expressed its concerns about the competitive structure of markets in some of these regions. In particular, it drew attention to 175 cases in which it believed it had found evidence that the major food retailers enjoyed effective 'monopoly' or 'duopoly' status in their local markets.

Having identified competitive situations which were a cause for concern and which 'should not be allowed to deteriorate', and expressed its belief that these were issues which 'cannot be addressed within the framework of the current planning regime' (Competition Commission: vol. 1, 154), the Commission considered possible remedies. These included: (a) requiring the major UK food retailers, as in the case of the FTC in the USA, to divest their monopoly or duopoly stores; (b) building competition issues and its regulatory authorities into a development consent process which, in the past under existing UK planning law, had been 'applicant blind'.

Table 8.2: *Regional shares (%) of grocery sales from stores of 1400 sq. metres (approx 15,000 sq. ft.) or more by major UK food retailers, 1998/99*

Region	Tesco	Sainsbury	Asda	Safeway	Morrison	Total
London	26.5	41.5	7.6	10.7	0.3	86.7
South East	36.2	31.0	10.7	11.5	0.1	89.5
East Anglia	42.1	27.5	9.4	9.5	0.0	88.5
South West	33.8	25.4	12.6	16.3	0.0	88.1
West Midlands	22.1	27.5	17.2	20.1	6.7	93.6
East Midlands	22.2	22.0	17.4	13.5	12.5	87.6
North West	21.7	17.8	31.5	8.6	12.8	92.3
Yorkshire and Humberside	17.8	17.4	24.3	7.8	27.8	95.2
North East	6.8	12.7	30.5	21.0	15.4	86.4
Wales	41.7	14.1	22.0	12.2	0.0	90.0
Scotland	20.1	8.9	27.0	29.5	0.0	85.5
Northern Ireland	43.1	21.4	0.0	20.0	0.0	84.6
United Kingdom	28.5	24.8	16.8	13.8	5.4	89.3

Source: adapted from Competition Commission (2000: vol. 1, 27).

Although neither of these possible remedies was, in practice, taken up by the Secretary of State for Trade and Industry (see Wrigley, 2001b), by virtue of the Commission's expression of its concerns about the trends in the competitive structure of local markets, and with the possibility of implementation of the Commission's proposals held in reserve, a message was being sent to the major UK food retailers that any major domestic consolidation moves via merger and acquisition would be met with the threat of significant regulatory action. Retail analysts at Deutsche Bank (2000b: 16), for example, estimated that 'were Sainsbury to acquire Safeway, we estimate that around 40 per cent of Safeway's selling area would be sold, at an unknown price, in order to obtain government approval'.

However, as in the US, potential regulatory action does not imply that merger/acquisition and divestment of this type will not occur at some point in the UK, and that the regional geography of the food retail industry will not be recast as a result. Indeed, the Leeds geography team (Poole et al., 2002) has recently attempted to examine the shifting geography of UK food retailing which would result from a number of merger/acquisition/divestment scenarios involving various combinations of the major firms or the major firms with the second-rank chains. Although extremely speculative, their 'what if' analysis is nevertheless highly germane in a context in which most of the leading firms reported to the Commission in the late 1990s that:

> they expected to see further consolidation in the shape of acquisition and merger activity over the next few years, either in a domestic or cross-border context (Competition Commission, 2000: vol. 1, 52)

and in which the entry of the world's largest retailer, Wal-Mart, into the UK via its acquisition of Asda in 1999, opened the UK market to the exigencies of the rapidly emerging globalization of retail capital. It is to this transformation of markets by a group of retail transnational corporations that we now turn.

The globalization of retail capital

During the late 1990s retail markets throughout the world began to be transformed by acquisition and merger-driven consolidation and by the rapid emergence within that process of an elite group of multi-national/transnational corporations (TNCs). The world's largest retailer, Wal-Mart, as we noted above, entered the UK in 1999 via the acquisition of Britain's third largest food retailer, Asda, and had previously entered Europe via acquisitions in Germany in 1997 and 1998. By the end of the 1990s Wal-Mart was well advanced along a path of internationalization which had seen it enter Mexico, Canada, Argentina, Brazil, Puerto Rico, China, Indonesia and South Korea, as well as Germany and the UK, taking it in the process from a firm which in 1995 drew less than 4 per cent of sales from outside its domestic market to one with 17 per cent of sales from its international operations. Other retailers, such as France's Casino and the UK's leading food retailer Tesco, were also moving rapidly along similar trajectories – in Tesco's case establishing important international operations in Central Europe (Hungary, Poland, the Czech Republic and Slovakia) and south-east/northern Asia (Thailand, South Korea, Taiwan and Malaysia). And at the same time established retail TNCs, such as Royal Ahold of the Netherlands and Carrefour of France, continued to add, at an ever increasing pace, to the global empires which they had begun to establish in the 1980s and 1990s – in Ahold's case producing a firm with over 75 per cent of its sales, employment and assets focused in its international operations.

Table 8.3 illustrates the pace of this merger- and acquisition-driven globalization in just one sector, food retailing, in selected countries in North/South America and Europe during a two-year period in the late 1990s. The same group of firms, Ahold, Carrefour, Promodès (prior to its merger with Carrefour), Casino, Delhaize and Wal-Mart, consistently feature as acquirers in this and similar tables which could be constructed for other parts of the world. And it is clear from such tables that the late 1990s were characterized by the rapid market entry and expansion in the emerging markets of Latin America, south-east/northern Asia and central and southern Europe of an elite group of international retailers with this small sub-set of firms (together with a few others such as the UK's Tesco and Kingfisher and France's Auchan) increasingly at its core. Table 8.4, for example, illustrates how very rapidly

Table 8.3: *Food retail sector merger and acquisition activity, Europe and North/South America: December 1997 to December 1999*

Country	Deal	Country	Deal
USA	Fred Meyer/Ralphs merger	Germany	Wal-Mart (US) acquires Wertkauf
	Fred Meyer/Quality Food Centers merge		Wal-Mart (US) acquires Interspar
	Albertson's acquires Buttrey, Seessel's		Metro acquires Allkauf
	Albertson's/American Stores merger		Metro acquires Kriegbaum
	Kroger/Fred Meyer merger		Tenglemann acquires Tip
	Safeway acquires Carr Gottstein		
	Safeway acquires Dominick's	Argentina	Ahold (Neth) acquires stake in Disco
	Ahold (Neth) acquires Giant Food		Casino (Fr) acquires stake in Libertad
	Ahold (Neth) acquires Pathmark[a]		Promodès (Fr) acquires stake in Norte
	Sainsbury (UK) acquires Star Markets		Promodès (Fr) acquires Tia
	Safeway acquires Randall's		Casino (Fr) acquires San Cayetano
	Delhaize (Bel) acquires Hannaford		Ahold (Neth)/Disco acquires Gonzalez, Supamer
Canada	Loblaw acquires Provigo	Brazil	Ahold (Neth) acquires stake in Bompreço
	Sobeys (Empire) acquires Oshawa		Carrefour (Fr) acquires Planaltão, Mineirão, Roncetti, Raihna
			Sonae (Port) acquires Real, Candia, Big, Mercadorama, Nacional, Coletão, Marmungar
France	Carrefour acquires Comptoirs Modernes		
	Casino and Cora create buying group		
	Carrefour/Promodès merger		Ahold (Neth)/Brompreço acquires PetiPreço
Netherlands	De Boer Unigro/Vendex Food merger (renamed Laurus)		Casino (Fr) acquires stake in Pão de Açucar
		Columbia	Casino (Fr) acquires stake in Exito
UK	Somerfield/Kwik Save merger	Chile	Ahold (Neth) acquires stake in Santa Isabel
	Wal-Mart (US) acquires Asda		Ahold (Neth)/Santa Isabel acquires Tops
Sweden/ Norway	Dagab/D-Gruppen merger	Uruguay	Casino (Fr) acquires stake in Disco
	ICA/Hakon merger		
	Ahold (Neth) acquires stake in ICA		

[a] Ahold withdrew from acquisition of Pathmark, December 1999, following FTC opposition.
Source: Wrigley (2000b).

Table 8.4: *Largest food retailers in Brazil and Argentina, 1999*

	Market share (%)
Brazil	
1. Carrefour	18
2. Pão de Açucar-CBD (Casino)	15
3. Brompreço (Ahold)	6
Argentina	
1. Carrefour + Norte/Tia (Promodès)	28
2. Disco (Ahold)	17

from the mid-1990s the retail industries of countries such as Brazil and Argentina fell under the control of the retail TNCs – in this case Carrefour/Promodès, Ahold and Casino – and the same process was being repeated by these and other retail TNCs in south-east/northern Asia and central/southern Europe.

In Reading 8.2, extracted from an early attempt to consider these developments in his chapter in *The Oxford handbook of economic geography* (2000), Neil Wrigley considers the economic logic which drove the retail TNCs to seek expansion in the emerging markets. Academic writing on the impacts in these markets of the entry of the retail TNCs is as yet in its infancy, but for an early example in the geographical literature see Tokatli and Boyaci (2001).

Reading 8.2 – Retail TNCs in emerging markets: leveraging scale, know-how and costs of capital

Emerging markets offered the attractions of potentially rapid economic development and rising levels of affluence, consumer spending, and retail sales, combined with extremely low levels of penetration of Western forms of large-store corporate retailing and associated distribution systems. Indeed, up to 90 per cent of retail sales in such markets were often in the hands of very small independent retailers or informal retail channels. The opportunity existed for the leading retailers to enter these markets, leveraging their scale, their lower costs of capital, and their superior distribution/logistics and IT systems operation know-how to obtain rapid revenue growth and high returns on investment.

Those firms (particularly Carrefour) which led the push into the emerging markets by the elite group of retail TNCs were initially able to achieve super-normal returns on their investment. Rapid organic growth was possible in markets in which competition to Western-style large-store corporate retailing was minimal. Licences to open new stores were easy to obtain (in marked contrast to the position in Western Europe), capital requirements for site acquisition and store construction were low, and the existing retailers in those markets typically had little purchasing scale and operated in an inefficient manner. Very rapid rates of revenue growth and market share gains were achieved (Carrefour, for example, had captured 19 per cent of the Brazilian food market by 1998) and exceptionally high returns on invested capital, in the 25 to 30 per cent range, were often recorded from operations which had very low levels of capital intensity. Indeed, Carrefour was required to make minimal recourse to shareholders' equity or debt financing to expand into emerging markets in the 1980s and early 1990s ...

The super-normal returns which firms like Carrefour were able to achieve in emerging markets in the early 1990s attracted, in turn, an influx of capital from other members of the elite group of international retailers – the case of Poland provides a classic example of the scale and rapidity of that influx. By the late 1990s, therefore, retail industries in the emerging markets had become far more capital intensive ... As a result, the asset turnover levels (a measure of how many dollars of annual sales are generated by a dollar of total assets) of the initial entrants fell steadily. For example, Carrefour's asset turnover which had been 13x in 1994 at a time when Wal-Mart in its pre-internationalization period was achieving merely 3.8x, had halved to just 6x by 1999 with Wal-Mart achieving a very similar level.

Nevertheless, in comparison to the position which the elite group of retail TNCs often faced in their core/mature markets, the attractions of the emerging markets remained essentially unchanged. Although more capital intensive than in the early 1990s, those markets continued to offer lower labour costs and lower capital investment per store requirements, whilst offering the potential of a rapid revenue/earnings growth and return on capital (when discounted for currency and political risk) increasingly difficult to achieve in the retailers' core markets. The Asian and Latin American economic crises of the late 1990s had relatively limited impact on the perception of those advantages by the retail TNCs. Indeed they were as likely to be viewed, particularly in the context of South East Asia, as an opportunity to acquire retail and real-estate assets at discounted prices. However, partnerships and joint ventures with local operators were often used by the retail TNCs in both South East Asia and Latin America during the late 1990s to attenuate some of the risks of market entry. (Ahold, for example, entered Latin America by acquiring 50 per cent of the voting shares of Brazilian retailer Brompreço and forming a 50/50 joint venture with Velox Holdings to acquire stakes in (subsequently full control of) Disco, the leading Argentinian food retailer, and Santa Isabel a retailer operating in Chile, Peru, Paraguay, and Ecuador.) In many cases, the retail TNCs subsequently increased their stakes in such partnerships – Promodès for example, after taking an initial 49 per cent stake in the leading Argentinian retailer Norte in 1998, subsequently raised that stake to 67 per cent after the acquisition and integration of a second Argentinian chain, Tia, in 1999.

Global retailing was characterized during the mid to late 1990s then by the efforts of an elite group of firms to leverage their increasing core-market scale, and the free cash flow for expansionary investment which those markets provided, in order to secure the longer-term higher growth opportunities offered by the emerging markets. Ahold's operations in Latin America with 292 stores in six countries and 11.5 per cent of its global selling space in the region by the end of 1998 is representative.

Extracted from: Neil Wrigley (2000b): The globalization of retail capital: themes for economic geography. Chapter 15 in Gordon Clark, Maryann Feldman and Meric Gertler (eds) *The Oxford handbook of economic geography*. Oxford: Oxford University Press, 292–313.

Positioning the retail TNCs

If the late 1990s were characterized by the rise of an elite group of retail TNCs and their rapid expansion in the emerging markets, what position do these firms now occupy within world retailing? Figure 8.5 from Wrigley (2000b) begins this task by placing the retail TNCs within a ranking of the world's top retailers in the late 1990s. The ranking is on the basis of annual retail sales and those sales have been adjusted to strip out any significant non-retailing activity, e.g. the element in Sears Roebuck's $41 billion turnover in 1998 which relates to its substantial credit card and appliance servicing operations. Sometimes, as in the case of Pinault Printemps-Redoute (the French distribution conglomerate) those non-retail sales (48 per cent of Pinault's total) were sufficient to remove that firm from the table.

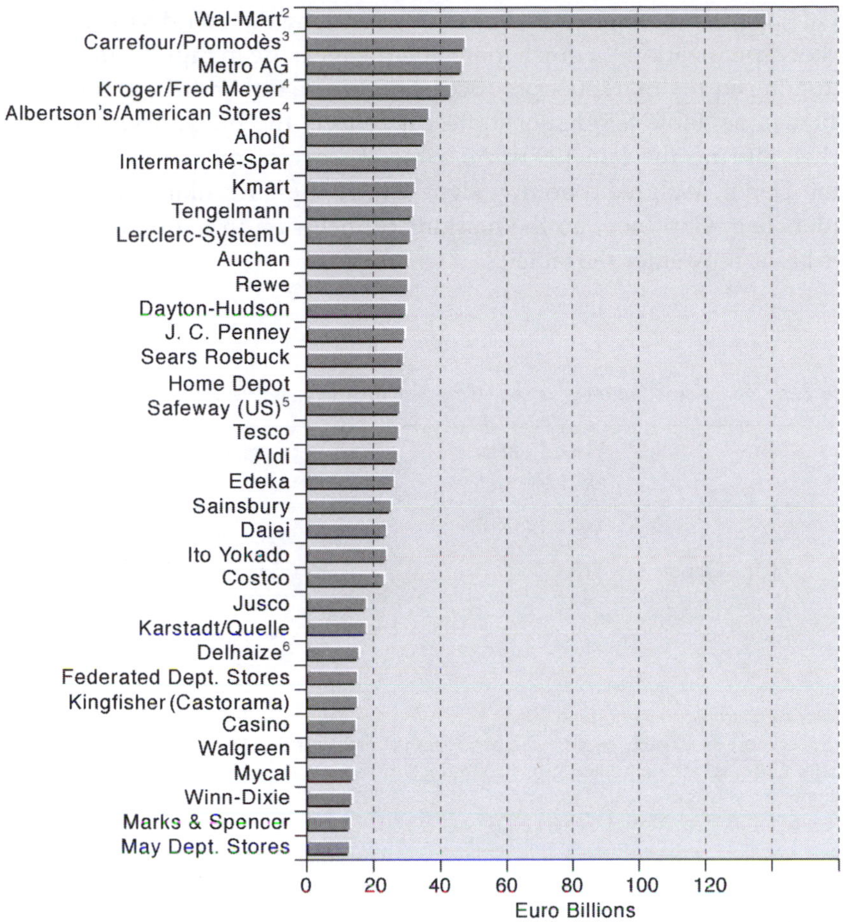

Figure 8.5: *World's top 35 retailers, mid-1999, ranked by pro forma 1998 retail sales*[1]

Notes: [1] *Any significant non-retailing revenue of firms stripped out of figures to provide true retail turnover.*

[2] *Includes sales of Asda plc following takeover announcement June 1999.*

[3] *Following merger announcement Aug. 1999.*

[4] *Mergers completed May 1999 (Kruger), July (Albertson's).*

[5] *Includes sales of acquisitions: Dominick's, Carr Gottstein and Randall's.*

[6] *Includes sales of Delhaize America companies: Good Lion, Kash n' Karry, and Hannaford (acquisition announced Aug. 1999).*

Source: Significantly adapted by author from Merrill Lynch (1999).

Revisiting the ranking of the world's largest retailers based on retail sales in 2000, Merrill Lynch (2001) find very little change. Indeed, 31 of the firms remain exactly the same – the marginal changes which occur see US retailers CVS, Lowe's and Best Buy entering the table to replace Winn-Dixie (US),

Mycal (Japan) and Marks & Spencer (UK). Ranking the world's top retailers by market capitalization – a much more volatile measure in rapidly and cyclically fluctuating equity markets – produces, however, a rather different ordering. In particular, as Table 8.5 demonstrates, it reflects the buoyancy of US equity markets during the late 1990s. As a result, several major US retailers (e.g. Home Depot, Walgreen) occupy significantly higher rankings and other US retailers (e.g. Gap Inc.), on the margins of inclusion in Figure 8.5 on a retail sales basis, now enter the table.

Table 8.5: *Top ten world retailers ranked by market capitalization, June 2001*[a]

Rank	Firm	Market capital. US $ billion	Rank	Firm	Market capital. US $ billion
1	Wal-Mart	230.2	6	Lowe's	28.0
2	Home Depot	118.4	7	Gap	27.7
3	Walgreen	41.8	8	Safeway (US)	26.6
4	Carrefour	38.3	9	Ahold	26.3
5	Target[b]	34.2	10	Tesco	25.0

[a] Market capitalization as of 5 June 2001.
[b] Target comprises Dayton Hudson, Target, Marshall Fields and is listed in Figure 8.5 as Dayton Hudson. Changed its name to Target to reflect increasing importance of that business in its portfolio.
Source: adapted from Merrill Lynch (2001).

On the basis of these rankings it can be seen that, although there can be no simple equating of the emerging retail TNCs with the world's largest retailers – some of the latter (Walgreen, Target, Kroger) remain entirely domestic US retail giants in the food, drug and general merchandise sectors – increasingly the retail TNCs and the world's largest retailers are becoming one and the same. Table 8.5, for example, includes four committed globalizers (Ahold, Carrefour, Wal-Mart, Tesco), one (Gap) with significant international operations in five countries outside the USA producing 10 per cent of its revenues, and another (Home Depot) just beginning to move outside the USA and expected by many retail analysts to play a strategically significant role in the retail globalization process over the next ten years. Not surprisingly, the view has increasingly been expressed by members of this leading group of retailers that, in food and general merchandise retailing in particular, the first decade of the twenty-first century will see the emergence of between four and six truly global mega-groupings. ABN-AMRO (1999), for example, reports the following statements by Ahold, Tesco and Carrefour:

> Before too long the world's food retail landscape will be ruled by between 4 and 6 groups (Ahold).
>
> There will be five or six major food retailers by 2005 (Tesco).
>
> In the future we will have local companies or global companies, but not much in between. Globalization will lead those companies who are not in the first team, or who are national retailers, to make alliances (Carrefour).

As a result, the struggle between these retailers is essentially centring during the early years of the twenty-first century upon which one can position themselves to dominate or become the corporate parents of these mega-groups, and upon which corporate model of globalized retail operation is likely to be the most appropriate and robust.

Table 8.6: *A characterization of alternative corporate models for retail globalization*

'Aggressively industrial'	versus	**'Intelligently federal'**
Low-format adaptation	versus	Multiple/flexible formats
Lack of partnerships/alliances in emerging markets	versus	Partnerships/alliances in emerging markets
Focus on economies of scale in purchasing, marketing and logistics	versus	Focus on back-end integration, accessing economies of skills as much as scale, best practice knowledge transfer
Centralized bureaucracy, export of key management and corporate culture from core	versus	Absorb, utilize/transfer best local management acquired
The global 'category killer' model	versus	The umbrella organization/corporate parent model

Source: Wrigley (2002b).

The contrast which is frequently drawn in the debates about alternative corporate models for retail globalization is essentially between (see Table 8.6) the so-called 'intelligently federal' model adopted by a firm such as Ahold, with its focus on local partnerships, best practice knowledge transfer, format adaptation, and back-end systems integration, with the much more centralized, exported corporate culture, low-format adaptation, 'global category killer' approach of Wal-Mart and, to a slightly lesser extent, Carrefour. Wrigley (2002b) and Fernie and Arnold (2002) debate some of these issues in relation to the emerging landscape of pan-European consolidation – a vital area in the future processes of retail globalization given the need for any potential global mega-group leader to have credible scale within a region which contains four of the top six retail markets in the world (Germany, France, the UK and Italy). Reading 8.3 by John Fernie and Steve Arnold, business school professors from Herriott Watt University, Scotland, and Queens University, Canada, respectively, considers Wal-Mart's entry into these European markets (for a more detailed assessment of Wal-Mart's entry into the UK see Burt and Sparks, 2001). Fernie and Arnold capture both the potential scale of Wal-Mart's globalization and also the considerable uncertainty which surrounded the export of Wal-Mart's corporate model to Europe during 2000/2001. At the same time, similar questions were being posed by retail analysts about Carrefour's model of retail globalization, particularly in relation to its suitability for some south-east Asian markets (ABN-AMRO, 2001).

Reading 8.3 – The uncertainties of Wal-Mart's entry into Europe

Wal-Mart is not only the world's largest retailer – it is nearly the globe's biggest corporation. Annual revenues of General Motors Corp. at the 2000 calendar year end were $183 billion. Total annual sales for Wal-Mart Stores Inc. at its January 31, 2001 fiscal year end were $191 billion. Furthermore, Chain Store Age forecast that Wal-Mart revenues in 2010 would approach $700 billion, the sheer scale of which is difficult to envisage. However, if international sales grow to only 27 per cent of total sales from the current 17 per cent, the 2010 proportion will be equivalent to all of Wal-Mart's 2001 sales.

To meet this forecasted growth, markets outside the US need to be developed ... Initially Wal-Mart moved to its geographically proximate neighbours, Mexico and Canada, before entering higher risk, more distant markets ... Europe was a logical target market for Wal-Mart because of its size, the expansion of the EU, the prospect of a common currency in the near future, and [because] no major retailer dominates the European market. However, many European retailers such as Carrefour and Ahold have a couple of decades more international experience than Wal-Mart ... Wal-Mart's strategy to date has been to acquire companies with the potential to be moulded to the 'Wal-Mart way' – hence their hypermarket acquisitions in Germany and the purchase of the value-orientated Asda in the UK ...

So what has Wal-Mart achieved to date? In Germany it has created a fierce price war ... This may work to Wal-Mart's advantage in the long run as it weakens competitors' profitability making them easier to acquire. In the short run, however, it has drained cash from the Wal-Mart empire at the rate of $200 million per year in 1999 and 2000 and plans to break even late in 2001 appear optimistic. US executives quickly superseded German management and attempted to impose Wal-Mart standards on the German operations [but analysts suggest] that the new management underestimated the scale of this task ...

In the UK Wal-Mart has made an impact on the market by changing the nature of retail competition ... to a greater price focus. In 1999/2000 the RPI increased by 3 per cent but food prices fell by 1.7 per cent ... While the market leader, Tesco, has been able to match Asda's EDLP strategy, the other key players in the market have had to revise their strategies ... Asda has been making ground on Sainsbury to challenge for the number two spot in the British grocery league table ... but where does Wal-Mart go from here? Its current expansion plans are fairly modest and it is going to be difficult to squeeze much more profitability from the existing store portfolio. Of more concern is the loss of key members of the management team who transformed Asda's fortunes in the 1990s ... Clearly there is some discord between Bentonville and Leeds about how to manage the British business ...

France is the next logical market for a Wal-Mart entry. It is the home of the hypermarket and has the highest level of sales through food retailers in Europe ... The problem for Wal-Mart is that strict planning legislation will preclude it from building large out of town stores so, like the UK and Germany, an acquisition or joint venture are the only possible market entry strategies available. Unfortunately for Wal-Mart, two of the best acquisition targets, Carrefour and Promodès, merged in September 1999 ... [it] is unlikely that the new Carrefour group will succumb to Wal-Mart ...

Wal-Mart needs to develop its international presence if it is to maintain its historical double-digit growth rates ... [But in Europe] it appears that Wal-Mart is at a crossroads. It only has a toehold in the market [and] needs other acquisitions to achieve sufficient scale economies. [But in both Germany and France] an acquisition is difficult because of the nature of ownership of targeted companies that are either family-owned, voluntary trading groups or co-operatives. Wal-Mart options appear limited. It could withdraw from the European market. Withdrawal, however, seems unlikely because it has only done this once before when local unrest forced it out of Indonesia. Wal-Mart has only been in the international arena for a decade and withdrawal would be an admission of defeat by an organization composed of extremely competitive individuals. Its other alternative is to make a mega-acquisition, an action consistent with the 2010 growth projections.

Extracted from: John Fernie and Steve Arnold (2002): Wal-Mart in Europe: Prospects for Germany, the UK and France. Paper in preparation for *International Journal of Retail and Distribution Management*, 30.

Conceptualizing the retail internationalization process

The difficulties Wal-Mart was encountering with its European operations four years after its entry into that market have frequently been mirrored in other retailers' experience of internationalization. In the same year, 2001, in which Fernie and Arnold were posing questions about Wal-Mart, UK retailers Marks & Spencer and Boots, for example, were pulling back in disarray from their international investments and operations in continental Europe/North America and Japan respectively – in the process reinforcing the long-held perception in the UK capital markets that the internationalization process was a graveyard for overambitious retailers. To what extent then do the conceptualizations of that process available within the academic literature adequately capture its difficulties and deal with the issues relating to appropriate and robust corporate globalization models outlined above?

Table 8.7 from Clarke and Rimmer (1997), drawing on early taxonomies of the retail internationalization process developed by Treadgold and Davies (1988), Treadgold (1990) and Dawson (1993), summarizes the range of so-

Table 8.7: *Factors traditionally cited as influencing retail internationalization*

'Push' factors	Facilitating factors	'Pull' factors
• Perceived/imminent saturation in domestic markets • Spreading of risk • Consolidation of buying power • Public policy constraints	• Use of surplus capital/ access to cheaper sources of capital • Entrepreneurial vision • Inducements from suppliers to enter new markets	• Unexploited markets • Higher profit potential • Consumer market segments not yet exploited • Access to new management • Reaction to manufacturer internationalization • Following existing customers abroad

Source: adapted from Clarke and Rimmer (1997: 364).

called 'push' and 'pull' motivations for retail internationalization which have traditionally been cited in the academic literature. Although useful at the broadest level – but see Williams (1992) and Alexander (1995) for critiques, and McGoldrick and Davies (1995), Alexander (1997) and Sternquist (1998) for wider summaries of the retail internationalization literature in the mid-1990s – the problem with the simple push/pull categorization of Table 8.7 is that it tends to focus attention on the '*implementation* of international expansion in strategic terms' (Whitehead, 1992: 74) rather than the problems of *sustaining* that expansion given the inevitable stresses the internationalization process poses for the firm. As a result, Clarke and Rimmer (1997) have argued that greater attention needs to be given than was the case in the early retail internationalization literature to the longitudinal and dynamic aspects of the process; in particular to understanding how the retail firm adapts its operations as it develops in international markets: that is to say, to 'the way a retail firm explores, invests, and then reflects on individual decisions it has made' (Clarke and Rimmer, 1997: 364) and learns from such experience. Adopting this perspective implies, therefore, a conceptualization of the retail internationalization process as being 'characterized inevitably by periods of retrenchment, strategic reassessment and discontinuity in which constant strategic renewal is necessary to long-term sustainability' (Wrigley, 2000a: 893).

Clarke and Rimmer (1997) and Wrigley (2000a) provide case studies of retail internationalization in which this perspective forms a central focus. Clarke and Rimmer consider Japanese department store chains entering the Australian market, whilst Wrigley considers the experience of the UK's second largest food retailer, Sainsbury, during a particular phase of Sainsbury's internationalization into the US market during the late 1990s. That phase was marked by Sainsbury's forced exit from its position in the leading Washington

DC/Baltimore food retail chain, Giant Food (see Wrigley, 1997b for background on Sainsbury's earlier acquisition of that stake in Giant) at the hands of rival food retailer Ahold, and Sainsbury's subsequent decision to remain in the US market and to acquire the Star Markets chain in the Boston area. The study highlights the often highly contested nature of the retail internationalization process, and the issues involved in sustaining international expansion during periods of strategic reassessment. In particular, it illustrates the tensions which can rapidly develop in the relationship between management of the firm and the capital markets if the internationalization process is perceived to threaten the strategic credibility of the firm.

Two other facets of retail internationalization which, by default, are also left unexplored in the traditions of the early literature on the topic summarized in Table 8.7 are:

(a) the consequences of the increasingly multi-channel nature of the retail TNCs;
(b) the consequences of increased global sourcing by those TNCs.

With respect to the former, we saw in Chapter 4 how the leading retailers, faced during the late 1990s with the challenge of the emerging pure-play 'e-tailers', responded by absorbing the new electronic channels to market as a complimentary part of their retail offer. Essentially they exploited the significant competitive advantage offered by the 'fulfilment' potential of their existing store networks and their distribution, inventory management, buying and branding expertise, becoming in the process multi-channel ('bricks and clicks') retailers to a greater or lesser extent. Multi-channel retailing, in its turn, offers the potential for retail internationalization along more than one channel simultaneously, and the development of hybrid forms of the process which might be termed 'matrix internationalization'. In other words, retail TNCs might operate very different corporate models of internationalization across different channels, experimenting with alternative market entry strategies and organizational structures. Early examples amongst the retail TNCs worthy of more detailed consideration include Ahold's acquisition of the longest-established on-line food retailer in the US, Peapod, in 2000 as a small part of Ahold's strategic push into the wider food distribution market in the US (Wrigley, 2001a). And, more specifically, the announcement by Tesco in June 2001 that it was to enter the US market in a low-capital-risk manner by investing just £16 million to take a 35 per cent stake in the on-line grocery business of leading US food retailer Safeway. Tesco was acknowledged by 2001 to be the world's largest and most successful internet grocer (albeit with on-line sales of a mere £300 million per year compared to its conventional sales of £22.8 billion). However, it was committed to a major conventional store-based internationalization programme, particularly in south-east/northern Asia and central Europe, which was absorbing its capital and management resources. As a result, the opportunity of a limited-risk but extremely high-potential internationalization via a non-store-based channel in which it was acknowledged to have competitive advantage, offered to Tesco an intriguing opportunity to advance its global ambitions.

Finally, with respect to increased global sourcing by retail TNCs, the traditional literature summarized in Table 8.7 is deficient in respect of its conceptualization of retail internationalization as a *reaction* to manufacturer internationalization. Although in the food and general merchandise sectors the major manufacturers (Proctor and Gamble, Unilever, etc.) did begin to change their organizational structures from a geographical region to a global production group basis in the late 1990s, most analysts of these industries do not believe that the huge wave of merger- and acquisition-driven retail globalization during that period came *in response* to a globally reorganizing supply base. For example, analysts at ABN-AMRO (1999: 16) state: 'we certainly do *not* believe food retail consolidation is a reaction to a consolidating supply base'.

Not least, the retail TNCs found during the late 1990s that some of the most important competitive advantages from global sourcing were, in practice, obtained from developing innovative *own-brand* global supply chains and networks – for example, the direct relationships which the UK food retailers such as Tesco forged with suppliers of high-value horticultural produce from Kenya and the Gambia (Barrett et al., 1999). Conceptualizing the nature and significance of these global own-brand supply chains within the wider debates on the globalization of retail capital – particularly in the context of a domestic regulatory environment in the UK which, following the Competition Commission's report, stipulates fairer treatment of suppliers, and pressures on the retail TNCs to conform to ethical trading initiatives – is a task recently begun by Alex Hughes (2001).

9

The configuration, manipulation and contestation of retail space

In his book *Fantasy city: pleasure and profit in the postmodern metropolis*, University of Toronto sociologist John Hannigan (1998: 89) charts the emergence of the 'mutual convergence and overlap of four consumer activity systems: shopping, dining, entertainment, education and culture' which in his view have given 'rise to three new hybrids which in the lexicon of the retail industry are known as shopertainment, eatertainment and edutainment'. This chapter is essentially concerned with the first of these which, according to Hannigan, emerged most forcefully in the theme park cities of the 1990s. But as Hannigan also observes, there were several earlier prototypes of '*shopertainment*' dating as far back as the 1890s to the rise of the great department stores. Reading 9.1 comprises an extended extract from Hannigan's text, and is used here as a template around which this chapter is structured. Our aim is to provide an account of the 'configuration and manipulation' of retail space from the close of the nineteenth century to the start of the twenty-first century, essentially considering the various ways in which the retail environment has been moulded by retailers in an attempt to sell commodities (Dowling, 1993). Along the way we are also concerned to illuminate the contestation of retail space – more specifically the geographies of exclusion evidenced in the private spaces of the shopping mall.

Reading 9.1 – Shopertainment

As far back as the 1890s, the great metropolitan department stores set out to attract downtown customers by providing free entertainment. For example, Siegel-Cooper, which opened at Sixth Avenue and 18th Street in New York in 1896, earned its reputation as 'the big store' by offering an orchestra, art shows, tearooms and 'spectacular extravaganzas' in its auditorium. One summer, the store mounted a six-week long, 'Carnival of Nations', which climaxed in the August with an exotic show, Phantasma, The Enchanted Bower, utilizing light and color effects to highlight a cast which included a Turkish harem, a parade of Turkish dancing girls, a 'genie of the lamp' and . . . 'Cleopatra of the Nile'. Not to be outdone, Marshall Fields opened its twelve-storey department store in Chicago in 1902 complete with six-string orchestras on various floors. In a similar fashion, McWhirters, a turn-of-the-century dry goods store in Brisbane, Australia, offered a fourth-floor tearoom where tired shoppers could enjoy a cool sea breeze and a charming view of the river and suburbs. The opening of McWhirters' new premises in August 1931, was promoted with a series of three-hour entertainments which included a dancing demonstration by Phyland Ray, Australia's leading adagio dancers, and a live revue advertising a leading brand of corset . . .

After the Second World War, suburban malls displaced downtown shopping districts as popular consumer destinations. At first, these shopping centers marketed themselves on the basis of easy automobile access and free parking. By the mid-1950s, however, mall developers rediscovered the appeal of turn-of-the-century department stores, transforming indoor spaces into theatrical 'sets' in which a form of retail drama could occur. The template for this new generation of enclosed malls was Southdale in Edina, Minnesota, a suburb of Minneapolis. Built by Victor Gruen, an Austrian urban architect who admired the covered pedestrian arcades in Europe, Southdale had as its focal point the 'Garden Court of Perpetual Spring', an atrium filled with orchids, azaleas, magnolias and palms which bloomed even in the midst of the deep freeze of Minnesota winters . . .

The West Edmonton Mall (WEM)'s . . . developers, the Ghermezian brothers, explicitly and ostentatiously set out to bring the world of the theme park to the environment of the shopping center. Among other things, WEM contains a 15-acre amusement park, a ten-acre water park, a full-size ice skating rink, the Fantasyland Hotel, a faux version of Bourbon Street in New Orleans, and a 2.5 acre artificial lagoon complete with a replica of Christopher Columbus' ship the *Santa Maria*, several mini submarines and electronically operated rubber sharks. Built in the 1980s, the West Edmonton Mall is a bizarre amalgamation of shopping, entertainment and social space . . . West Edmonton Mall also radically changed the shopping centre formula, boosting the footage dedicated to entertainment up to 40 per cent, the largest proportion up to that time in a suburban mall. Seven years later, the Ghermezians succeeded in cloning WEM with their first American project, the Mall of America in Bloomington. The centrepiece of the Mall of America is 'Camp Snoopy', an amusement park.

In the theme park cities of the 1990s shopping, fantasy and fun have further bonded in a number of ways. As Margaret Crawford has observed, the two activities have become part of the same loop: shopping has become intensely entertaining and this in turn encourages more shopping. Furthermore, theme parks themselves have begun to function as 'disguised market places'. This convergence is described as 'shopertainment' . . .

One form of shopertainment is the themed retail experience known as 'experiential retailing'. This is represented by Nike Town, a retail theatre showcase in New York. Opened in November 1996, on the site of the former Les Galeries Lafayette, Nike's flagship store is 'a fantasy environment, one part nostalgia to two parts high-tech, and it exists to bedazzle the customer, to give its merchandise sex appeal and establish Nike as the essence not just of athletic wear but also of our culture and way of life' . . . According to its creative director, John Hoke III, the store is designed like a ship in a bottle . . . The bottle in question is a simulated, old style gymnasium made to look as if it was built in the 1930s or 1940s. This sense of age is created on the exterior through an arched limestone and sandstone façade with the numbers PS 6453 added, a reference to a time when boxers trained in the gym of the local public school. Inside, the old gym theme is continued with aged brick detail, wooden sports flooring, wireglass windows, gym clocks, wrestling mats and 'authentic' bleachers reclaimed from a gym on Long Island . . . The retail element of the store is more muted: one can buy Nike products at Nike Town but the store exists primarily to promote brand recognition.

Extracted from: John Hannigan (1998): *Fantasy city: pleasure and profit in the postmodern metropolis.* London and New York: Routledge, 90–2.

The configuration and manipulation of retail space is an intensely geographical phenomenon. Developers and designers of the retail built environment have consistently exploited the power of place and an intuitive understanding of the structuration of space in order to facilitate consumption (Goss, 1993). At its most simplistic this process comprises attempts to attract target consumers and to keep them on the premises for as long as possible, on the assumption that the amount of spending is related *directly* to the amount of time spent at centres or in stores (Goss, 1993; Hopkins, 1990; Reynolds, 1990). But retailers' strategies in this area have become progressively more sophisticated, so that by the late 1990s, for example, Pine and Gilmore's *Experience economy* highlights the use of what they term 'retail theatre' in order to entice customers. At Nike Town, in Chicago, for example, 'the store was built as theatre, where ... customers are the audience participating in the production' (1999: 63). The concept of Nike Town, and other retailers propagating 'retail theatre' is examined in detail towards the end of this chapter. Here, though, and following the framework etched out by Hannigan in Reading 9.1, we begin with an analysis of the great metropolitan department stores and their role in the configuration and manipulation of retail space.

A new shopping experience in a new environment

In his *The department store: a social history* (1995), Bill Lancaster in presenting the first historical survey of the British department store, illuminates the fundamental importance of the Parisian family the Boucicauts and their store the Bon Marché which introduced 'the notion of 'democratic luxury' ... and exhibition-style displays [which] fundamentally changed the nature of retailing and brought the department store to maturity'. For Lancaster (1995: 17), 'the major innovation of the Bon Marché and its Parisian competitors was the creation of a new shopping experience in a new environment, a development as profound as the twentieth-century shopping mall'. What is fascinating about Lancaster's social history, however, is the connection drawn between Boucicaut's store and the Grand Exhibition held in Paris in 1855. Reading 9.2 highlights this linkage and in so doing draws attention to the fundamental importance of the spatial layout of the store as a motor for increased consumption. Lancaster's account is mainly a historical one. In a much more explicitly *geographical* framework, but dealing with the same theme, is Nick Blomley's 1996 reading of Zola's *Au bonheur des dames* ('The ladies' paradise'). As we discussed in Chapter 1 (Reading 1.4), Zola's novel, published in 1883 and set in Paris, is reputedly based on the many hours Zola spent researching his subject at the Parisian department stores, the Bon Marché and the Louvre. Reading 9.3 extracted from Blomley's paper should be read alongside Reading 9.2. As Blomley (1996: 240) shows, Mouret, the hero of

Zola's tale, 'an aggressive, unscrupulous and predatory merchant' is modelled perhaps on Boucicaut, the founder of the Bon Marché, and is a 'genius of store layout and design'.

Reading 9.2 – The *grand magasin*

A legend of retailing is that of the Paris draper, Aristide Boucicaut, having lost his way in the middle of the 1855 Exposition. Instead of being confused Boucicaut found himself enraptured by the spectacle of the goods on view and delighted in the surprises that met his every turn. It is difficult to confirm this story but the connection between the Paris exhibition and the development of the *grand magasin* is undeniable . . . The historian of the Bon Marché has described the new store as 'Dazzling and sensuous . . . a permanent fair'. The Bon Marché acknowledged its debt to the expositions by becoming a regular exhibitor, even participating in the international fairs of Chicago and St Louis. This process of imitation reached full circle when the journalist, Maurice Talmeyr, described the 1900 Paris exposition as a 'sort of Louvre or Bon Marché . . .

What was significant about 1855 was the combination of price tags, sumptuous display and the ability to browse, explore and dream of potential ownership . . . Boucicaut realised that spectacle and browsing were integral to the success of the exposition and these elements had tremendous potential for retailing . . .

The sheer size of the new emporium was a key to success. Its numerous departments and floors facilitated a vast traffic of customers and potential customers. The Bon Marché offered a new type of liberty. Anyone could enter, browse in departments, wander from floor to floor, without spending a centime . . . Customers in the Bon Marché were encouraged to wander by means of stunning visual displays. Goods draped across gallery rails sucked crowds up to the upper floors. Boucicaut was anxious to recreate the experience of the exhibition. He knew only too well that a customer who entered the Bon Marché to buy an umbrella would soon be back for something else. Even browsers would leave the store with new desires in their minds.

Extracted from: Bill Lancaster (1995): *The department store: a social history.* London: Leicester University Press, 17–18.

Reading 9.3 – 'How to sell'

The production and configuration of space within the emergent retail economy is a constant concern . . . Geographies unfold throughout the novel. It is quickly made clear that questions of store design and layout are pivotal to Mouret's success. At the beginning of the novel, the reader is introduced to the spatial configuration of the store itself, as we walk through the store with Mouret on his round of inspection. This is quickly followed by the description of the use of display techniques during a sale. Mouret is a genius at layout and display, anticipating many contemporary discussions on strike zones, customer traffic and product exposure. While the traditional retailers foreswear showy advertisement, Mouret crafts an assertive and aggressive visual economy . . . He has an obsession with colour, for example: the white sale, in which a profusion of white goods are draped everywhere, is characterized as 'vivified, colossal, burning' . . . He criticises one of his salesmen who arranges silks to 'please the eyes'. 'Don't be afraid', he insists, 'blind them. Look! red, green,

> yellow'... He delights 'in a tumbling of stuffs, as if they had fallen from the crowded shelves by chance, making them glow with the most ardent colours, lighting each other up by the contrast, declaring that the customers ought to have sore eyes on going out of the shop'.
>
> These brash and intoxicating displays, in which women lose themselves, are, however, insufficient. Mouret's success also rests on the fact that he was 'an unrivalled master... in the interior arrangement of the shops, He realizes the dangers of 'dead space' insisting that every corner of the store must be busy. By careful positioning of displays and restrictions on circulation he directs and manages the crowd. Bargains are placed at the entrance to encourage a crush 'so that people in the street should mistake it for a riot'... and be tempted in. He is seized with flashes of spatial insight: just before a sale he realizes that the 'orderly' and 'logical' layout of the store and its departments is 'wrong and stupid' and insists that it be repositioned. The problem, as he puts it, is that it would have 'localised the crowd. A woman would have come in, gone straight to the department she wished ... then retired, without having lost herself for a moment.' There is a need to disorder space so as to disorder and 'lose' the consumer.
>
> **Extracted from Nicholas Blomley (1996): 'I'd like to dress her all over': masculinity, power and retail space. In Neil Wrigley and Michelle Lowe (eds) *Retailing, consumption and capital: towards the new retail geography*. Harlow: Addison Wesley Longman, 238–56.**

In the late nineteenth and twentieth centuries the kinds of techniques employed by Bon Marché and described in Zola's fictional *Ladies' paradise* were systematically replicated on both sides of the Atlantic. In London, and in other big cities in England, stores 'set out to attract shoppers from a distance by offering auxiliary, non-selling services such as restaurants, banking facilities and exhibitions' (Adburgham, 1989: 231). William Whiteley, for example, dubbed himself 'the Universal Provider' and added estate agency, hairdressing and tearooms to his emporium in Bayswater. Although 'very different from the bourgeois spectacle of the Paris *grands magasins* ... [Whiteleys] also served the similar function of making shopping more than just shopping' (Lancaster, 1995: 22). But it was in the United States that the most phenomenal growth of stores on a par with those in Paris occurred. Chapter 11, in the final part of our volume, examines these stores in detail, drawing on the work of several North American scholars (Benson, 1986; Domosh, 1996a; Dowling, 1991 and 1993; Leach, 1984) on stores such as A.T. Stewart's and Wanamaker's in New York and Woodward's in Vancouver, as well as on Mica Nava's important research on Selfridge's in London (Nava, 1997 and 1998) which was founded in 1909 by Gordon Selfridge, a self-made American entrepreneur (see Reading 11.1). Selfridge had previously been a key figure in the transformation of Marshall Field's Chicago store into 'one of America's most innovative department stores':

> This was more than a store, combining as it did all the spectacle and drama of expositions and museums. The five thousand people who daily used the store's dining facilities enjoyed the finest restaurant settings in Chicago, and many looked upon the store and its services as a haven of order in the merciless laissez-faire world of Chicago's Loop (Lancaster, 1995: 61–7).

At his Oxford Street store, Selfridge employed techniques of 'showmanship' ranging from exhibiting all the pictures rejected by the Royal Academy for their Summer Show, to displaying Louis Bleriot's plane fresh from its ground-breaking first flight across the English Channel (Honeycombe, 1984) in order to allure customers. During the same period, a conventional wisdom about the placement of the store's departments began to crystallize.

> Commentators noted that such factors as 'chance' and 'architectural convenience' as determinants of store layout were on their way to a well-deserved extinction; the new watchword was 'deliberation'. Basements sold bargain goods, groceries and (less often) housewares. Street floors offered cosmetics, notions, gloves, hosiery, jewellery and other small wares – glamour and impulse items to waylay women on their way to the upper floors – and clothing and furnishings for men who were presumed too timid to venture farther into the store ... Upper floors featured furniture, appliances, carpeting and housewares, items that were usually the object of a special shopping trip.
>
> Within floors ... store managers tried both to maximise the sales potential of well-trafficked areas and to rehabilitate lost corners with irresistible merchandise ... The bargain basement was a common device for turning an unattractive area to a revenue producer, but some managers were more creative. A Brooklyn store president boasted that he used fashion fads to draw customers to dead corners; he was particularly proud of his music department tucked under a basement staircase – customers were drawn, as if by sirens, to the sound of music (Benson, 1988: 44–6).

Returning to Hannigan's template, however, we now move on to consider the post-war shopping malls and the various ways in which the retail spaces of those malls drew on the experience and example of the department stores in their configuration and manipulation of retail space.

Mall magic

As Hannigan (1998) notes in Reading 9.1, the prototype suburban shopping centres of the immediate post-war years initially displaced downtown shopping districts via an emphasis on easy access, automobile parking and an orderly urban experience in contrast to the chaotic downtown (see Chapter 7). But it was not long before such innovations seemed passé and therefore rather more dramatic measures were needed to attract consumers to the newly developing malls. Southdale in Minneapolis, which opened in 1956, is universally regarded as the world's first enclosed (fully covered) mall. Chapter

12 tackles in detail the story of Southdale as part and parcel of a more detailed study of the history and development of the shopping mall. Suffice it to say here that Southdale was a clear innovator with regard to mall layout. As Kowinski (1985) comments, 'when Southdale was being planned, the normal shopping centre was a long strip, all on one level with at best one department store' (Reading 12.1). But Southdale's Austrian architect, Victor Gruen, had been influenced by his experiences of the covered pedestrian arcades in Europe and as a result modelled his Southdale Centre on the principles of enclosure and vertical dimension, both of which ensured the constant circulation of customers only too pleased to be sheltered from the deep freeze of Minnesota winters.

From Gruen's Southdale it was but a short step to a whole armoury of mall design techniques utilized by mall developers. As Crawford (1992: 7) states 'between 1960 and 1980 . . . the basic regional mall paradigm was perfected and systematically replicated'. And hand in hand with these developments went a series of 'familiar tricks of mall design':

> Limited entrances, escalators placed only at one end of corridors, fountains and benches carefully positioned to entice shoppers into stores – control the flow of consumers through the numbingly repetitive corridor of shops. The orderly processions of goods along endless aisles continuously stimulate the desire to buy . . . The jargon used by mall management demonstrates not only their awareness of these side effects but also their . . . attempts to capitalize on them. The Gruen Transfer (named after architect Victor Gruen) designates the moment when a 'destination buyer', with a specific purchase in mind, is transformed into an impulse shopper, a crucial point immediately visible in the shift from a determined stride to an erratic and meandering gait (Crawford, 1992: 13– 4).

From the 1950s onwards, planned shopping centres of North America with their 'carefully selected tenants and a floor plan designed to manipulate consumer flows and maximise sales' (Jones and Simmons, 1990: 121) became commonplace. Figure 9.1 illustrates some examples of typical mall layouts and lists the common principles usually incorporated into mall design. But it was in the 1980s and 1990s that shopping mall design entered an entirely new phase. Like the great metropolitan department stores before them, this latest generation of shopping centres quickly became tourist attractions in their own right and 'the mall' was never to be the same again. It is to these new centres, once again following Hannigan's framework, that we now turn.

At the West Edmonton 'Mega' Mall (WEM) in Alberta, Canada 'developers the Ghermezian brothers, explicitly and ostentatiously set out to bring the world of the theme park to the environment of the shopping centre' (Hannigan, 1998: 91). Chapter 12 deals with WEM in some detail, and also considers the work of Hopkins (1990), Shields (1989) and Goss (1993), all of whom have emphasized the various ways in which WEM stimulates a sense of

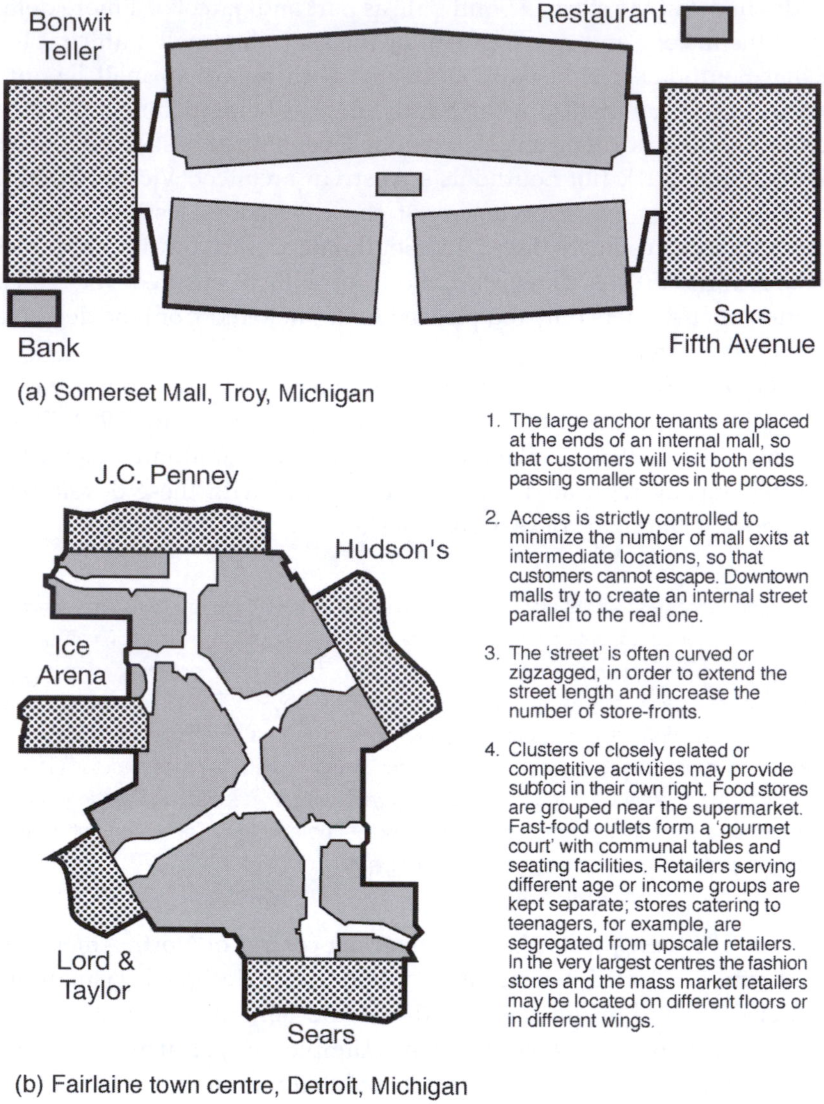

Bonwit
Teller
Restaurant
Bank
Saks
Fifth Avenue
(a) Somerset Mall, Troy, Michigan
J.C. Penney
Hudson's
Ice
Arena
Lord &
Taylor
Sears
(b) Fairlaine town centre, Detroit, Michigan
1. The large anchor tenants are placed at the ends of an internal mall, so that customers will visit both ends passing smaller stores in the process.
2. Access is strictly controlled to minimize the number of mall exits at intermediate locations, so that customers cannot escape. Downtown malls try to create an internal street parallel to the real one.
3. The 'street' is often curved or zigzagged, in order to extend the street length and increase the number of store-fronts.
4. Clusters of closely related or competitive activities may provide subfoci in their own right. Food stores are grouped near the supermarket. Fast-food outlets form a 'gourmet court' with communal tables and seating facilities. Retailers serving different age or income groups are kept separate; stores catering to teenagers, for example, are segregated from upscale retailers. In the very largest centres the fashion stores and the mass market retailers may be located on different floors or in different wings.

Figure 9.1: *Examples of mall layouts*
Source: Adapted from Jones and Simmons (1990: 124–6).

'elsewhereness' – 'the overt manipulation of time and/or space to simulate or evoke experiences of other places' (Hopkins, 1990: 2) – in its clientele (see Reading 12.2, in particular). But Chapter 12 also mentions in passing the Mall of America in Bloomington, Minneapolis. Here we choose to focus on the Mall of America and some recent work on this centre which is not detailed in later chapters.

The Mall of America

> The largest mall in the United States is the Mall of America, located in Bloomington, Minnesota, outside Minneapolis. This 'megamall', which opened for business on August 21, 1992, is in many ways a separate small city itself ... The Mall of America covers seventy-eight acres with over 4 million square feet of floor area, including 2.5 million square feet of actual retailing space. It has over four hundred speciality shops, four large department stores, a fourteen-screen movie theater, nightclubs, bars, nine acres of family entertainment, over twenty-two restaurants, and twenty-three more fast food outlets. But that is not all. At the center of this three-storey complex, beneath an immense hyperspace of skylights, mall developers located a seven-acre theme park that is run by Knotts Berry Farms of southern California (Gottdeiner, 1998: 39).

For Jon Goss the Mall of America (MOA) – 'the largest themed retail and entertainment complex in the US', is the 'apotheosis of the modern mall' (1999: 45). Reading 9.4 is taken from Goss's extended 'critical reading' of the Mall published in the *Annals of the Association of American Geographers*. In this reading Goss, a geographer based at the University of Hawaii, who has written extensively on a range of consumer landscapes, including pieces on West Edmonton Mall (Goss, 1993) and festival marketplaces (Goss, 1996 – see Reading 7.4), provides a 'critical participatory' view of the Mall. Here he positions himself between what he terms a 'disdainful' and a 'celebratory' approach 'taking pleasure [*jouissance*] in the play of reality and fantasy, while critically examining how things actually seem' (1999: 49).

Reading 9.4 – The apotheosis of the modern mall

Like the arcade, only at greater speed and intensity, The Mall is a kinesthetic space, and contemporary flaneurs are literally moved to aggregate, shop and celebrate: they are drawn across thresholds and along paths by the use of contrast in color and light, focal attractions and linear design elements; carried away by escalators and elevators; directed by spatialized narratives in the form of waymarkers and sequential interpretive texts; whirled around and whipped along on fairground rides; and reflected endlessly in mirrors and glass surfaces. The aesthetics of motion are repeated in the decor, in colorful flags, banners, mobiles, and water features and in modes of transport in The Mall's innumerable attractions and retail concepts ...

The speculative activity of shopping has been effectively combined with mobilities of tourism since at least the nineteenth century, when the retail built environment became a tourist destination, and tourist resorts, shopping centers ... The Mall, however is the acme of this process. The number-one tourist attraction in the US, it operates its own Tourism Department and contains six Mall of America Gift Stores. The Mall is divided into districts based on tourist–retail destinations: West Market evokes a bustling European marketplace under a train station-style roof, with street furniture, vendors, carts and traditional stores; North Garden reproduces a European landscaped garden with terraces, gazebos, and fine shops; South Avenue is based upon luxury hotels with

> carpeted walks and sophisticated stores; and East Broadway is modelled on an upbeat American city with its chrome, neon lights and contemporary fashions. The transit station seems like an airport, and the parking areas are named after the fourteen leading tourism states in the US . . .
>
> Retailers in The Mall exploit travel themes to sell accessories, adventure clothing, and shoes. Perhaps the best example of this is Timbuktu Station, a women's clothing store whose products are displayed in antique suitcases and shipping cases decorated with whimsical literary quotations about travel. Concept restaurants and fast-food outlets are positioned as imaginary destinations in the consumer–tourist world. Rainforest Café, for example, is described as an 'enchanted place for fun far away, that's just beyond your doorstep'. Diners are issued 'passports' for a meal that is a 'safari' and are exhorted to complete their 'trip' by buying a souvenir stuffed-animal toy in the adjacent 'Retail Village'.
>
> **Extracted from: Jon Goss (1999): 'Once-upon-a-time in the commodity world: an official guide to the Mall of America. *Annals of the Associations of American Geographers*, 89, 45– 75.**

Goss centres his analysis explicitly on the connection between retailing and tourism which ensures in the case of the MOA (as indeed it did in the case of the great department stores) the ultimate in terms of spatial manipulation – the clever combination of a retail and a tourist destination attracting consumers and capturing them at The Mall for as long as possible. Essentially what Goss is describing here is the evolution of a sophisticated 'themed environment', the 'underlying referent [of which] is the ideology of pedestrian city life'. At MOA 'The grand themed environment . . . functions as a sign vehicle that aids its role as a container of many commercial enterprises because it is also attractive as a desirable destination itself' (Gottdeiner, 1998: 40–2). We elaborate on the concept of the 'themed environment' in the final section of this chapter. But Goss also focuses here on the more mundane 'spatial' tricks of retail design – contrasts in colour and light, escalators and elevators, waymarkers, mirrors and glass surfaces, that we have reflected on earlier in this chapter, all of which act, together with the more sophisticated 'imagery' evidenced here, as important stimuli to conspicuous consumption. Finally in his paper, though, Goss draws attention to the privatization of public space in the retail built environment that goes hand in hand with the configuration and manipulation of retail space which he and several other authors have described. Essentially, the logical conclusion of the kinds of 'spatial manipulation' discussed here, in particular the construction of 'themed environments' in enclosed places, is the exclusion of certain individuals and the transposition of what were previously public activities – shopping and strolling – into private environments. It is to this thorny issue, regarding the private control of public space, that we now turn.

How public is a mall?

From the late 1980s onwards, human geography developed an explicit concern with the privatization of public space (see, for example, Harvey, 1989a; Zukin, 1991). Nowhere was this concern more obvious, however, than in the example of the shopping mall. Indeed, for Christopherson (1994: 417) the shopping centre or mall is the predecessor of *all* privatized urban environments, the quintessential features of which are 'separation from the larger urban environment, limited pedestrian access, multi-level functionally integrated spaces through which users are channelled via walkways and high levels of security'. According to Crawford (1992: 22):

> the enclosed mall compressed and intensified space. Glass-enclosed elevators and zigzagging escalators added dynamic vertical and diagonal movement to the basic horizontal plan of the mall. Architects manipulated space and light to achieve the density and bustle of a city downtown – to create essentially a *fantasy urbanism* devoid of the city's negative aspects: weather, traffic, and poor people.

Once again it is the West Edmonton Mall (WEM) which has attracted the most attention in this regard, with Shields (1989), Hopkins (1990) and Goss (1993) commenting variously on the ways in which WEM constructs a so-called civic space, privately controlled and tightly policed, essentially reclaiming 'the street' for the middle-class imagination (see also Chapter 12). But this 'privatization' issue is even more pronounced in the case of city centre (as opposed to suburban) malls where the mall's private spaces begin to function effectively as city streets – in the process, subjecting key parts of downtown, as Frieden and Sagalyn (1989: 236) suggest, to the 'discipline' of suburban malls' (see also Chapter 7). Reading 9.5 by Margaret Crawford and extracted from an edited collection by Michael Sorkin which deals more broadly with 'the American city and the end of public space', neatly summarizes many of these issues, focusing in particular on the surveillance present in shopping centres and on the ironic construction of a colourful 'street life' in the mall's 'simulated city'.

Reading 9.5 – Public life in a pleasure dome

As the mall incorporated more and more of the city inside its walls, the nascent conflict between private and public space became acute. Supreme court decisions confirmed an Oregon mall's legal right to be defined as a private space, allowing bans on any activity the owners deemed detrimental to consumption . . . Many malls now clarify the extent of their public role by posting signs that read: 'Areas in this mall used by the public are not public ways, but are for the use of the tenants and the public transacting business with them. Permission to use said areas may be revoked at any time', thus 'protecting' their customers from potentially disturbing petitions or pickets. According to the manager of Greengate Mall in Pennsylvania, 'We simply don't want anything to interfere with the shopper's freedom to not be bothered and have fun' . . .

> If mall decor and design are not explicit enough to tell young blacks or the homeless that they are not welcome, more literal warnings can be issued. Since statistics show that shopping-mall crimes, from shoplifting and purse-snatching to car theft and kidnapping, have measurably increased, the assurance of safety implied by the mall's sealed space is no longer adequate. At the WEM, the mall's security headquarters, central Dispatch, is prominently showcased. Behind a glass wall, a high-tech command post lined with banks of closed-circuit televisions and computers is constantly monitored by uniformed members of the mall's security force. This electronic Panopticon surveys every corner of the mall, making patrons aware of its omnipresence and theatricalizing routine security activities into a spectacle of reassurance and deterrence. But the ambiguous attractions of a lively street life, although excluded from the WEM by a strictly enforced code of behaviour, are not wholly absent. Rather, they are vicariously acknowledged, at a nostalgic distance to be sure, by Bourbon Street's collection of mannequins, 'depicting the streetpeople of New Orleans'. Frozen in permanent poses of abandon, drunks, prostitutes and panhandlers act out transgressions forbidden in the mall's simulated city'.
>
> **Extracted from: Margaret Crawford (1992): The world in a shopping mall. In Michael Sorkin (ed.) *Variations on a theme park: the new American city and the end of public space*. New York: Noonday Press, 3–30.**

It would be quite wrong, however, to suggest that the privatization of public space evidenced in both suburban and downtown malls goes ahead in a completely uncontested fashion. Crawford (1992: 22) draws attention to the 'nascent conflict between private and public space' as 'the mall incorporated more and more of the city inside its walls', and Frieden and Sagalyn (1989: 227–33) document some of the legal struggles in the US during the 1970s concerning the limits imposed by mall developers and managers over the public uses of their private spaces. However, it is other authors, whose work we review in Chapter 12, who have drawn direct attention to the *contestation* of retail space. Rob Shields (1989: 160), for example, is particularly concerned with 'the carnivalization of the mall by its users [which] provides the only means at hand to balance its 'commercial terror':

> Like Urry's 'post-tourist', who knows that mass tourism is a game played for status, the West Edmonton Mall has its 'post-shoppers' who, as *flâneurs*, play at being consumers in complex, self-conscious mockery . . . the users, both young and old, are not just resigned victims, but actively subvert the ambitions of the mall developers by developing the insulation value of the stance of the jaded, world-weary *flâneur*, asserting their independence in a multitude of ways apart from consuming (see Reading 12.6 for an extended extract).

In similar fashion, Hopkins (1990: 7) refers to 'individuals who see shopping centres as a public place and go there for the purpose of socializing and not to carry out traditional economic activity', whilst Goss (1993: 435) comments on *mall rats* who claim public space by sitting on the floor of the mall or 'mallingering' (see also Kowinski, 1985). Taking this issue further, Shields (1994: 211) details the various ways in which refugee Somalis managed to appropriate part of the private mall space of the Rideau Centre in Ottawa for gathering:

> securing a privileged status as loiterers in the cafés and food courts of the . . . Centre. The primary (unconscious) tactic was to quietly dress the part of better-off mall users and to enact a chic European café culture in the lower, more popular, 'teen-oriented' areas and levels of the malls . . . The Somalis' long presence in groups of two and three (mostly male but not always) over drawn-out cups of coffee stabilized the sometimes boisterous, younger shoppers attracted to the jean shops and hunting knife boutiques.

In this latter article, Shields is concerned to illuminate what he terms 'the logic of the mall' and to demonstrate how such environments 'once entered, enfold and engulf us in their ordering time-space logic of interior spaces and store opening hours' (Shields, 1994: 203). But, as we have already suggested, it is not only shopping malls that are developed with a specific 'spatial logic' in mind. The designers of store interiors, in particular, are in the business of constructing ever more complex means of 'enfolding and engulfing' their patrons. Returning to Hannigan's (1998) template, we now turn to what he terms 'experiential retailing', the creation of themed retail environments which represent (at least currently) the most sophisticated form of 'shopertainment'.

Experiential retailing

In earlier sections of this chapter we have discussed the various ways in which layout and design of stores have long been used to facilitate consumption. But as Hannigan observes, by the early 1990s, many retail spaces had become stale – 'consumers embraced the concept of "value retail" – brand name goods at prices lower than those offered by department and speciality stores. Tired of the hassle of fighting traffic on the freeways or muggers in the parking lot, millions of consumers looked to other alternatives – at home catalogue shopping, on-line computer services, cable television shopping channels'. In short, 'a new in-and-out style of shopping was adopted' (1998: 89). For retailers faced with this conundrum, the answer to their problems lay in providing once again an *entertaining shopping experience*. Just as the great department stores had done at the close of the nineteenth century, so by the close of the twentieth century many retail stores began to provide *social* experiences way beyond their sheer 'economic' role. In Pine and Gilmore's terms, the concept of 'retail theatre' was reborn, an essential component of which in its contemporary form is the creation of an *experiential brand image*, 'emphasizing the experience customers can have surrounding the purchase, use or ownership of a good' (1999: 17). Pine and Gilmore's text details a number of examples of 'retail theatre'. At its most simplistic this concept involves an understanding of 'the shopping experience':

> One retailer that does understand the experience of shopping is Leonard Riggio. When the Barnes and Noble CEO began to expand the chain of bookstores into superstores, he hit on the simple theme of 'theatre'. Riggio realized that people visited bookstores for the same reason they go to the theatre: for the *social* experience. So he changed everything about the stores to express this theme: the architecture, the way salespeople acted, the decor and furnishings. And of course he added cafes as an 'intermission' from mingling, browsing and buying (Pine and Gilmore, 1999: 46–7).

But experiential retailing has become increasingly sophisticated. Reading 9.6 by John F. Sherry, Jr., a US business school professor from Northwestern University, details the process by which Nike's flagship stores aim to build brand recognition and to stimulate buying of Nike products at other retail outlets. Nike Town, Chicago (NTC), was specifically 'built as theater, where . . . consumers are the audience participating in the production'. Fascinatingly here, Sherry highlights the *cultural geography* of NTC, focusing on the importance of 'location, location, location' to NTC's success. In this case, then, the configuration and manipulation of retail space extends beyond the immediate confines of the store to include its placing in the broader retail environment. (See also Crewe and Lowe, 1995).

Reading 9.6 – Nike Town Chicago (NTC) and the emplaced brandscape

'Just do it'. As this tagline . . . has been embroidered into the fabric of adcult, so has the building expanded our notion of alternative translations of retail space. With the exception of pricing strategy, every designed element of the servicescape encourages impulsive behaviour and invites instant gratification . . .

NTC, the second in a series of seven company stores launched to date, opened in Chicago in the summer of 1992. Designed in-house by Gordon Thompson, the 68,000 square-foot store boasts three selling floors and eighty feet of frontage on the 'magnificent Mile' of Michigan Avenue. The store is designed to deliver a 'landmark experience', comparable to 'enter[ing] Wrigley Field or hop[ping] on a ride at Disneyland'. The NTC 'retail theater' concept is intended to combine 'the fun of Disneyland and FAO Schwarz, the museum quality of the Smithsonian Institution and the merchandising of Ralph Lauren with the sights and sounds associated with MTV'. The store comprises eighteen pavilions that display products related to twenty sports . . .

A significant measure of the experience NTC afford consumers derives from its prestigious location on 'the Boule Mich', a celebrated stretch of North Michigan Avenue . . . NTC is flanked by a range of upscale retail outlets and galleries. The art-and-commerce ambience of this setting is not reflected solely in discrete and distinctive offerings by specialized shops. Rather, the effect is heightened by the kind of hybridized merchandising that gives NTC its own particular appeal, and the architectural diversity for which Chicago is renowned . . .

As one looks up and down the boulevard, this urban marketplace resembles nothing so much as a canyon of consumption, its glass and concrete walls reigning over a river of pedestrian and vehicular traffic. As a transparently designed canyon, its cultural ecology is characterized by spectacle and desire. The energy and pace of this urban setting contribute to the immediacy of NTC's external presence and mirror the phenomenal realms contained within the building.

Extracted from: John F. Sherry, Jr. (1998): The soul of the company store, Nike Town Chicago and the emplaced brandscape. In John. F. Sherry, Jr. (ed.) *Servicescapes: the concept of place in contemporary markets.* Lincolnwood, IL: NTC Business Books, 109–46.

We elaborate the concept of 'entertainment retail' further in Chapter 11, where we give attention on the crucial role of *design* in the retail built environment. Here, though, we would like to conclude by taking this chapter full circle back to the department store, in this case focusing on the application of 'experiential retailing' techniques to an ailing US department store industry during the 1990s.

In the broader article from which Reading 9.6 is extracted, John. F. Sherry, Jr. draws attention to the 'theatrical embrace' of the Nike Town concept which in Katz's (1994) view is reminiscent of a '1939 World's Fair'. Hence, in the same fashion as Boucicaut's Bon Marché is allegedly intimately connected with the 1855 Grand Exhibition, so Nike Town as a 'brand-building 3-D commercial' is evocative of the World's Fair. In many senses then, and as we argue later in Chapter 11, late twentieth-century flagship stores like Nike Town have effectively assumed the role of the late nineteenth-century department store (see also Lowe and Wrigley, 1996). However, the department store industry itself in the 1990s, in both the USA and (to a lesser extent) in the UK, was in deep crisis. Steve Wood has documented the restructuring of the US department store industry during this period (Wood, 2001a, b and c), and in so doing has analysed the ways in which department stores have sought to restructure their store environments, merchandise selection and marketing image, positioning themselves as 'retailer brands'. The configuration and manipulation of the internal spaces within the store was seen as a key component of successful restructuring with, for example, open selling environments (removing the counter between the sales consultant and the customer) being introduced. At the same time an explicit attempt was made to reintroduce 'fun' to the spaces of the US department store – an element often conspicuous by its absence since the mid-1970s. Somewhat ironically, then, during the early years of the twenty-first century and in the light of competition from elsewhere in the retail industry, US department stores have begun aggressively to pursue the concept of 'retail entertainment', taking their cue from speciality stores such as Old Navy, whose 'industrial chic' – concrete partitions, pumping music, bright lights and electronic messages – appealed to the important youth market. In 1999, for example, Macy's (now under the ownership of the Federated department store chain) unveiled its 'Macy's Sport' department at its Union Square, New York, flagship store:

> A three storey vertical billboard marked the 34th Street entrance to the 15,000 square foot department and a barrier on the billboard highlighted a different Macy sport vendor each week. The space was configured to be elliptical . . . lighting features complemented this shape . . . and [were] used to highlight graphics. Each sport had its own area of the department . . . way finding signs were of sports imagery [and], in common with Old Navy, Macy Sport offered flooring of Asbestolith, a material designed to resemble concrete (Wood, 2001a: 277–8).

'Shopertainment', as we have seen throughout this chapter, has become progressively more refined as retailers have become more keenly aware of the importance of the configuration and manipulation of their spaces in order to sell commodities. Having engaged closely with these aspects of retail space in this chapter, we now move on to consider consumption spaces and places more explicitly in the final section of our book.

PART 4

Consumption places and spaces

10
The street

In this section of the book we consider in detail geographies of consumption spaces and places, focusing explicitly on the street, the store, the mall and the home. We aim to show how these four quintessential consumption sites are not just passive surfaces upon which the business of consumption takes place but instead are 'actively produced, represented and contested' (Blomley, 1996: 239). Within this framework, two specific sub-themes emerge. The first centres on the gendered nature of these spaces and places. The second relates to the issues considered in Chapter 9 – the various ways in which the geographical organization of such spaces facilitates consumption.

First, however, we turn to 'the street' – the original centre of urban retail life and the retail space most often at the heart of our everyday experience.

Street life and street culture

Between November 1994 and February 1995 an exhibition entitled 'Streetstyle' was held at the Victoria and Albert Museum in London. Throughout the exhibition – and throughout Polhemus's volume *Streetstyle: from sidewalk to catwalk* published to coincide with it – the street is 'a focus of modern life' (Polhemus, 1994: 6), a key site of identity construction.

> All human life is here. Mingling. Checking each other out. Doing their thing ... the excitement and the sense of 'this is where it's at' is so tangible that you want to reach out and grab handfuls of it. Instead of just passing through ... en route to somewhere else, we want to linger here. It is a destination as well as a thoroughfare' (Polhemus, 1994: 7).

For 1940s zooties, for bikers and teddy boys, for mods and rockers, for skinheads and goths, for indie kids, and for 1990s technos and cyberpunks, the street is a place to see and be seen and 'styles which start life on the street corner have a way of ending up on the backs of top models on the world's most prestigious fashion catwalks' (Polhemus, 1994: 8).

As we shall see, this view of the street as 'a structured and skilful space' (Glennie and Thrift, 1996b: 227) pervades much geographical literature on this topic. The street as a site of display, of strolling and of seeing, dates back to the key nineteenth-century urban figure of the *flâneur,* but there are many resonances of this nineteenth-century urban geography of display in what we term contemporary 'streets of style' such as Old Compton Street or Frith Street in Soho or New Bond Street in London's West End.

Of course, the street in the nineteenth century was not just a site for the lone bourgeois promenader or *flâneur*; rather 'urban streets belonged in large part to labouring class people . . . For poor men, women and children the streets were workplaces and playgrounds' (Glennie and Thrift, 1996b: 227). And today's streets also bear the hallmarks of these 'socialities'. Contemporary street markets and street carnivals display many similar qualities. As Zukin has argued, 'ordinary shopping districts frequented by ordinary people are important sites for negotiating the street-level practices of urban public culture in all large cities . . . A commercial street is nearly always the "heart" of the modern city' (Zukin, 1995: 191). For Zukin, then, 'streets and fleamarkets as "ghetto shopping centres" are just as important as the malls for constructing identity and difference' (Zukin, 1995: 253). And, as Mort suggests, there have always been 'a number of more heterotopic forms of city life; a series of other worlds and spaces which jostled for attention on the boulevards' (Mort, 1996: 163). These other worlds and spaces will be the subject of our attention later in this chapter. Here though we begin with an investigation of the role of the *flâneur* in the nineteenth-century shopping street, the street as a site of display and streets of style.

The *flâneur* and the street as a site of display

For many commentators the *flâneur*, 'an independent but impecunious single man who strolled the city's streets . . . on the lookout for the new, the exciting and the unfamiliar' (Zukin, 1998: 8) is the key figure of the nineteenth-century shopping street. Reading 10.1 by Mike Featherstone, a cultural theorist writing in a theme issue of *Urban Studies* (1998) on 'Urban consumption' captures the place of the *flâneur* in nineteenth-century urban life. In addition, the reading draws attention to the fact that the *flâneur*'s gaze in his native territory, the street, was an unequivocally masculine one – 'there is no corresponding literary figure of the *flâneuse*' (Zukin, 1998: 8). The department store – as the first in a series of attempts to move the street into the interior – brought about the feminization of the *flâneur*, a process which will be discussed in more detail in Chapter 11.

Reading 10.1 – The *flâneur* in the city

The *flâneur* has his origin in the Paris of the early nineteenth century in which around 30 arcades were constructed between 1800 and 1850 which provided enclosed spaces for people to stroll and look, to idle and dawdle – as we find in the oft-quoted example of the *flâneur* who showed his indifference to the pace of modern life by taking a turtle for a walk. This tension between looking and idling points to a number of other tensions. On the one hand, the *flâneur* is the idler or waster; on the other hand, he is the observer or detective, the suspicious person who is always looking, noting and classifying: the person who as Benjamin put it 'goes botanising on the asphalt'. The *flâneur* seeks an immersion in the sensations of the city, he seeks to 'bathe in the crowd'; to become lost in feelings, to succumb to the pull of random desires and the pleasures of scopophilia . . . The *flâneur*'s craft entailed a hermeneutic of seeing which appealed to the growing urban educated middle-class reading public. It was a hermeneutic which made the city exotic and sought to follow the key maxim of romanticism: 'to make the strange familiar and the familiar strange'. The task was to see the city anew, as if for the first time . . .

If the *flâneur* was evidently a male social type, then the rise of the department stores can be seen as involving a process of the feminisation of the *flâneur*. The department store can be seen as a further attempt to move the street into the interior – something we see in the twentieth century with the rise of shopping centres, malls and resort hotels. Whereas for the poor the street was an interior, the place where the homeless and street people have to live, for the rich the interior became made into a street as all the variety of goods, shops and exotic urban experience became assembled in artificial settings. In the department store, people who once roamed in the city now roam amongst merchandise.

Extracted from: Mike Featherstone (1998): The flâneur, the city and virtual public life. *Urban Studies*, 35, 909–25.

The clearest attempt to revive the spirit of the *flâneur* in the late twentieth century must be in the work of Mort (1995, 1996, 1998). In various studies of cultures of consumption in London from the 1950s to the 1980s, Mort foregrounds the figure of the urban/metropolitan *flâneur*. Importantly, Mort's writing gives the persona of the *flâneur* a new twist – in 1980s London the *flâneur* is a homosexual man 'cruising the streets with a clear agenda' (Mort, 1996: 176). Regardless of his sexual tendencies, however, the image of the *flâneur* gives us valuable perspective on geographies of display in contemporary shopping streets. Reading 10.2 by Frank Mort, a cultural historian whose work is widely known in human geography, emphasizes the importance of image and identity to the renaissance of Soho in the 1980s and the place of the *flâneur* in the new 'streets of style'.

Reading 10.2 – Boulevards for the fashionable and famous

Cosmopolitan, bohemian, wildly trendy – Soho is London's very own rive gauche. This square mile of style is a kingdom unto itself: the land of the brasserie lunch and the after-hours watering hole, the land of accessories and attitude, where fashion relentlessly struggles to become style and image is simply everything; the glittering heart of media land where the worlds of art, journalism, film, advertising and theatre blend into one glamorous heady cocktail . . . How does one become part of this bohemian world? How can you . . . cut a dash as a get-ahead young turk effortlessly oozing that quintessential Soho style – style that you have only glimpsed in the pages of *The Face* . . . There is a whole unspoken language to be learnt, a code of behaviour and dress, a long list of do's and don'ts . . .

Alastair Little's celebrated restaurant in Frith Street appeared in 1985. Less intimate, but more imposing, was Braganza's sited further along Frith Street, which had started the previous year. For this venture architects and interior designers had been specially commissioned to give each of the three floors a particular atmosphere. Opening its doors at the same time, the Soho Brasserie, in Old Compton Street, was one of Soho's first deliberate emulations of a Parisian café bar. Furnished with the ubiquitous chrome and stainless steel interior, marble-topped tables and authentic foreign waiters, the café offered a 'continental' ambience. Ordering a Kir, or perhaps a *salade de fruits de mer*, customers could seat themselves at window tables opening directly onto Soho's main thoroughfare. From this vantage point they could watch the boulevard like latter-day *flâneurs* . . .

Christopher New, whose first clothes shop opened in Dean Street in 1985, pointed out that many retailers moved to Soho not simply on account of lower rents, but because the district was 'considerably more interesting'. New felt that the area had a genuine village feel . . . There were eye-catching items displayed at American Retro in Old Compton Street where owner Sue Tahran brought together an abundance of classic design items: from Braun alarm clocks and zippo lighters . . . Or for personal organizers there might be a visit to Just Facts, in Broadwick Street, reputedly the only shop in the world to stock the entire Filofax range. Distinctive 'eye wear' could be bought at Eye Tech in Brewer Street, designer clothes picked up at Workers for Freedom or the Academy.

Extracted from: Frank Mort (1996): *Cultures of consumption: masculinities and social space in late twentieth-century Britain*. London and New York: Routledge, 157–63.

Contemporary streets of style

Mort's cultural history in which the street – specifically Frith Street, Old Compton Street and Dean Street in Soho – plays a particular role in the gestation of new urban landscapes of consumption, reads in stark contrast to more 'economic' views of contemporary London 'streets of style' – specifically concerning the importance of Bond Street to the urban retail scene. As many readers will doubtless be aware, Bond Street, in Mayfair, which suffered from the retail recession of the late 1980s/early 1990s subsequently revived strongly and became a key site for major flagship stores discussed in detail in Chapter 11. Bond Street's recent growth has been in parallel with expansions in Knightsbridge (in which Sloane Street is the dominant street for fashion designer stores). Both streets have experienced above average growth since 1992 so that 'in terms of floorspace as well as store numbers the dominance of Bond Street and Knightsbridge has grown markedly in the last few years and, with developments in the pipeline, this dominance is set to increase' (Hillier Parker, 1996: 5). Table 10.1 shows that of over 50 foreign fashion designer outlets in central London in 1997 no less than 43 were sited in these two key locations.

Of course, Bond Street has always been one of the pre-eminent locations of London's luxury shops (Glennie, 1998) but in the mid- to late 1990s this

Table 10.1: *Foreign fashion designer outlets in central London, 1997*

Fascia	Country of origin	Location of store	District	Date of opening
Anna Molinari Blumarine	Italy	Bond Street	Mayfair	1995
Armani – Emporio	Italy	Long Acre	Covent Garden	1992
Armani – Emporio	Italy	Brompton Road	Knightsbridge	1990
Armani – Emporio	Italy	Bond Street	Mayfair	1993
Armani – Giorgio	Italy	Sloane Street	Knightsbridge	1995
Celine	France	Brompton Road	Knightsbridge	1969
Chanel	France	Sloane Street	Knightsbridge	1988
Chanel	France	Bond Street	Mayfair	1981
Christian Dior	France	Sloane Street	Knightsbridge	1993
Christian Lacroix	France	Bond Street	Mayfair	1994
Christian Lacroix	France	Sloane Street	Knightsbridge	1992
Comme des Garcons	Japan	Davies Street	Mayfair	1988
Dolce & Gabbana	Italy	Sloane Street	Knightsbridge	1994
Donna Karan	US	Bond Street	Mayfair	1996
Donna Karan – DKNY	US	Bond Street	Mayfair	1994
Emanuel Ungaro	Italy	Sloane Street	Knightsbridge	1991
Fendi	France	Conduit Street	Mayfair	1988
GianFranco Ferre	Italy	Sloane Street	Knightsbridge	1995
GianFranco Ferre	Italy	Brook Street	Mayfair	1987
GianFranco Ferre	Italy	Brompton Road	Knightsbridge	1988
Gianni Versace	Italy	Bond Street	Mayfair	1995
Gianni Versace – Istante	Italy	Sloane Street	Knightsbridge	1995
Gianni Versace – Versus	Italy	Brompton Road	Knightsbridge	1987
Gucci	Italy	Bond Street	Mayfair	1967
Gucci	Italy	Sloane Street	Knightsbridge	1990
Guy Laroche	France	Bond Street	Mayfair	1993
Hermes	France	Bond Street	Mayfair	1974
Hermes	France	Sloane Street	Knightsbridge	1987
Hugo Boss	Germany	Regent Street	Mayfair	1994
Iceberg	Italy	Brompton Road	Knightsbridge	1992
Issey Miyake	Japan	Brompton Road	Knightsbridge	1991
Jean Paul Gaultier	France	Draycott Avenue	Kensington	1991
Kenzo	France	Sloane Street	Knightsbridge	1993
Lanvin	France	Brompton Road	Knightsbridge	1987
Lanvin	France	Bond Street	Mayfair	1987
Loewe	Spain	Bond Street	Mayfair	1978
Louis Feraud	France	Bond Street	Mayfair	1988
Max Mara	Italy	Bond Street	Mayfair	1991
Max Mara	Italy	Sloane Street	Knightsbridge	1988
Oscar de la Renta	US	Savile Row	Mayfair	1988
Paul Costelloe	Ireland	Brompton Road	Knightsbridge	1994
Paul Costelloe	Ireland	Pelham Street	Kensington	1996
Polo Ralph Lauren	US	Bond Street	Mayfair	1981
Prada	Italy	Sloane Street	Knightsbridge	1994
Romeo Gigli	Italy	South Molton Street	Mayfair	1987
Romeo Gigli	Italy	South Molton Street	Mayfair	1986
Sonia Rykiel	France	South Molton Street	Mayfair	1986
Thierry Mugler	France	Bond Street	Mayfair	1995
Valentino	Italy	Bond Street	Mayfair	1985
Valentino	Italy	Sloane Street	Knightsbridge	1982
Yves Saint Laurent	France	Bond Street	Mayfair	1971
Yves Saint Laurent	France	Sloane Street	Knightsbridge	1966

Source: Corporate Intelligence UK Retail Report No. 76, December 1996. p. 135, based on a Hillier Parker Research Survey.

dominance was given new emphasis as large numbers of foreign (especially US) fashion designer stores were attracted there. Figure 10.1 shows the position in 1997. Set in the context of the rising profile of London as a fashion centre, this concentration of designers on such a high-rent, high-profile thoroughfare was seen to be due to economic imperatives such as the growing prosperity of top earners in the UK and the growing wealth and disposable income in the capital. Openings on Bond Street in the mid-1990s included Donna Karen, Armani, Ralph Lauren, Calvin Klein and Tommy Hilfiger, whilst indigenous UK designers with less financial muscle were often relegated to locations in less prominent streets in Covent Garden and South Kensington.

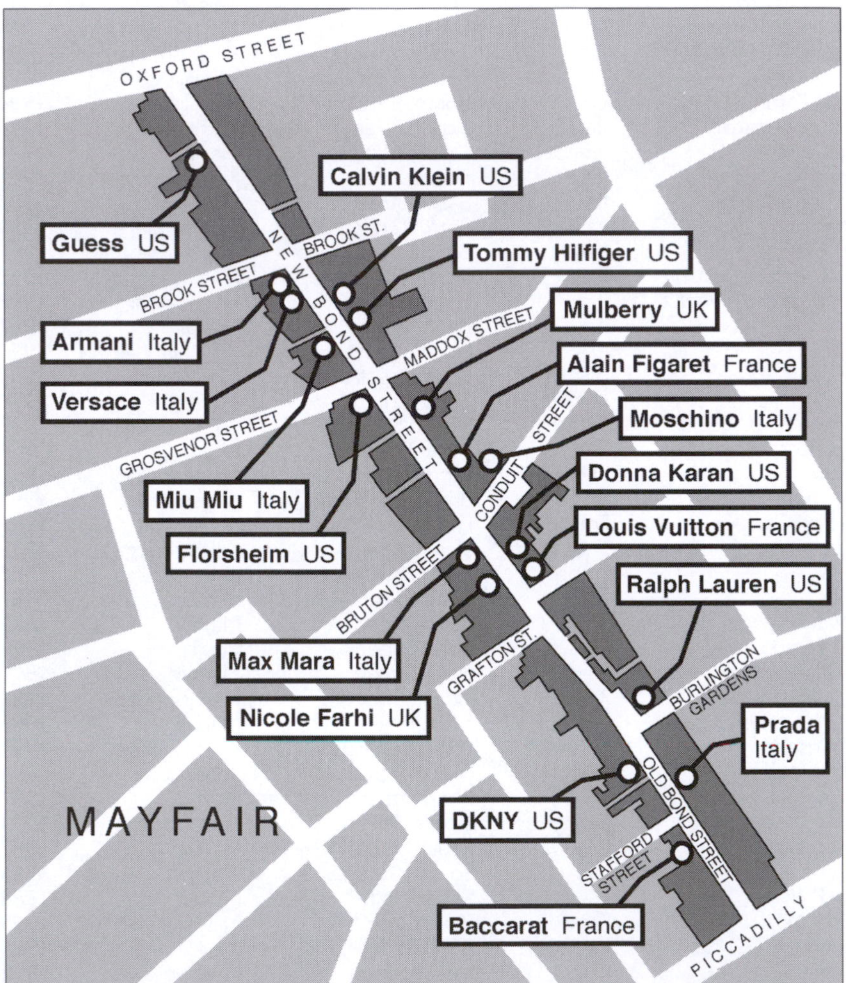

Figure 10.1: *Fashionable lettings in Bond Street 1996/97*
Source: Adapted from Sunday Times *15 February 1998 and* Sunday Telegraph *19 January 1997.*

But it is possible to offer an explanation for the place of Bond Street within the microgeography of fashion and design in 1990s Britain which, like Mort's work, stresses 'cultural' as well as 'economic' factors, in particular the importance of the cultural cachet of a Bond Street location to many of the designer stores under discussion here. Reading 10.3, taken from a summary of an article by John Fernie, Christopher Moore, Alex Lawrie and Alan Hallsworth, provides a fascinating perspective on the Bond Street phenomenon and captures, in particular, the importance of branding to the success of this particular 'street of style' (see also Moore, 2000).

Reading 10.3 – The branded street

London has rediscovered its position as a fashion centre over the period since 1985. This reflects the attractiveness of the city to international travellers and the growth in wealth and disposable income among local residents. UK fashion houses face the challenge of the big international brands from Europe and the USA. Either because of ownership, franchising or sheer international power these established fashion brands were able to force their way into the better London locations . . .

Just as who buys the exclusive brand is important, so is the location of the outlet. The image associated with a particular location has an effect on the brand . . . Within London, Fernie et al. remark on the distinction between 'primary' locations for fashion outlets (Knightsbridge, Mayfair) and 'secondary' locations (South Kensington, Covent Garden). The strong international brands dominate the former locations while local brands concentrate on the latter. This concentration reflects the ability of the big fashion brands to afford very expensive leases (Fernie et al. report figures of £6 million to £10 million for leases and development on some sites).

One interesting observation from the research here is the concentration of investment by international fashion brands. Despite the ability of their very wealthy customers to travel, the fashion investment has focused on just two streets – Bond Street and Sloane Street. Yet again the tendency of businesses in the same line to congregate is demonstrated. Just as, in the past, Bond Street for jewellers, Savile Row for tailors and Threadneedle or Lombard Street for banking represented this trend, the desire of top fashion brands to locate in one or two streets repeats this classic locational pattern.

Why does this occur? It can't just be a matter of convenience. After all it's only 400 yards from Bond Street to Regent Street or Oxford Street (central London's two main shopping streets). Fernie et al. hint at the reason when they say that investment activity 'has occurred in two prestigious areas of London'. The key word here is prestigious. What the fashion retailers are exploiting is the brand of the location. A Bond Street address carries a cachet not enjoyed by neighbouring streets . . . it cannot be said that there is anything peculiar about Bond Street. It's not the main street nor is it pedestrianized. It doesn't have any particular attractions along its length and it isn't close to the main tourist attractions in London. Yet, for some reason, it is seen as a key 'branded' street to international fashion houses.

Extracted from: Executive summary of John Fernie et al. (1997): The internationalization of the high fashion brand: the case of central London. *Journal of Product and Brand Management*, 6, 151–62.

Outside London, in provincial cities of the UK, Crewe and Lowe (1995) consider the emergence of 'differentiated spaces of consumption' i.e. how retail spaces become invested with particular identities. They suggest that 'identity-based' location preferences have led to 'pioneering clothing retailers' such as Hobbs and Jigsaw locating in centres such as Guildford, Nottingham, Cheltenham and Glasgow and further that *within* these centres the identity and image of particular streets – for example Little Clarendon Street in Oxford or the Promenade in Cheltenham make them prime (if somewhat offbeat) locations for companies keen to differentiate themselves on grounds of style and exclusivity. Like New Bond Street in London, Oxford's Little Clarendon Street or Cheltenham's Promenade appear to have specific cultures and images attached to them – they are effectively 'branded streets' and this branding enables them to attract and maintain upmarket retailers (see also the discussion of North Michigan Avenue, Chicago in Chapter 9).

The street and 'sociality'

So far our consideration of the street as a consumption site has been a somewhat elitist one. Our focus on the historical figure of the *flâneur* as well as his contemporary equivalent, together with our emphasis on exclusive 'branded streets' or 'streets of style' in late twentieth-century London and elsewhere can be seen as a partial view. As we hinted in the introduction to this chapter, the street as a consumption site in the nineteenth century functioned not only as a space for the solitary activities of the *flâneur*, but the street in that and earlier periods also acted as 'a kind of classroom ... in which people learned about commodities, styles and their uses and meanings (Glennie and Thrift, 1996b: 227). Shopping then was a social activity in which consumers' interaction within 'the throng' influenced their behaviour (see also Chapter 11). Glennie and Thrift (1996b: 226–7) define sociality as:

> the basic everyday ways in which people relate to one another and maintain an atmosphere of normality, even in the midst of antagonisms based on gender, race, class or other social fractures ... it consists of a 'contact' or tactile community built up from the solidarity and reciprocity of everyday life ... Moreover, visually rich public sites provide an intensive level of stimuli and imaginative cues. In the past, the street, market places and public gatherings of all sorts have been important sites.

Such socialities are also present in today's streets (as well as in other retail and consumption spaces discussed in Chapters 11 and 12) and it is to these perspectives on 'ordinary' people in 'ordinary' shopping settings that this chapter will now turn. In particular we wish to emphasize the fact that 'the shop' – often the centre of our retail geographies and the focus in the form of 'the store' of Chapter 11 – 'was [and remains] but one channel among many through which goods might be bought' (Glennie and Thrift, 1996b: 228).

In various historical geographies of consumption, Glennie and Thrift (1992, 1996a and b) emphasize the importance of a perspective which recognizes that 'modern' consumption dates back to the seventeenth and eighteenth centuries and beyond. In this vein they indicate:

> What is remarkable is how far back in time the history of retail shops and shopping stretches in England; the scale on which shops and shopping operate; and how much of this history of shops and shopping is based on a reflexive relationship between consumers, shopkeepers, and the *sites* of consumption (understood as streets, markets, shops, galleries, and so on), sites which act as an active context rather than a passive backdrop (Glennie and Thrift, 1996a: 26).

Within these various accounts the street as a key consumption space plays a major role – 'Very large numbers of purchasers bought goods in open markets and fairs, or directly from artisan producers, or from hawkers and chapman. Consumers of almost all social strata routinely acquired goods in face-to-face interaction with vendors in public settings, rather than in specialist shops' (Glennie and Thrift, 1996b: 228). In the eighteenth and nineteenth centuries, as Glennie (1998) demonstrates, stimulated by demographic growth, higher household incomes and narrowing in household self-provisioning (Benson and Shaw, 1992), scores of new retail markets were established especially in rapidly growing industrial towns (Shaw, 1985; Scola, 1982). Figure 10.2, for

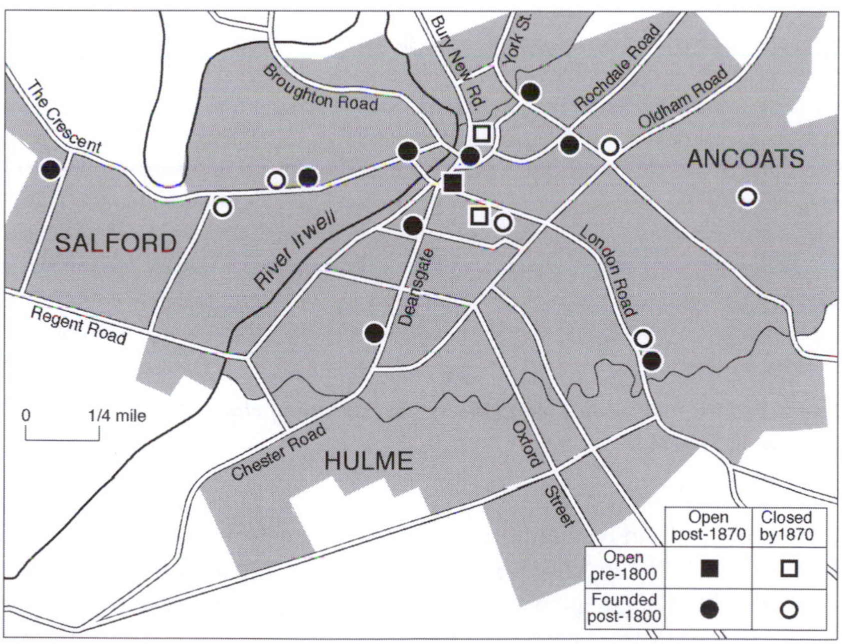

Figure 10.2: *Markets in early nineteenth-century Manchester and Salford*
Source: Redrawn from Glennie (1998), after Scola (1982).

example, shows markets established in the industrial cities of Manchester and Salford in the nineteenth century. During this period many shopkeepers also held stalls in weekly markets and itinerant casualized street trading was important to retail geographies.

In the contemporary period the social nature of shopping in the street has been largely preserved. Reading 10.4 by Sharon Zukin, an American sociologist who has published extensively on urban cultures, captures very eloquently the diversity of cultures on shopping streets and the vitality of urban markets in New York City.

Reading 10.4 – Urban markets and urban cultures – twentieth-century New York City

While retail shopping [in the United States] has been changing, new waves of immigration from Asia, Africa and Latin America have been bringing new consumers and entrepreneurs to urban markets ... They create new spaces of ethnic identity (Little Havanas, Odessas and Cambodias and suburban Chinatowns) and place new people in existing ethnic divisions of labour. Korean and (East) Indian shopkeepers often replace Jewish merchants in low-income neighbourhoods, and buy from Jewish and Italian wholesalers before establishing their own suppliers' networks. Immigrants from all parts of the world swell the ranks of street vendors. West Africans sell knock-off designer scarves and Rolex watches on Fulton Street in downtown Brooklyn and on Fifth Avenue in midtown Manhattan. They sell African art that resembles museum art on 53rd Street, down the block from the museum of Modern Art, and 'tourist' art on 125th Street in Harlem. Chinese sell frozen shrimp and dried mushrooms on the sidewalk outside grocery shops on Canal Street. Central Americans dispense hot dogs from vending carts all over the city.

Hardly tolerated by local store owners on ethnic shopping streets – many of whom are, themselves, Italian, Chinese, African American, or 'Arab', usually Syrian Jews – immigrant street peddlers recreate a bit of the experience of Third World street markets and stalls. They also engage in less sanctioned, informal markets. They join a street economy in legal and illegal goods already flourishing in poor areas of the city. Some sell stolen or pilfered goods. Poor Russian immigrants stand around on Brighton Beach Avenue with shopping bags of their household goods and personal possessions, hoping to barter or sell. Individual blocks and whole shopping streets are in flux between stable ethnic identities, diversity and change.

Extracted from: Sharon Zukin (1995): *The culture of cities*. Oxford: Blackwell, 210–11.

But it is not only in conventional street market settings that 'sociality' is an integral part of 'going shopping'. Some of the most novel work in this framework is that of Gregson and Crewe who have studied contemporary car boot sales in the UK from a variety of perspectives. As well as concentrating on the regulation of these alternative 'spaces of exchange' (Gregson et al., 1997) and on the gendered nature of consumption within them (Gregson and Crewe,

1998), Gregson and Crewe emphasize the importance of what they term 'fun and sociality' to such activities as well as the various ways in which car boot fairs 'encapsulate the carnivalesque' (Gregson and Crewe, 1997, 1998).

Whilst not strictly meeting the criteria of this chapter – car boot sales/fairs do not generally take place in 'street' settings – we suggest that such marginal spaces of contemporary consumption are effectively (as indeed Gregson and Crewe themselves argue) today's equivalents of historic markets, albeit held in largely different venues, and as a result our coverage of them here is, we suggest, wholly justified. Reading 10.5 – a composite of some of Gregson and Crewe's work – profiles car boot sales, discusses their 'intrinsically social character for both vendors and buyers alike' (Crewe and Gregson 1998: 42) and draws out the festival aspects of these phenomenon.

Reading 10.5 – Car boot sales – markets and carnival

Car boot sales ... first appeared in Britain in the late 1970s and early 1980s. Situated primarily, although not exclusively, in fields, car parks, and/or open spaces on the urban fringe, and organised both by private sector promoters and by institutions such as schools and hospitals for fund-raising, these events involve the exchange of, for the most part, used household and personal goods. As such they connect with other similar spaces of secondhand exchange, notably jumble sales, flea markets, swap meets and garage sales ... Sellers drive to a site where they pay a flat-rate fee to the promoter or organizer for a 'pitch' ... on which they park their car and from which they sell their goods ...

So great has been the proliferation of car boot sales in Britain through the 1990s that it is currently estimated that over 1500 sales occur every week, attracting over one million people either to 'stall out' or to revel wholeheartedly as buyers in a punishing dawn-to-afternoon ritual which sometimes takes in several sales, and which makes car boot sales Britain's fastest growing leisure pursuit and one of the largest.[1]

In contrast to the reliability of conventional retail environments ... car boot sales provide participants with the opportunity to stroll, gaze, browse and rummage for the unexpected and to haggle for a bargain ... car boot sales, then, are dirty, cluttered and unpredictable but as such they are also exciting, challenging and fun ...

At the same time, this sense of fun and excitement is not just something experienced at the level of the individual. Rather, much of the pleasure gained from participation stems from its intrinsically social character for both vendors and buyers alike. The importance of this element of sociality within exchange can be traced back to the historic role of markets as sites of communication and social exchange. However, while sociality has for some time been recognized in relation to the buyers' experience, that this sociality extends to sellers has not been fully acknowledged ... Vendors find the freedom, flexibility and easy-going entrepreneurialism of the car boot sale attractive ... Barriers to entry are low, social exchanges are important and work becomes indistinguishable from leisure ... For buyers too, car boot sale consumption is a commercial activity in which sociality, tactile interaction and tribal solidarity as part of the crowd figure centrally ... Just as in the case of traditional markets, the exploration of the alternative landscape of the car boot sale is explained not only by the desire to acquire products but also by the imaginative, sensory experiences which it promises.[2]

> We read the car boot sale phenomenon in the Bakhtinian sense of carnival, as lived festival ... Indeed we see remarkable resonances between the car boot sale and Bakhtin's representations of medieval market place festival ... It is in its oppositional positioning in relation to the world of conventional retailing, in its transgression of the rules which structure formal exchange, that the car boot sale most closely conforms with 'carnival'. Here ... we have a space of exchange in which the accepted conventions of exchange ... are quite literally turned upside down ... the disciplined social order of fixed prices, the non-contestable, non-negotiable social relations of retailer–salesperson–consumer and the trading regulations designed to 'protect the consumer' are all suspended as the 'consumer' transforms into this hybrid entity-vendor, buyer, stroller, gazer, and even entertainer. Thus, here is a space of exchange constructed by 'consumers', the subordinates to retail capital and its representatives, and a space which captures the notion of festive life.[3]
>
> **Extracted from: [1] Nicky Gregson et al. (1997): Excluded spaces of regulation: car boot sales as an enterprise culture out of control. *Environment and Planning A*, 29, 1717–37; [2] Louise Crewe and Nicky Gregson (1998): Tales of the unexpected: exploring car boot sales as marginal spaces of contemporary consumption. *Transactions of the Institute of British Geographers* NS 23, 39–53; [3] Nicky Gregson and Louise Crewe (1997): The bargain, the knowledge, and the spectacle: making sense of consumption in the space of the car boot sale. *Environment and Planning D: Society and Space*, 15, 87–112.**

The street as a site of resistance

Of course it is perhaps a little simplistic to view the car boot sale as a space of exchange 'constructed by consumers' and to compare this firmly with the 'high street, the mall and the out-of-town retail park' (Gregson and Crewe, 1997: 109). The street as a consumption space (unlike stores, or various types of mall discussed in Chapters 11 and 12) has the potential to act as an important site of resistance to dominant cultural norms. In contrast to the privately regulated public space of malls, the sheer accessibility of the street – like the relative accessibility of the car boot fair – enables the activities of various groups in this public space. Latter day street carnivals such as the West Indian–American Day Carnival Parade in Brooklyn (Zukin, 1995) or the Notting Hill festivities in west London stand in sharp contrast to more regulated and manicured 'events' that often coincide with public holidays in malled environments. Interestingly, though, such street festivals themselves have sometimes coalesced around various landscapes of consumption. To return to the work of Mort (1996: 164) on Soho:

> By February 1993 Old Compton Street had become the venue for a different type of spectacle – Soho's 'Queer Valentine Carnival'. In festive atmosphere nearly two thousand 'lesbians, gay men and their friends' gathered for an afternoon party. The high point of the day was a street parade ... As the parade turned into Old Compton Street, activist Peter

> Tatchell conducted a renaming ceremony. Hence forward, he proclaimed, the thoroughfare was to be known as 'Queer Street' . . . Such manifestations of the carnivalesque were not entirely spontaneous. They marked the deliberate attempt to fuse together a new upsurge of radical sexual politics with the celebratory style of the street festival . . . Yet Soho's carnival involved something more than an exercise in sexual politics. It was also a testament to the growing commercialization of homosexuality. Every time the Valentine parade stopped on its way through the area, it drew attention to the diverse network of consumer culture which was now established. Bars and clubs, cafés and shops held out the promise of a homosexual life, shaped by the market.

In this case, then, the 'resistance of the street' is centred around the promotion of alternative forms of consumption. Of course, none of this is new, London's Carnaby Street developed a specific identity for itself in the 1960s focused on fashion. As Polhemus (1994: 61) comments:

> A few years earlier . . . [Carnaby Street] had been a fairly typical Soho back street with newsagents, greasy spoon cafés and a few traditional tailors. In the early sixties it was transformed into a haven for early mods who discovered that menswear entrepreneurs like John Stephen had an eye for detail which matched their own. Then, as swinging London became the centrepoint of the western world's popular culture, an endless stream of new 'boutiques' opened on Carnaby Street, selling anything and everything that might catch the eye.

In a similar vein, 'Worlds End' – the 'wrong end' of Kings Road which had 'never enjoyed the trendy prosperity of the rest of this famous street' became the 1970s centre of the 'proto-punk revolution' (Polhemus, 1994: 90). Mort's 'tales of Soho', then, are the latest in a long line of cultural movements centred around resistance to dominant consumption norms of various eras. And the street, as this chapter has shown, has always offered the potential for such uprisings. Somewhat alarmingly, though, the late twentieth century brought about an attempt to capture the vitality and spirit of the street within themed (and importantly, controlled) mall environments.

The street as mall

Universal Studio's City Walk in Los Angeles or the 'Fremont Street Experience' in Las Vegas, both designed by the Jerde Partnership International (Jerde, 1998), recreate the diversity of the street in the form of a mall. As Scott and Soja (1996: 454) comment: 'City Walk, described by its imagineers as an "idealized reality, L.A. style," is an attempt to deliver the unkept promise of Los Angeles.' A $100 million addition to MCA's 'Entertainment City', City Walk aims to capture the 'real' feel of an L.A. street with boutiqued facades borrowed from Melrose Avenue, 3D billboards (with moving parts) copied from Sunset Strip, and a faux Venice Beach, complete with sand, artificially induced waves, and strolling troubadours. Even history has been fabricated, with buildings painted 'as if they had been occupied before' and candy wrappers embedded into the terrazzo flooring to give 'a simulated patina of use'. A 'new and improved Los Angeles' is needed, so say the project's market researchers, because 'reality has become too much of a hassle'. City Walk borrows heavily from the architectural styles of various shopping areas in Los Angeles such as Venice, Melrose Avenue and Third Street, Santa Monica, and has promoted itself as a 'safe' alternative to traditional L.A. streets especially for family shopping. Significantly though, this 'mall as street' has failed to capture some of the vital features and liveliness of the street portrayed earlier in this chapter – 'No amount of lush subtropical planting can disguise the fact that City Walk doesn't encourage loitering, and people generally don't – they criss-cross its streets from shop to shop without quite looking at home' (Beckett, 1994: 12). The *flâneur* would certainly not be welcome here! But Jerde's 'streets' are the ultimate antidote to the problematics of an unkept 'street life' and with the privatization of public space that has occurred in so many cities (see Chapter 9), the style and purpose of such developments seem set to continue. Of course stores and malls offer at the outset much more 'controlled' consumption spaces and it is to these sites that we will now turn.

11
The store

On Thanksgiving Day each year New York City plays host to the colourful and elaborate Macy's Thanksgiving Day Parade. The parade through the streets of Manhattan is one of the most important events in the city's calendar; dating from 1924, it remains a symbol of the wider role of the department store as an institution at the heart of New York urban life and aims to create an almost emotional bond between the store and its customers. Despite its bankruptcy in 1991, and its subsequent incorporation as merely one of several divisions of the giant Federated Department Stores (see Chapter 2), Macy's – as New York City's and the world's largest department store – retains a certain aura and a unique place in the hearts and minds of consumers in New York and beyond.

This chapter charts the role of the store, more broadly, as one of the classic consumption spaces. Beginning with an analysis of the great department store of the late nineteenth and early twentieth centuries which – like their latter day equivalents the shopping malls (see Chapter 12) – served an important purpose as entertainment and tourist attractions alongside their merchandising role, we move on to consider stores of the 1920s and 1950s, followed by late twentieth-century flagship stores. We highlight the various ways in which concepts pioneered by the likes of Macy's are still important mechanisms by which these consumption spaces function today. In particular, we emphasize the importance of design in fashioning the interior spaces of the store from the nineteenth century to the present day. Unusually here, though, we begin with a reading from Mica Nava, a cultural theorist whose pioneering work on consumption and identity is highly regarded by historical, contemporary human and feminist geographers. Reading 11.1, extracted from an article which more generally engages with post-colonial theorizations of orientalism and imperial ideologies, highlights many themes regarding the store as a consumption space which are taken up in more detail later in the chapter.

Reading 11.1 – Selfridge's – monument to modernity

Selfridge's was founded in 1909 by Gordon Selfridge, a self-made American entrepreneur . . . [Selfridge] had risen from stock boy to partner of Marshall Fields of Chicago, the largest and most magnificent department store in the United States, had been denied a senior partnership by Marshall Field, and had come to Britain in 1907, partly out of pique, determined to make his mark and revolutionize British retailing practices. He was already an established innovator and modernizer and the London enterprise, the biggest purpose-built store in the world and the first in Britain, was indeed to have a massive commercial and cultural resonance. The launch in 1909 was preceded by the largest advertising campaign in British history and during the first week alone the store attracted an astonishing one and a quarter million visitors – 'a cosmopolitan crowd', according to *The Star* (29 March 1909). Its architecture, plate-glass windows, roof gardens, lifts and electric lighting quickly established it as a technological and aesthetic 'monument to modernity' but it was a modernizing cultural force in a number of less obvious ways as well.

From the beginning Selfridge intended the store to become a 'social centre', a place where people could browse and meet; the first advertising campaign urged customers to 'spend the day at Selfridge's' and stressed that there was 'no obligation to buy'. In this way it contributed to the expansion of public leisure space, particularly for women. By reiterating in many of its adverts that its prices were 'the lowest – always' and by introducing from the outset the American innovation of sales and bargain basements, it expanded the class spectrum of its targeted customers and helped enfranchise the middle to lower-middle classes into the world of customer citizenship. As an employer, Selfridge operated exceptionally progressive and generous practices for his day and expected his staff to identify with and benefit from the commercial project. He was a supporter of women's suffrage, advertised regularly in the feminist press and made clear his respect for the astuteness and economic power of women customers.

Selfridge's style of publicity, although considered vulgarly American by rivals, was imaginative and unprecedented, and in addition to full-page advertising the store provided entertainments and exhibitions not directly related to sales in order to attract the public. As part of the routine promotion of the store, Selfridge published a daily 500-word syndicated column in several national and London newspapers in which he or a member of his staff commented on events and moods of cultural interest both inside and outside the store.

Extracted from: Mica Nava (1998): The cosmopolitanism of commerce and the allure of difference: Selfridge's, the Russian ballet and the tango 1911–1914. *International Journal of Cultural Studies*, 1, 163–96.

The great department stores

The great department stores of the late nineteenth and early twentieth centuries emerged as what Nava terms one of the 'key iconic aspects of modern urban society' (Nava, 1997: 56). As Reading 11.1 indicates, a central feature of stores like Selfridge's was their monumental architecture and sheer physical scale, which alongside their often cathedral-like ornamental interiors and illuminated window displays made them 'the most visible, urban representations of consumer culture and the economics of mass production and selling' (Domosh, 1996a: 257). But it was not only the quasi-civic architecture of these new consumption spaces which rendered them unique. Indeed, as Reading 11.1 also shows, a central feature of the new department stores was their role as important entertainment centres and tourist attractions. Selfridge's, for example, 'was considered one of the great show sights of London, like

Westminster Abbey' whilst 'Macy's restaurant in New York could accommodate 2500 in one sitting' (Nava, 1997: 69). Stores in general provided vast ranges of facilities including children's areas, restaurants and tea rooms, reading rooms and libraries, roof gardens, banks and travel agents. Like their twentieth-century counterparts the shopping malls, the nineteenth-century department stores quickly stamped their identity on the urban scene and acted as important focal points for community affairs.

From their inception, the stores served as particularly welcoming spaces for women. Indeed, and counter to Wolff's much quoted treatise on the 'invisible *flâneuse*' (see Chapter 10), department stores quickly developed a role as everyday arenas for ordinary women, with historical accounts indicating that visits to them took place very frequently – often several times a week (Abelson, 1989; Domosh, 1996a; Nava, 1997). In the relative safety of these new environments, nineteenth-century women were able to participate in the public sphere, and the sociality the stores promoted often played an important part in the development of women's political consciousness. Gordon Selfridge was known to be a supporter of women's suffrage, whilst Wanamaker's, the American store, 'gave all women employees time off during working hours to march in suffrage parades' (Nava, 1997: 73).

Reading 11.2 by Mona Domosh, an American geographer who has published widely on the nineteenth-century urban landscapes of New York and Boston, captures the importance of the late nineteenth-century retail district of New York and of Stewart's department stores on Broadway to the emerging cultural geography of the city. This reading parallels Reading 11.1 by Mica Nava on Selfridge's, but also deals with the nineteenth-century store as an important site of display as well as a social destination for women.

Reading 11.2 – The new domestic downtown

In late nineteenth-century New York, the retailing district centred on Fifth Avenue between Union and Madison Squares, extending west to Sixth Avenue and east to Broadway. It was an area of ornamental architecture and grand boulevards, of restaurants and bars, and of small boutiques and large department stores. It was, above all, an urban landscape designed specifically for consumption. This new retailing area was not only functionally different from its predecessors, in that its land use was dedicated only to retailing, but was also architecturally distinct. The new department stores that dominated this area were dramatically larger than earlier dry goods stores, and differed from their predecessors in the degree of ornamentation, the attention to decorative detail and display of goods, the concern with internal organization of departments, and the catering to the personal needs of shoppers, most of whom were women.

New York's first department store, Stewart's, opened on Broadway in 1846. Its four storeys, all dedicated to retailing, were unprecedented in the city, as was its white marble façade, a feature that made the building perhaps the most conspicuous in a city constructed almost exclusively of brick or wood. Its Italianate style, and its internal design centring on a dome with accompanying rotunda, were direct references to the public structures of the city, particularly the City Hall, which was located just to the south of the store. The store's owner, A.T. Stewart, was acutely aware of the importance of the appearance of his store, since he understood that in the new culture of consumption, the actual place of purchase was as important as the goods consumed, a notion that in the 1990s we take for granted in our image-laden shopping malls and festival settings. Yet in the mid-nineteenth century this was a novel idea ... In the new department stores ... the actual act of purchasing became insignificant compared to the activity of shopping ... Thus, Stewart was intent on designing a building that would provide an appropriate setting for consumption ...

The store's large interior was arranged to maximise the display of goods for comparative shopping ... The central hall of the store was decorated with frescoes on the walls and ceilings, an ornate chandelier in the centre, and large mirrors imported from Paris ... The rich interior displays included mahogany countertops, marble shelves, and an elegant ladies' parlour complete with large mirrors ...

Stewart's later store, completed in 1862 further uptown on Broadway, continued the idea of the department store as a stage of consumption ... Using cast iron for the façade, Stewart again created a visual sensation. This six-storey iron and glass palazzo became a site not to be missed by visitors to the city ... The entire store revolved around a skylight, that provided direct light to the large main floor. The five storeys that rose above were arranged as encircling balconies, thus providing natural light to all floors, and enabling shoppers to observe each other as they strolled through the enormous building ...

Department stores ... were designed specifically for women, appointed with elegant lounges and spaces to cater for women's needs. By the end of the century, the domestic was more fully incorporated, and department stores began to function almost as home parlours, with tea rooms, restaurants, art galleries and grand architectural displays.

Extracted from: Mona Domosh (1996a): The feminised retail landscape: gender ideology and consumer culture in nineteenth-century New York City. Chapter 14 in Neil Wrigley and Michelle Lowe (eds) *Retailing, consumption and capital: towards the new retail geography*. Harlow: Addison Wesley Longman, 257–70.

The central place of women in the newly emergent spaces of the downtown department stores is expressed most eloquently by Émile Zola in his aptly titled text *Au bonheur des dames* ('The ladies' paradise') (Zola, 1992), modelled on the Parisian store Bon Marché. As we demonstrate in Reading 1.4, Nick Blomley provides multiple readings of *Au bonheur*. In an article which focuses on Zola's novel as a signpost towards a 'new retail geography', Blomley also tackles 'the implication of retail spaces in the reproduction of femininity' (Blomley, 1996: 238). As such his account should be read alongside those of Nava and Domosh (discussed above), but also in parallel with work by Dowling (1993) on the construction of place and femininity at Woodward's department store in Vancouver in the mid-twentieth century (see below), and by Reekie (1992) on McWhirters in Brisbane. This latter article contrasts the gendered nature of the original McWhirters store which emphasized the interconnection between consumption and femininity, with the less obviously gendered space of the new McWhirters Marketplace (a festival market of the type discussed in Chapter 12).

But Blomley's chapter not only reads Zola's novel as an important treatise on the construction of femininity in the nineteenth-century city. Indeed, and somewhat uniquely, Blomley's work also views *Au bonheur* as providing an insight into masculinity, for as he argues 'the consumption spaces within and beyond the store . . . [as much as] they are feminized, offering a site for female *flânerie*, are also constructed by men' (Blomley, 1996: 254). Reading 11.3 briefly summarizes those places of masculinity and what Blomley (1996: 249) calls 'the geographies of men' in Zola's novel.

Reading 11.3 – Masculinities, space and power

At first glance, masculinity seems far removed from *Au bonheur*. Department stores have frequently been cast as feminine spaces. To Benson (1986), they were the 'Adamless Eden', in which women appeared omnipresent even triumphant. Certainly, *Au bonheur*, as its name suggests, appears feminized. Not only does Mouret [the founder and manager of the store] offer endless diversions targeted at women (such as a children's section, writing rooms and other amenities), but also the pleasures of consumption are deemed peculiarly feminine. At the height of the sale, for example, the women reign supreme, the salesmen becoming their slaves 'of whom they disposed with a sovereign's tyranny' (Zola, 1992: 236). The women revel in their new bourgeois wealth and the relative independence that it gives them, ignoring the concerns of their husbands. Male consumers appear notable by their absence in the store. If they are present, they are like children in tow 'buried beneath the overflow of bosoms, casting anxious glances around them' (Zola, 1992: 214).

However important is the construction of femininity, it must not be forgotten that the store is also a masculine space. Not only are the majority of workers men, but also the geographies of the store itself are powerfully masculine. The spatial logic of the store and its spreading reach across Paris and the world is structured not simply according to the imperatives of capital accumulation, but also in relation to specific conceptions of 'manly' sexuality and identity . . . The masculine space of the store is at once a cathedral to femininity, and a prison . . . this ostensibly Adamless Eden is constructed under a sexualized and voyeuristic male gaze . . . The commodities offer themselves up to women who, in turn, offer themselves up to an invisible but omnipresent masculine observer . . . This sexualization continues outside the store. Modernity and the expanding store are coded as masculine – virile, assertive and expansive . . . [But while Zola] offers an account of a triumphant masculinity, in which masculine and commercial imperatives dovetail, a closer reading reveals that the masculinity of the novel is, in fact, multiple, fractured and far from certain . . . on several occasions, Zola hints at the implication of these fractured masculinities in the geography of retail capital accumulation.

Extracted from Nick Blomley (1996): 'I'd like to dress her all over': masculinity, power and retail space. Chapter 13 in Neil Wrigley and Michelle Lowe (eds) *Retailing, consumption and capital: towards the new retail geography*. Harlow: Addison Wesley Longman, 238–56.

We will return to the notion of the construction of masculinities in specific spaces of consumption via our consideration of the various ways in which such masculinities were represented through practices of selling, and specifically via design interiors in the 1980s, later in this chapter. Here, though, we continue our discussion of the store by focusing on important developments that took place in the 1920s to 1950s. Such an assessment centres on the key role that shifts in the economics and layout of stores played in broader cultural transformations of the period, but also necessitates an examination of the cultural construction of 'Mrs and Mr Consumer'.

Shopping in the modern way – the store in the 1940s and 1950s

Just as the great department stores of the late nineteenth and early twentieth centuries were central institutions in the drive to modernity (and specifically in allowing women to participate in public life), so in the 1920s to 1950s – at first in North America and subsequently in Britain – changes internal to stores (often the simple move from counter service to self-service formats), proved culturally important symbols of the drive to a 'modern' society.

The first self-service food stores were opened in the years 1912–16 in the USA, with that developed by Clarence Saunders of the Piggly Wiggly chain in Memphis, Tennessee in 1916 being particularly well known (Bowlby, 1997). However, it was in the 1920s and 1930s that self-service retailing developed strongly across the USA (see Chapter 4). Self-service stores, like department stores before them, were viewed as attractive and comfortable places for the consumer to enjoy, where looking and touching was encouraged by the open display, and where, critically, as in Selfridge's (Reading 11.1), there was 'no obligation to buy'. As Bowlby (1997: 99) suggests, 'in this interim, provisional space, between the entrance and the checkout' the consumer really could have everything. This was 'the dreamlike face of self-service' and 'in this regard, the rationale of the self-service store is much like that of the department store, equally devoted to the careful construction of a fantasy experience for women'.

Immediately after the Second World War, many other western countries looked to the USA as a symbol of the modern society. In particular, 'shopping in the modern way' became a critically important cultural export. Our next two readings from Robyn Dowling, a Canadian geographer, and Frank Mort, draw out aspects of the transformations of this immediate post-war period, and focus specifically on the construction of femininity and masculinity within the stores of the 1940s and 1950s. In the broader article from which Reading 11.4 is extracted, Dowling discusses the construction of femininity and the creation of place at Woodward's department store in Vancouver. Here, she analyses the food floor at Woodward's and shows how 'shopping in the modern way' was an important component of the construction of femininity during this period.

Reading 11.4 – Mrs Consumer: Woodward's food floor and femininity

Consistent with the modern environment Woodward's created, the type and diversity of commodities available on the food floor and the associated femininity were intricately modern . . . With packaged goods acquired in an easy manner, it was thought that women's domestic chores would be reduced considerably. A mixture of familiar and rational femininity was being constructed. Caring for the family remained a primary component of domestic labour but this was to be achieved in a scientific manner, requiring the use of ready made commodities . . .

The mode of shopping at Woodward's was primarily scientific, rational and new . . . Efficient food merchandising involved a particular way of buying. The modern mode of shopping on the food floor consisted of self service . . . By self service is meant a mode of acquiring commodities where the shopper browses and chooses the commodities herself and then takes them to a cashier to pay for them. In contrast, a more traditional (and more labor intensive) mode of shopping involved asking a sales clerk for a particular commodity . . .

After the 1940s, the modern mode of shopping prevailed on the food floor, with customers roaming the aisles searching through the many commodities to find the one they wanted . . . The aisles were wide, and commodities were well organized, both facilitating a modern mode of shopping that consisted of methodically searching out desired commodities . . . The modern Vancouver woman chose from a wide variety of name brand commodities, superbly organized and presented, herself with a modern shopping cart and took her purchases home with her . . .

Shopping at Woodward's was seen as enabling the women of Vancouver to better cook for and nurture their families. More subtly, food floor shoppers . . . had to be taught how to be modern. The Woodward's food floor was not only a site . . . where the women of Vancouver bought their food, it was also a place where they would learn how to cook, learn what new and modern foods were available and hence how to be a better wife and mother. Learning about commodities was achieved through frequent special promotions or displays within the food floor encompassing such diverse topics as 'Brazilian Coffee Week' and an Australian Food Fair . . . Such displays were augmented by specific educational events. For instance, in May 1954 a display kitchen was built . . . for a food floor home economist service. Here, the Woodward's shopper could learn how to cook the modern foods she had bought.

Extracted from Robyn Dowling (1993): Femininity, place and commodities: a retail case study. *Antipode*, 25, 295–319.

Reading 11.5 by Mort centres on Burton's – flagship of British retailing – in the 1950s. The retail branch network of Burton's had developed rapidly in the preceding decades, and Burton's stores contributed in various ways to the modernization of the retail system (Alexander, 1997). More specifically, Burton's became an iconic symbol of 1950s British masculinity.

Reading 11.5 – Mr Consumer and the tailor of taste

Viewed through the prism of cultural memory, Burton's has the power to evoke distinctive images of Britain in the 1950s. Positive reminiscences nostalgically recall the 'tailor of taste' as a stable landmark in town centres across the country. A focal point on the high street, Burton's window was always the most brightly lit spot ... the atmosphere of a Burton's shop was unmistakable men talking to men ... customers standing on the silent, polished wooden floor. Legs akimbo, arms up. Keep it perfectly still. This reassuring image of collective cultural conformity – of a shared masculine culture fixed by retailing – was also reinforced by the size and scale of the Burton's empire. The company's precise market share is not recorded, but from the 1930s Burton's was the largest manufacturer and retailer of menswear in Britain ... Even as late as 1961 Burton's remained the biggest brand name in men's suits ...

From the 1930s through to the mid-1950s Burton's techniques of salesmanship centred on one clearly identifiable icon of masculinity. This was their image of the Gentleman. The emblem not only dominated the company's advertising ... it also shaped the approach to customers at the level of the shop floor. Burton's gentlemanly codes of masculinity were formal and fixed. The firm's ideal type was of full adult manhood, indeterminate age, but secure in position. Burton's gentleman might appear in different milieus – about town as well as on the way to the office – but his standard pose was always upright, if not stiff ...

Burton's approach to retailing was premised on the exclusion of women and all associations of feminine culture. Selling was understood as man's business. Women were kept away from customers. Though the branch cashier and the cleaners were women, they were never visible. The ostensible reason for this marginalisation of women was the delicate issue of measuring the male body. It was deemed unsuitable for a 'lady' to measure a man. But there were also broader cultural assumptions at work. Until the mid-1950s, Burton's projected a rational and implicitly masculine understanding of the purchasing process. The customer was cast as a variant of economic man, calculating and in control. Significantly the term 'shopping', with its more feminine and potentially more chaotic connotations, was never used by the firm.

Extracted from Frank Mort (1996): *Cultures of consumption: masculinities and social space in late twentieth-century Britain.* London: Routledge, 135–8.

We have seen then how the great department stores of the late nineteenth and early twentieth centuries, as well as stores in the 1940s and 1950s, were important markers of shifts taking place in broader society at these times. The sheer architectural presence – be it of Stewart's marble-fronted or cast-iron stores on Broadway in New York, or Burton's flagship store Burton House, which by 1939 sat proudly on the corner of New Oxford Street and Tottenham Court Road, in London (Alexander, 1997) – served as prominent reminders of the role of retailing in public life and, as we have indicated, the new experiences that these spaces allowed were symptomatic of significant cultural changes. Such interconnections are still visible today. Indeed, in the contemporary era, the store retains a certain cultural resonance and its linkages with important developments in wider society make it a fascinating prism through which to view urban life. It is to stores of the 1990s – beginning with the flagship stores of our large cities – that we will now turn.

Late twentieth-century flagship stores

Despite a good deal of discussion regarding the potential role of mail order catalogues, television shopping channels and the Internet (see Chapter 13), the essential 'sociality' which defines shopping in the street or mall (Chapters 10 and 12) and which played an important role in stores of earlier eras, retains an vital role in the store of today. Indeed, it has been suggested that in the contemporary era – in cultures starved of public experiences – the store functions as an essential space of public entertainment and social interaction (Goldberger, 1997). The ultimate gatherings of such new spaces of consumption take place in contemporary 'streets of style' discussed in Chapter 10. In such locations, retailing has taken on a similar role to that which it played in the case of the great department stores of the late nineteenth and early twentieth centuries – a time when such stores 'set themselves up not merely as merchants but also as veritable bazaars, bringing women out of their houses and into the public realm for the first time' (Goldberger, 1997: 46).

Collections of stores such those in Bond Street or Madison Avenue draw upon the notion of 'lifestyle shopping' pioneered by the likes of Habitat in the 1960s. Habitat developed 'the 'lifestyle' organization of products coordinated through selling techniques and addressed to a specific group of consumers' (Nixon, 1996: 49). From early beginnings in Fulham Road, London, in 1964, Habitat's founder Terence Conran 'sold an aspirational and affordable image of domestic living to a young, post-war generation looking for a way to differentiate themselves from the traditional pre-war values and taste of their parents' (Gardner and Sheppard, 1989: 77). Building on concepts utilized slightly earlier by individual store proprietors like Mary Quant and John Stephen in the mid- to late 1950s, Habitat's modernist store designs and cleverly displayed room layouts spoke to the new generation of consumers and provided a retail destination which was wildly different from those of its predecessors.

It was over 20 years later, in 1986, that Polo/Ralph Lauren opened the first 'lifestyle' store on Madison Avenue in New York at 72nd Street. Lauren's 'flagship' was quickly followed by others keen to establish a foothold on this high-fashion boulevard (see Figure 11.1). Capitalizing on concepts developed by Habitat and others, the new designer stores merchandise 'lifestyles' not just apparel, and are arguably becoming mini department stores in their own right (Lowe and Wrigley, 1996) – 'expanding their lines from clothes to sheets, candles, furniture, dishes and potpourri' (Handelman, 1997: 50). The 'flagship' store concept is also central to the revival of London's New Bond Street (see Figure 10.1). Companies keen to establish a clear and coherent brand identity

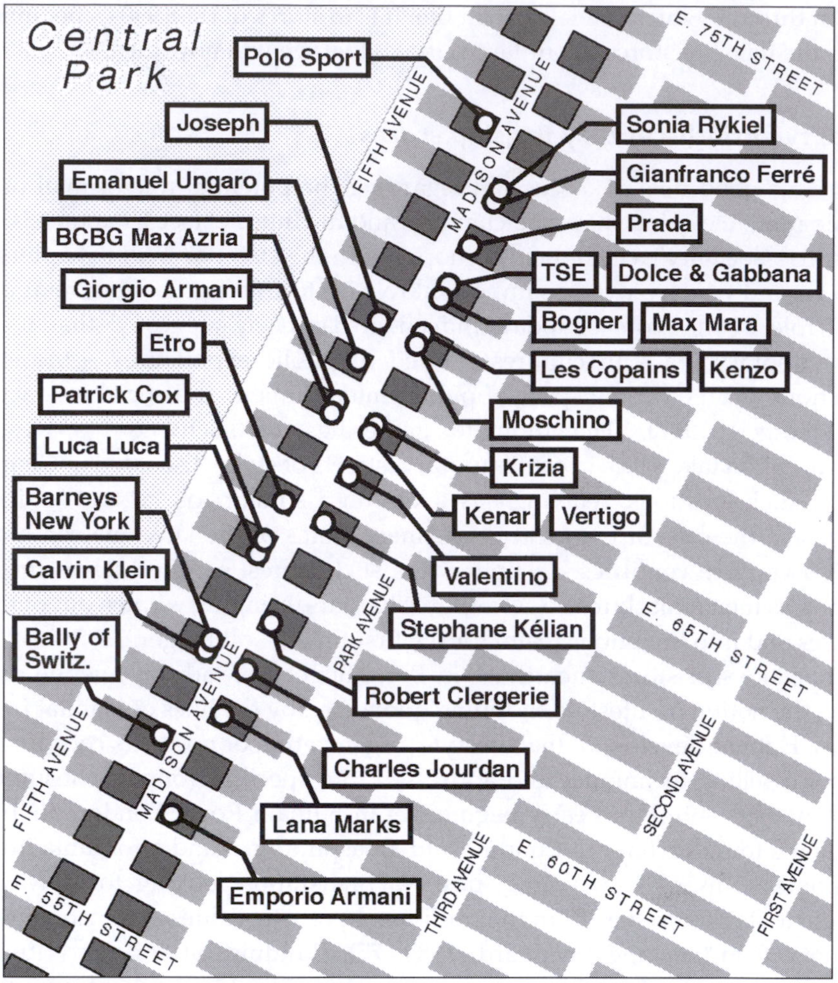

Figure 11.1: *Fashionable lettings in Madison Avenue 1993–96*

for their products have taken up leases on this 'street of style' at vastly inflated prices. At Donna Karen, for example, DKNY mineral water, New York bagels and New England cheesecake allow the contemporary consumer to lose themselves in the ultimate 'own-label' experience, whilst Nicole Farhi's fashionable café – Nicole's – draws in 'ladies' and gentlemen who lunch in a manner that is hard to distinguish from the attractions of Gordon Selfridge's nineteenth-century extravaganza.

The idea of the flagship store is taken to its extreme in the notion of 'entertainment retail' discussed in detail in Chapter 9. At Calvin Klein's Madison Avenue store this concept appears quite subtly with Klein's minimalist interiors being combined with various art exhibits in order to enhance the store's appeal. But Main Street/High Street retailers have taken this idea further such that contemporary retailing – at least in certain places – has many of the characteristics of the theme park. As Zukin (1998: 833) suggests, 'most of the prototype entertainment retail stores have opened in the largest cities . . . where they have become new landmarks on the urban scene . . . Like the old department stores, entertainment retail stores enjoy favourable coverage in local newspapers for their "enchantment" of the urban landscape.'

Retailers which have launched 'entertainment retail' stores include:

> Diesel which with its in-store D.J. and retro-chic coffee bar is designed to make you feel as if you are in a hip nightclub . . . [and] Recreational Equipment in Seattle [which] contains a 65-foot-high freestanding artificial rock for climbing, a glass enclosed wet stall for testing rain gear, a vented area for testing camp stoves and an outdoor trail for mountain biking' (Goldberger, 1997: 46).

Such novel consumption spaces capture the consumer's imagination by inviting them to participate in simulated forms of non-shopping entertainment. NikeTown – with several locations across the USA, including a store at the Forum Shops in Las Vegas (see Chapter 12), and in the former Galeries Lafayette building on East 57th Street in New York, establishes a new retail culture via its spectacular interiors. At East 57th Street NikeTown's façade is a carbon copy of a New York high school building complete with sports-arena-style turnstiles. Once inside the store consumers can take part in a range of athletic pursuits as well as hear athletes describe their experiences or register to get involved in specific community projects such an environmental clean-ups (see also Chapter 9). In this way, and in the manner of Macy's Thanksgiving Day Parade, stores like NikeTown offer a connection to a broader public culture. Nike is thus established 'as the essence not just of athletic wear but also of our culture and way of life' (Goldberger, 1997: 45). Of course this way of life is only open and available to certain sections of society such that the 'sociality' of the contemporary store is 'dependent on visual coherence and security guards, a collective memory of commercial culture rather than either tolerance or moral solidarity' (Zukin, 1998: 834). Moreover, the new flagship stores – like the great department stores before them – are only too aware of the merchandising potential of these new schema. Taken to the extreme in the rapidly multiplying bookstore/coffee bars which began in the USA with Borders and Barnes and Noble, but which in the 1990s were rolled out in the UK in the new formats of Waterstones and Books Etc., such spaces put together the growing popularity of 'designer' coffee brands such as Starbucks (See Reading 1.1, Chapter 1) with the relaxed atmosphere of the historic library and provide the ultimate extension of what retail designers and marketeers term 'dwell time'. Indeed, many Barnes and Noble and Borders superstores now serve the function historically occupied by the department store – as 'anchor' at the shopping mall (Panek, 1997). In the UK, Waterstones in Glasgow has a Café Internet as well as a computer-free Costa Café and a roving coffee cart that is moved around the store. Part and parcel of retail design, the bookstore/coffee shop concept is the latest in a long line of elaborate techniques designed to keep the consumer in the store for longer (see Chapter 9). But retail design – a concept which was brought to the fore of retail innovation by Habitat and others in the 1960s – often has a number of more subtle functions as well. It is to the central place of design in the contemporary store that we will now turn.

Designing the interior spaces of the store

We have already hinted at the crucial role of design in the retail built environment and specifically in the spaces of the store with reference, in particular, to the flagship stores of the late twentieth century. But interior design in retailing holds such a pivotal position in what Dowling (1993: 298) terms 'place-making' that it is essential to give this concept broader consideration here. Throughout the last century, from the time of the great department stores of the late nineteenth and early twentieth centuries, the store has been 'a resource which retailers can use to make themselves and their commodities distinctive' (Dowling, 1993: 298). But whilst individuals like Gordon Selfridge or A.T. Stewart employed a range of architectural and other techniques in order to make their retail spaces popular everyday and tourist destinations, and whilst pioneers of the art of window dressing such as L. Frank Baum, the famous author of the *Wizard of Oz,* began to use window displays in the department stores of New York City during the same period (Lancaster, 1995), it was not until the inter-war years that the notion of 'display' – and its critical role in retail design – began to gain a foothold in the retail mainstream. At Burton's, for example, the mid-1950s ushered in a new concern with image and style and that concern was at the root of the ascendancy of a new breed of commercial expert – the display specialist. Indeed, 'by 1964 Burton's display department had a staff of over four hundred, who were kept up to date via a central advisory bureau in Leeds' (Mort, 1996: 140).

Retail display then – and specifically the art of window dressing – became a prominent means by which store owners marketed their stock. But it was not until several decades later – in the 1980s – that the specific function of retail design began to pervade all aspects of retail culture. As Gardner and Sheppard (1989: 69) argued:

> Design has become the buzzword of the 1980s . . . The public results are obvious even to those with only a rudimentary awareness of their surroundings – the look of the high street is new, and seemingly in a permanent state of renewal. It bristles with new graphic identities and brand names; more invitingly extravagant interiors in timber, chrome glass and stainless steel seem to appear every week.

Of course, and as we have seen earlier in this chapter, retail pioneers like Conran of Habitat fame had given strength to the principles of retail design as key markers of retail exclusivity and brand identity in the 1960s, but in the 1980s a new breed of retailer and a new role for retail design was in the ascendancy. By that time, retail design 'functioned to represent [a store's] . . . "identity" and to address that "identity" to a specific segmented set of consumers'. Essentially, design was viewed as a 'strategic business resource' that could be used by retailers 'first to differentiate; secondly, to focus or segment operations; thirdly to reposition stores [and] fourthly to represent stores as brands, fixing their image' (Nixon, 1996: 50–1). The strategic significance of design to the operations of the store is best captured by the example of George Davies' Next, which from small beginnings in 1982, as retailer 'to smart, better-off women in the 25–35 age band', came to occupy a pivotal position in Britain's 1980s high street revolution (Mort, 1996: 122).

Reading 11.6 by Sean Nixon, a British sociologist, captures the significance of retail display techniques to the success of Next in the 1980s, and in particular focuses on the way in which specific masculinities were represented at the point of sale via the clever use of design and display. In addition, Nixon hints at the importance of the image of the sales staff as part and parcel of retail interiors (see Chapter 5 on retail employment).

Reading 11.6 – Next and geographies of display

Next's retailing strategy was focused around the presentation . . . of a limited and co-ordinated garment collection; what George Davies called a 'tightly edited range'. This garment range was targeted to a specific group of consumers – 25 year old plus ABC1 consumers or, more specifically (and initially), 'young working women who were weary of the fast fashion in the High Street boutiques but not weary enough for the staider styles of the Department stores'. The menswear range was merchandised effectively to address and shape an 'upmarket' but affordable middle market in menswear. Next then . . . identified an underserviced segment of the clothing market and based their retailing strategies on effective servicing of the segment. The development of Next – through a proliferation of its retailing forms (Next Too, Next Essentials, Next Accessories, Next Café, Department X, Next Directory, Next Interiors, Next, Next for Men, Next Lingerie) – aimed for a 'complete' Next store . . . that would extend and maintain the 'exclusivity' of Next into a wider range of goods and services. Absolutely pivotal to both the initial formulation and the expansion of the business was the construction and careful regulation of Next's 'image' . . . [As George Davies suggested] . . . 'Image was uppermost in my mind at the outset, and I knew that image must start with the shop fitments.'

Davies drew inspiration from the early Benetton stores in London. He was impressed by the way Benetton broke with established conventions of display: 'Unlike everyone else, they weren't using window dummies to display their merchandise – they simply draped it over black plastic fittings.'

Next followed this innovation as part of its approach to the retail interior; an approach that aimed to be – in Davies' favourite phrase – 'upmarket from the shop fittings to the sales girls, who were dressed in Next merchandise and looked wonderful'. The design formats of the stores varied across the different retailing formats whilst retaining a strong corporate coherence ...

Central to the design of the Next menswear interiors ... was the use of space and materials. The frontage of the stores gave the first indication of this: a large window set in a dark matt-grey from beneath the trademark signage – Next – in lowercase lettering. The window displays – framed by this frontage – were similarly uncluttered. A combination of garments was displayed on abstract mannequins, often backed by large display or showcards that gave written accounts of the merchandise range. The display cards – featuring details of the clothes being worn as well as accompanying copy – played off the themes of space, colour and line in the shop through their layout and lettering. Inside, the lighting, colouring and organization of space were distinctive. Here were the features that formed a coherent design vocabulary: bleached wood pigeonholes and dresser units; downlighting spotlights; gently spiralled staircases with matt black banisters. The 'edited' collection of clothes was displayed in various ways. Around the sides of the shop slatted wooden units displayed a few folded jumpers next to hangers with three jackets; socks were folded in pigeon-holes or individual shoes perched on bleached wood units. A dresser unit commanded the central space of the shop, placed upon a classic woven carpet. Such features acted as centripetal counterpoints to the displays of clothes set against the walls and encouraged customers to circulate around the shop. The overall feel of these shops was spaciousness and a cool modernity that invited echoes of cruise-liner aesthetics together with strong references to traditional gentleman's outfitters ... In Great Newport Street, London [Davies evoked] 'the lost environments of Edwardian England: old fashioned cricket bats; an ancient looking canoe; authentic cigarette cases and lighters set out in glass-topped cabinets; and an antique dressing mirror. Here was a set of objects that most directly connoted the visual style and ambience of Merchant and Ivory's film adaptation of Forster's novels: clothes and accessories for your 'room with a view'.

Extracted from Sean Nixon (1996): *Hard looks: masculinities, spectatorship and contemporary consumption.* London: UCL Press, 52–6.

In *Hard looks,* Nixon combines his close focus on the Next retail chain with parallel readings of designer and mass market menswear. But Nixon's study is not only significant in drawing out the crucial role of retail design in retail marketing but also, and critically, because, as Reading 11.6 indicates, Nixon highlights the various ways in which store design can influence the manipulation of space in the retail built environment. Such spatial manipulation has already been examined in detail in Chapter 9. There we demonstrated how the kinds of design and display techniques utilized by Next and others in the 1980s have also been important in supermarkets and shopping malls. Our next chapter 'The Mall' revisits these latter consumption spaces and once again stresses the overarching importance of design in the retail environment.

12
The mall

In his 1960 publication *Shopping towns USA* Victor Gruen, acknowledged as a pioneering architect of the shopping mall in North America, looks to ancient Greece with its *stoa* (merchants' building) centrally placed within the *agora* or city square for the basic principles of town square design. According to Gruen, the growth of suburban America in the 1950s (encouraged by the automobile) led to the rapid movement away from this framework. In the absence of alternatives, planned shopping centres provided:

> the needed place and opportunity for participation in modern community life that the ancient Greek *agora*, the medieval market place and ... town squares provided in the past. That the shopping centre can fulfil this ... urgent need of suburbanites for the amenities of urban living, is convincingly proved in a large number of centres. In such centres, pedestrian areas are filled with teeming life not only during normal shopping hours, but on Sundays and holidays when people windowshop, promenade, relax in the garden courts, view exhibits and patronize the restaurants (Gruen and Smith, 1960: 22).

Fascinatingly, over four decades on, one of the leading examples of shopping mall design in contemporary USA, 'The Forum Shops' in Las Vegas, also looks to ancient Greece and Rome for its design principles. A recreation of the streets of Rome from 300 BC to AD 300, The Forum Shops has as its centrepiece a spectacular fountain which features four Greek and Roman gods which regularly come to life and perform an eight-minute production of the 'Gods of Olympus'. The star attraction at the Forum is the painted ceiling or 'sky' which is lit in such a way that a 24-hour cycle – sunrise to sunset – takes just three hours. The ultimate in fashioning 'the mall as street' – bringing the exterior (the sky) to the interior of the mall – the Forum is, like Gruen's early shopping centres, also 'filled with teeming life' and is providing a model for future mall development in North America and elsewhere.

This chapter begins by examining the history and development of the shopping mall; it then considers North America's 'mega-malls' followed by their counterparts in the UK the regional shopping centres. Attention is paid to the people who inhabit these new consumption environments. Finally, the chapter charts the emergence of a number of new landscapes of consumption – from speciality centres and festival marketplaces to factory outlet centres and airport shopping malls.

Constructing suburban utopias – history and development of the shopping mall

Planned shopping districts existed in the United States from the early decades of the twentieth century. Market Square, which opened in 1916 in the Chicago suburb of Lake Forest, was the first and was followed by Highland Park Shopping Village completed in 1931 in Dallas, where importantly 'storefronts were turned away from the public street and inward around a central area – a special courtyard where cars couldn't go' (Kowinski, 1985: 105). But it was not until the 1950s that shopping centres – classically constructed with a large department store at one end, two parallel rows of shops, and a pedestrian area in the middle – developed rapidly. Victor Gruen Associates as architects were responsible for a large number of America's shopping centres or *malls* as they became known. From the outset Gruen's centres were designed to serve the civic, cultural and social needs of new suburban communities and were intended to be developed alongside apartments, office buildings, theatres, etc.

Southdale, Minneapolis, which opened in 1956, was the world's first enclosed (fully covered) mall. Air-conditioned and temperature-controlled, Southdale avoided the vagaries of the Minnesota climate (very cold winters, very hot summers) in which Victor Gruen, its architect, had calculated that there were only 126 'ideal weather shopping days' per annum. In this context, 1950s advertisements for the centre promised 'Tomorrow's Main Street Today' and Southdale was marketed as 'A whole new shopping world in itself'. Located at Edina, Minneapolis, the centre provided the prototype for the majority of malls built since. Reading 12.1 by Wilbur Kowinski, an American journalist, recounts the story of the Southdale Mall and demonstrates how the centre provided an important model for subsequent mall development.

Reading 12.1 – The invention of the mall

Southdale . . . is where the mall was invented. Although it now seems to be a superior but not unusual mall – with its soaring Garden Court, its Interior Systems furniture shop, B. Dalton Booksellers, Berman Buckskin, Children's Barber, Chrome Concepts, Fanny Farmer and Eat & Run – in 1956 Southdale represented innovation, creative problem-solving, and aesthetic daring, as well as shopping-center heresy.

When Southdale was being planned, the normal shopping center was a long strip, all on one level, with at best one department store. Gospel was that nobody was going to walk or even ride up to a second level in a drive-in shopping center; if they couldn't reach a store by practically parking in front of it, car-dependent people wouldn't go.

Finally a few daring developers tried a two-level center, and it worked. The Rouse Company did it with the Mordawmin Shopping Center in Baltimore and the Dayton-Hudson Company with Northland in Detroit . . . At Northland, the two-level open mall proved itself. By a handy coincidence, one chain store had a shop on each level and they did identical business, which seemed to mean that people weren't entirely averse to getting out of their cars and walking around a shopping center, or riding an escalator to see

> what was going on upstairs. But while others contemplated what exactly this might mean, Northland's designer was hurrying on to the next step, the truly fateful one, at Southdale. His name was Victor Gruen . . .
>
> [The] problem, in Edina, Minnesota, was the weather. It not only got very cold and snowy in the winter . . . it also got baking hot in the summer. Gruen's answer to the problem turned out to be the most fateful advance made at Southdale.
>
> The solution was, of course, complete enclosure. Gruen saw it immediately and went to the Dayton-Hudson hierarchy with his proposal. He told them about the covered pedestrian arcades in Europe, especially the Galleria Vittorio Emanuele in Milan, Italy, with its arcades rising four storeys to a glass barrel vault and a central glass cupola 160 feet high . . . Enclosure could also make the central court a much more dramatic place, as became apparent in the planning stages of Southdale . . . For the first time a shopping center would have a real vertical dimension, with the Central Garden Court soaring to a high ceiling, and the two levels of the mall visible to each other.
>
> **Extracted from: Wilbur Kowinski (1985): *The malling of America: an inside look at the great consumer paradise*. New York: Morrow, 117–19.**

The Southdale 'model' was adapted and utilized for shopping centres throughout America. Significantly, however, Gruen and his associates became disillusioned with the way in which their ideas were diluted by other developers. As Cesar Pelli, a design partner, lamented, 'Malls have not become true community centers. At Southdale it was realised that people will come in great numbers with just a few public activities' (Kowinski, 1985: 123). Notwithstanding this, however, advances made at Southdale still form the basic framework of shopping centre design. Importantly, the concept of enclosure gave mall architects the possibility of fashioning an interior space sheltered from the outside world and its elements. Forty years later, in the 1990s, the Mall of America, also in Minnesota, and reputedly the world's largest mall, extended on a gigantic scale some of the concepts pioneered at Southdale.

North America's mega-malls

During what Margaret Crawford (1992: 7) refers to as the 'golden years' of the 'malling of America' – the period 1960 to 1980 – when 'the basic regional mall paradigm was perfected and systematically replicated, [and] developers methodically surveyed, divided and appropriated suburban cornfields and orange groves to create a new landscape of consumption', almost 30,000 malls were constructed. In time, as regional malls approached saturation point, super-regional malls began to appear 'at freeway interchanges – such as the Galleria outside Houston, South Coast Plaza in Orange County, and Tyson's Corner near Washington, DC – [and] became catalysts for new suburban mini cities' (Crawford, 1992: 24 – see also Garreau, 1991). Ed Soja (1996: 11) paints a fascinating picture of such a mini city in Orange Country with at its heart what he describes as the 'curiously insubstantial' South Coast Plaza, 'California's largest and most profitable shopping mall, with nearly three million square feet of space and almost ten thousand parking places, Nordstrom's, May Co., Sears, Bullocks, Saks Fifth Avenue, Robinson's, the Broadway, over two hundred other stores and boutiques, and nearly half a billion dollars of taxable retail sales in 1986'. However, it is two 'mega-malls' completed in the late 1980s and early 1990s that have attracted increasing amounts of attention. The largest of these, the Mall of America, is located on the former site of the Met Stadium – home of Minnesota's professional baseball and football teams in Bloomington, Minneapolis. Built by Melvin Simon Associates, one of the largest developers of shopping centres in the USA, the mall opened in August 1992 and includes four nationally recognized department stores – Bloomingdale's, Macy's, Nordstrom and Sears, over 400 other stores and a seven-acre amusement park (Goss, 1999). But it is another giant North American mall development – West Edmonton Mall constructed in the 1980s – that has so far attracted the greatest degree of attention by geographers (Hopkins, 1990; Johnson, 1991). The importance of West Edmonton Mall was that it was explicitly 'the shopping mall as entertainment centre (and) tourist attraction' (Jones and Simmons, 1990: 223). Figure 12.1 shows the major magnets within the mall (department stores and leisure facilities) in its early years.

Reading 12.2 by Jeffrey Hopkins, a Canadian geographer, provides a picture of the West Edmonton 'Mega'-Mall (WEM) and hints at the way in which geographers have explored this pioneering consumption space. Hopkins' intention in the broader article from which this reading is extracted is to provide a 'reading' of WEM which emphasizes the various ways in which the mall stimulates a sense of 'elsewhereness' – 'the overt manipulation of time and/or space to simulate or evoke experiences of other places' (Hopkins, 1990: 2) – in its clientele. As such Hopkins' article should be read alongside those of Shields (1989) and Goss (1993) who also focus on the manipulation of space within the retail built environment, a theme which has been explored in detail in Chapter 9.

Reading 12.2 – The West Edmonton mega-mall

A novel mixture of retail, commercial, recreational and entertainment facilities literally under one 950-million-dollar roof . . . Occupying a 110-acre site about six miles west of the centre of Edmonton, Alberta, this two-storey 5,200,000-square-foot structure, completed in 1986, contains over six hundred stores and services . . . including four major department stores, four junior department stores, two auto showrooms and 11,775 parking stalls . . . WEM employs approximately 18,000 people . . . contains almost 23 per cent of Edmonton's total retail space, takes in 42 per cent of all consumer dollars spent in

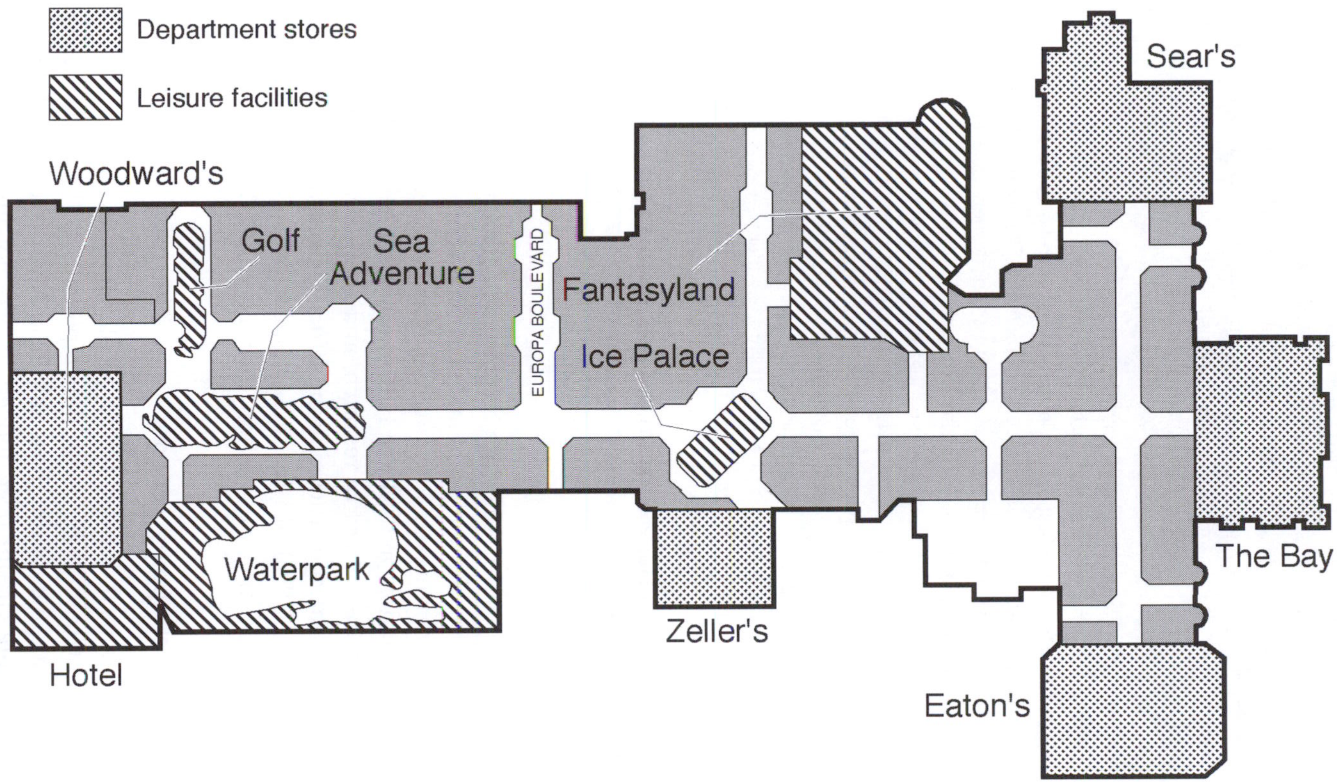

Figure 12.1: *The retail and leisure mix in the West Edmonton Mall – main floor*
Source: Redrawn from Jones and Simmons (1990).

shopping centres in the Edmonton area . . . and accounts for more than 1 per cent of *all* retail sales in Canada . . .

While the exterior of WEM is a standard frame structure of brick with flat and galleria roofs surrounded by four-lane roads and two-level parking garages, the interior is distinct. Mall attractions include a seven-acre waterpark, a National Hockey League-size skating rink, an 18-hole miniature golf course, a Fantasyland with a triple-loop roller coaster, a four-acre sea aquarium with four Atlantic bottlenose dolphins, four fully operational submarines, and a scaled replica of Christopher Columbus's *Santa Maria*. There are also numerous exotic flora, fauna and fountains, a 'New Orleans' – themed streetscape, eight statues, 19 movie theatres and a Fantasyland Hotel with 360 rooms, 120 of which are modelled according to one of six themes: Arabian, Coach, Hollywood, Polynesian, Roman or Truck . . .

This mega-mall typifies 'disneyfication' and 'imagineering', and reflects the trend in North America towards specialized, self-contained built environments. The annexation of much of a city's retail/social life into a corporate, self-contained, 'disneyfied' built environment, however, is unparalleled. Disney's parks are isolated entities, world's fairs are temporary and amusement parks tend to be seasonal. The mega-mall is 'spectacle' integrated into the everyday and open year round . . . but a statement that WEM is merely a 'disneyfied' or 'imagineered' shopping mall is a static analogy, one that is metaphorically correct but reveals nothing about process(es). Built environments which provide such 'specialised experiences' go beyond two-dimensional illusion by providing three-dimensional theatre in which patrons are both spectators and actors (sightseers themselves are part of the show). WEM is neither merely disneyfied, nor Disneyland simply imagineered, both are part of a much larger set of processes, one of which is simulated elsewhereness.

Extracted from: Jeffrey Hopkins (1990): West Edmonton Mall: landscape of myths and elsewhereness. *Canadian Geographer*, 34, 2–17.

But it is not only the North American 'mega-malls' which have attracted this kind of critical attention. Britain's regional shopping centres, in particular the Metrocentre in the north-east, Meadowhall outside Sheffield, Lakeside at Thurrock, east of London, and Merry Hill in the West Midlands – all 1980s centres – have been the subject of research, and Brent Cross – Britain's first regional shopping centre constructed in the 1970s – has also been examined in detail by Miller et al. (1998). It is to these smaller British equivalents of the North American centres that we will now turn (Lowe, 2000b).

Britain's regional shopping centres

In the UK, in the context of more restricted planning regulation and higher land values (see Chapter 7), mall development, at least on anything approaching the scale and type of those at Mall of America or West Edmonton, has been much more circumspect. Indeed, it was not until 1976 that Brent Cross, Britain's first purpose-built regional shopping centre was opened, and it was another decade before there were any more such centres. Brent Cross

> was unusual at the time in that it was the only shopping centre to have been built on a previously undeveloped site outside of an established shopping area and not being part of a new town. When it opened . . . [it] had 82 tenants with 800,000 square feet of retail space on a site of 52 acres. There were 3500 free car parking spaces and over 4000 employees. Besides two anchor stores (John Lewis and Fenwick), the centre included a Waitrose supermarket and branches of Boots, C&A, Marks & Spencer and W.H. Smith (Miller et al., 1998: 32).

Significantly, Brent Cross was built in an area without major existing shopping facilities and with the approval of local authorities (Guy, 1994b: 300).

In the early 1980s, in the first term of the Thatcher administration, a deregulationist stance led to a relatively favourable environment for proposed regional shopping centres and a wave of planning applications were submitted (Schiller, 1986). More specifically, the enterprise zone experiment created the possibility of large tracts of land on which unencumbered development could take place. The Metrocentre in Gateshead (opened in 1986) was constructed on part of the Tyneside enterprise zone, whilst Merry Hill in the West Midlands was sited on two adjoining enterprise zones (Dudley and Round Oak). In both of these cases early developments on the enterprise zones comprised retail warehousing and these were later supplemented by more 'up-market' enclosed regional malls. At Meadowhall in Sheffield a regional shopping centre was planned at the outset and this was also the case at Lakeside (these centres both opened in 1990).

Much of the literature on Britain's regional shopping centres has concentrated on planning dimensions (Guy, 1994a and b), specifically their impact on neighbouring towns and cities. But regional shopping centres have also provided foci for two other types of research: first, work emphasizing the crucial role of the regional entrepreneur in the development of these centres; and, second, research which has examined the experience of shoppers within them. Here we focus on three readings on Britain's regional shopping centres which deal with these issues.

Reading 12.3 is a composite of extracts from a study of the retail impact of Meadowhall – a regional shopping centre of more than 1,250,000 square feet three miles north-east of the centre of Sheffield – by Jonathan Reynolds and Elizabeth Howard of the Oxford Institute of Retail Management (OXIRM), and a paper by Gwyn Rowley which examines the decay in Sheffield's CBD following the opening of the Meadowhall complex. Both of these studies were undertaken in the early 1990s, in the preliminary stages of Meadowhall's development.

Reading 12.3 – The early impact of Meadowhall on Sheffield

Any assessment of Meadowhall inevitably involves consideration of its effects, both economic and social, on the city centre. However, this analysis should be seen in the context of the combined, and to some extent complementary roles which the city centre and Meadowhall perform . . .The changes in Sheffield's role have not been straightforward, however, and the 'impact' of Meadowhall in the conventional sense of trade diversion is difficult to quantify . . .

Sheffield's role began to change long before Meadowhall was developed, even though the pace of change has accelerated rapidly in the last five years . . . The limitations of Sheffield's [shopping centre] layout – its very elongated nature which unusually is weak at or near the central point . . . offering the dual focus points ofThe Moor and Fargate [which] present a confusing shopping 'offer' for shoppers and retailers – and the failure to create strength and retail depth at the core have meant the city has failed to improve as others have done . . .

Our survey shows Sheffield's share of non-food shopping trips has fallen by 38 per cent . . . Therefore, in terms of its relative decline, Sheffield appears to have lost substantially more of its market share than originally forecast in planning for Meadowhall. Of course, not all this decline will be directly attributable to Meadowhall itself . . . competing centres, including retail parks [have also] increased their market share . . .

However, given its size, proximity and the degree of overlap between the centres, together with the corresponding decline of the city centre . . . We consider Meadowhall to have been a significant factor in Sheffield's further decline between 1990–1994. Meadowhall's 'impact' appears to have been higher than that predicted [because] . . . regional household expenditure has not continued to grow as it did in the 1980s: thus there has been no 'cushion' for existing centres. Meadowhall has opened in a static or slow growing market. [In addition] the relative attraction of Meadowhall over Sheffield is greater than that of the Metrocentre over Newcastle. The latter was – and is – a much stronger centre than Sheffield, with a much better range of modern shopping including significant new additions during the 1970s (Eldon Square) and subsequently.[1]

Since Meadowhall opened in 1990, trade in the city centre appears to have fallen by more than one third, and the entire northern third of the CBD, the Haymarket/ CastleMarkets area, has been devastated. Streets are lined with boarded-up shops, whitewashed windows and down-market discounters. Several charity shops now occupy what were prime retail sites until the late 1980s . . . While the central section of the CBD around Fargate retains a certain ebullience there are signs of decay there as well. The Marks & Spencer store is running a definite second to the company's Meadowhall branch and ranges of goods are consequently relatively limited.The local John Lewis Store, Cole Brothers, still survived in 1992 but a number of smaller retailers . . . had closed down and moved to Meadowhall . . .The southern section of the CBD, around the Moor, is seeking a specific market niche in what can be termed a bazaar economy with cheap street-market stalls along the pedestrianized Moor . . . the general air is of a non-CBD local shopping district . . . Marks & Spencer closed its store on the Moor when it opened its Meadowhall branch.[2]

Extracted from: [1] OXIRM (1994): *Sheffield retail study*. Oxford: Oxford Institute of Retail Management); [2] Gwyn Rowley (1993): Prospects for the central business district. Chapter 6 in Rosemary Bromley and Colin Thomas (eds) *Retail change: contemporary issues*. London: UCL Press, 110–25.

Reading 12.4 by Michelle Lowe discusses the 'regional entrepreneurs' behind the Metrocentre and Meadowhall before focusing explicitly on the Richardson brothers' activities at Merry Hill. This extract, from the early 1990s, hints at substantial leisure developments due to be constructed alongside the retail centres at Meadowhall and Merry Hill. Such a mixture of retail and leisure was modelled on the success of parallel developments in North America, but, in reality, as Reynolds (1993: 80) notes, 'Planning the leisure component within the UK regional centres . . . in line with US principles was . . . fraught with difficulty.' Indeed, at the Metrocentre, the leisure element was substantially scaled down 'and leisure as such failed to appear at all in Sheffield's Meadowhall Centre or Thurrock's Lakeside'. Figure 12.2 shows the layout of the Meadowhall Centre and illustrates how leisure was largely omitted from this scheme. At Merry Hill, 'waterfront' developments currently comprise office buildings, a hotel and some restaurants, pubs and a nightclub but other leisure facilities have yet to materialize (Lowe, 1998).

Reading 12.4 – Regional entrepreneurs and regional shopping centres in the 1980s

Regional entrepreneurs became flag bearers for Mrs Thatcher's vision of regional regeneration. Not surprisingly, they have become the targets for a great deal of media attention. John Hall, for example, the most famous of these characters, the man behind the massive Metrocentre at Gateshead, played a pivotal role in Mrs Thatcher's cleverly managed 1987 election campaign. His centre, situated on a site on the banks of the River Tyne, provided an image of a Britain – and perhaps more significantly from the point of view of vote-winning, a view of a North-East – pulling itself up by its bootstraps. Hall is currently developing a major leisure/living complex around his home, Wynyard Hall, near Billingham in Cleveland.

John Hall, albeit the best known, is only one of an increasing number of individuals who have emerged in some of Britain's regions as leaders of a new form of regional regeneration. Within the same retailing/leisure framework Eddie Healey has instigated the development of Meadowhall, a new shopping and leisure experience close to the M1 on Sheffield's outskirts . . . At Merry Hill in the West Midlands . . . Don and Roy Richardson, twin brothers, have developed an out-of-town regional shopping centre . . . The Merry Hill development rose like a phoenix from the ashes of industrial decline and is currently into Phase Five of its operations.

The Richardson family background is in heavy truck distribution in the West Midlands region. Their business bought and sold trucks and lorries nationwide, and was very successful. In addition, the family had also been involved in the acquisition and redevelopment of steelworks sites in the Black Country area. In many ways, then, these brothers were and still are archetypal 'local heroes'.

As well-known local characters and businessmen, the brothers were approached by Dudley Metropolitan Borough Council who wanted them to become involved in the Dudley enterprise zone in order to 'create activity' . . . The centrepiece of the Richardsons' involvement in the . . . enterprise zone . . . is the Merry Hill Centre. By 1989 Merry Hill retail

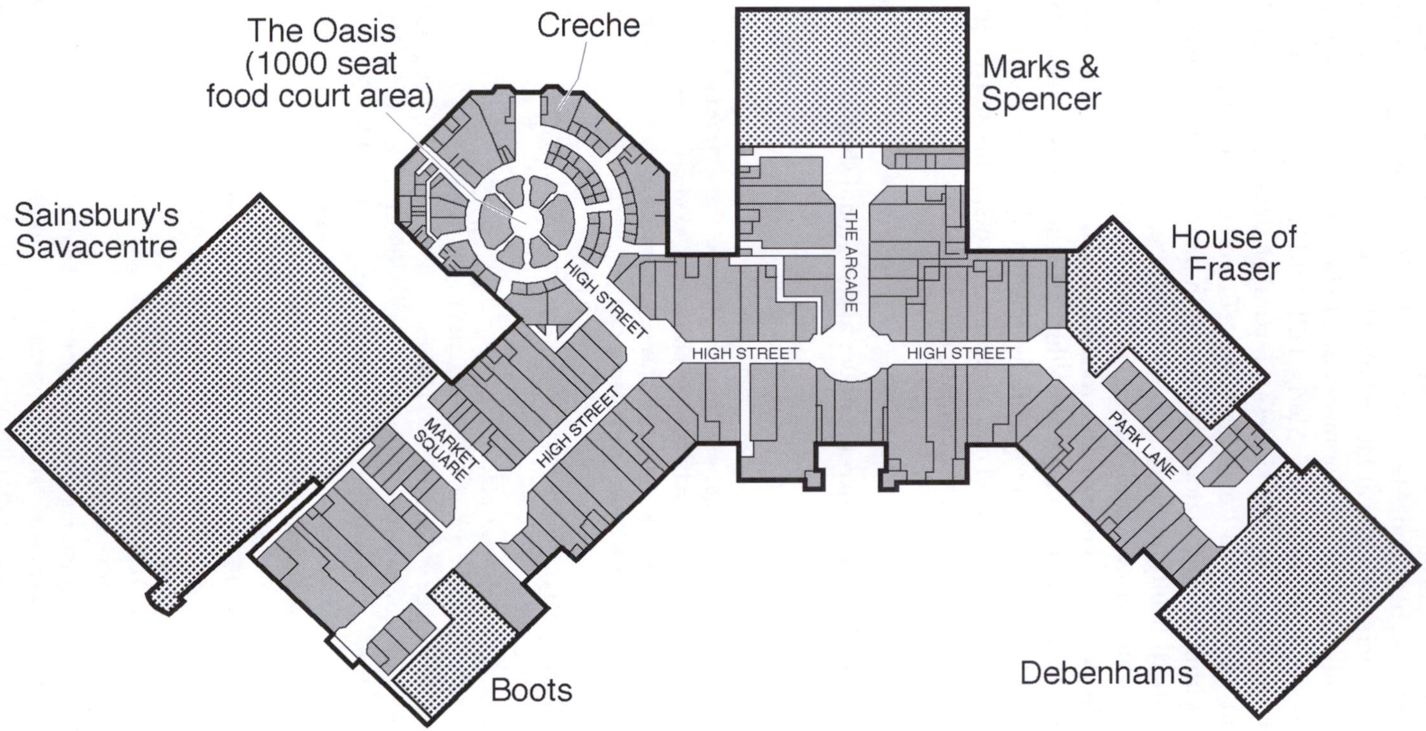

Figure 12.2: *The Meadowhall Centre, Sheffield, illustrating how leisure was largely omitted from Britain's regional shopping centres*
Source: Redrawn from Guy (1994).

had reached Phase Five. Phases One, Two and Three respectively comprised retail warehousing . . . MFI, Toys R Us, Comet, Macdonalds and Children's World . . . In later stages . . . there was a shift in the style of development . . . Phase Four was the first phase of upmarket retail, a double-story connecting mall, and this has subsequently been followed by Phase Five, a retail fashion-based mall of over one million square feet incorporating . . . major stores . . . Sainsburys, Debenhams, Next, BHS [and Marks & Spencer] . . .

Not content with their success in the retail field . . . the Richardson brothers have topped the whole Merry Hill Scheme with a monorail system threading through the current complex and scheduled to link Merry Hill retail with further Richardson developments currently under construction in the vicinity . . . There is a planned waterfront leisure and business scheme on an adjoining site . . . 'Merry Hill: The Waterfront' will be a major leisure complex. The latest proposals include at 'Waterfront East' office development, and at 'Waterfront West' two hotels, conference and exhibition centre, a fun park, waterworld, ice rink and other sporting facilities . . .

The Richardsons will not rest with the transformation of the West Midlands region which they have orchestrated by their Merry Hill Scheme. Their latest project involves one of Britain's most famous industrial landmarks. The former tyre factory, Fort Dunlop, alongside the M6 in Birmingham, is to become the centrepiece of a multi-million pound scheme creating thousands of new jobs . . . It is clearly the case that the Richardson brothers . . . [are] 'regional entrepreneurs' . . . these individuals are exemplary of the curious intermingling of regional culture, regional identity and regional regeneration which is occurring during an era largely dominated by increased globalisation.

Extracted from: Michelle Lowe (1993): Local hero! An examination of the role of the regional entrepreneur in the regeneration of Britain's regions. Chapter 10 in Gerry Kearns and Chris Philo (eds) *Selling places: the city as cultural capital, past and present*. Oxford: Pergamon Press, 214–21.

Reading 12.5 is taken from research by Daniel Miller, Peter Jackson, Nigel Thrift, Michael Rowlands (like Miller an anthropologist) and Beverley Holbrook which explores the experience of shopping at Brent Cross from the perspective of 'ordinary consumers' and uses a variety of methodological approaches (from a questionnaire survey to focus groups and ethnography). Here we take an extract which places their work in context and suggests that shopping (in such malls and elsewhere) provides an active and independent component of identity construction.

Reading 12.5 – The mother of UK shopping malls: shopping and identity

In 1997, Brent Cross Shopping Centre celebrated its twenty-first birthday; it had officially come of age. Described in *The Independent* newspaper as 'the mother of malls' . . . Brent Cross was Britain's first purpose-built regional shopping centre. Since its opening in 1976, Brent Cross has been overtaken by several bigger and more spectacular malls (Merry Hill,

Meadowhall and Lakeside among them), but its claims to have been in some sense 'the first' have rarely been challenged. As such, it has attracted both praise and blame. Accused of destroying the traditional English high street, wrecking the environment and replacing freely accessible public places with sanitised and tightly controlled private space, shopping centres such as Brent Cross (and their latter-day successor, shopping malls) have nonetheless proved extremely popular and financially successful places. At £300 a square foot, rents in Brent Cross are as high as in London's West End, offering consumers a safe and climate-controlled alternative to the perceived dangers and unpredictability of city-centre shopping. Over the years, Brent Cross has become an accepted part of many consumers' weekly (and in some cases daily) routine, no longer seen as a spectacular symbol of modernity and progress. Brent Cross has responded to the competition from more recent developments undertaking a multi-million pound face lift, letting in more daylight, increasing the amount of free parking space and generally sprucing up its appearance, aiming to attract a younger generation of shoppers as well as those who grew up with the centre . . .

What we were seeking to establish through this research was less the 'meaning of the mall' and rather more the diverse and often contested meaning of . . . shopping places and their relationship to the identity of those who shopped there. Rejecting the ungrounded theorising that has tended to dominate recent work on consumption . . . we ground . . . our understanding of contemporary consumption in the lives of 'ordinary consumers' and let . . . their voices be heard through transcriptions from our focus groups and ethnographic observations . . .

[We argue] that shopping does not merely reproduce identities that are forged elsewhere but provides an active and independent component of identity construction . . . One of our key findings [is] that shopping is an investment in social relationships, often within a relatively narrowly defined household or domestic context, as much as it is an economic activity, devoted to the acquisition of particular commodities. While our respondents rarely shopped as whole families . . . we found social relations within the family to be the dominant context of contemporary consumption . . .

The carefully controlled environment of shopping centres and malls provides consumers with a haven from the perceived dangers of high street shopping and the risk of unplanned encounters with various (often racialised) others. As such, the privatisation of space within shopping centres and malls provides a solution to the now widespread fear of public space, with closed circuit television and other visible means of improved personal security adding to the sense of risk-free shopping . . .

Consumers' fears about the increasingly 'artificial' nature of contemporary shopping represents the other side of their desire for a carefully managed and crime-free shopping environment. Again, we argue that these feelings are as much about the social context of shopping as they are about the physical setting of the shopping centre or mall. When people yearn for a return to 'personal service' or support current trends for opening up enclosed shopping centres to 'natural light', we suggest that they are at least as concerned about the increasing artificiality of their social relationships (and in particular the perceived materialism of their children) as they are about the physical environment itself.

Extracted from: Daniel Miller et al. (1998): *Shopping, place and identity*. London: Routledge, viii–xi.

The work of Miller, Thrift, Jackson, Rowlands and Holbrook is considered to be novel to the extent that it pays attention to the *people* who shop in malls – 'what shoppers do and what they understand as shopping'. Whilst earlier studies of malls have given a good deal of attention to the people within these shopping places, these literatures have concentrated primarily on the public's use of these spaces but have ignored 'the cultural practices of shopping' per se. It is to an assessment of this work that we will now turn.

Peopling the mall

Our starting point in this section is the fact that suburban malls have often been somewhat romantically viewed as new egalitarian public spaces. As Langman (1991, quoted in Shields, 1992: 5) suggests,

> whatever one's status or job in the world of work or even without job, there is an equality of just being there and looking at the shows of decor, goods and other people. Malls appear democratic and open to all, rich or poor, young or old . . . This is the realm where the goods of the good life promised in the magazine ads and television commercials can be found.

In this framework, there has been considerable attention paid to the activities of mall-rats – 'people who seem to do nothing else but hang out at the mall, all day, everyday' (Kowinski, 1985: 33). Mall-rats are usually (but not exclusively) adolescents for whom 'malls have become a primary hang out and site of such truly significant life events as first use of a charge card, driving a car (to the mall) and losing one's virginity in the parking lot' (Langman, 1992: 58). Like 'the street' discussed in Chapter 10, the mall is viewed in these accounts as offering people 'a third place beyond home and work/school, a venue where . . . [they] can congregate, commune and "see and be seen"' (Goss, 1993: 25). The mega-malls of North America and Britain's regional shopping centres specifically market themselves as 'open to all' and provide a range of activities from old-age pensioners' tea dances to teenage school break activities in order to support this image. Indeed, in North America many malls even throw open their doors early in the morning in order to give space to the multitude of 'mall walkers' who like to practise their keep-fit activities in the safety and security of these centres (Goss, 1993). Customers at the mall then – of all ages – 'become performers in their own drama' (Chaney, 1990: 63) as a new indoor *flânerie* – reminiscent of the 'sociality of the street' (see Chapter 10) takes shapes.

In reality, of course, the key word here is *customers* and any *flânerie* that does happen is strictly controlled. As Chaney (1990: 64) notes 'people walk here . . . as if their conduct might be called into question at any moment'. Malls are effectively spaces for the white middle classes – 'they reclaim, for the middle-class imagination, "The Street" – an idealized social space free, by virtue of private property, planning and strict control, from the inconvenience of the weather and the danger and pollution of the automobile, but most importantly from the terror of crime associated with today's urban environment' (Goss, 1993: 24) (see Chapter 9).

Reading 12.6 by Rob Shields, a sociologist whose pioneering work on consumption spaces (Shields, 1992) is well known in geography, is taken from *Environment and Planning D: Society and Space*, a leading journal of critical social and cultural geography. In this extract Shields contrasts the spaces of the mall – and the activities these allow – with those of the street. But Shields also draws attention to the fact that not all users of malls are 'passive dupes', rather 'post-shoppers' at malls play at being consumers and are thus able to exploit the possibilities of these places for their own purposes.

Reading 12.6 – Malls, *flânerie* and post-shoppers

Being in the tightly policed, semiprivate interior of a mall is quite different from being 'on the street'. 'No loitering', as the signs in the mall say. Certain types of comportment are expected . . . in malls, business deals are struck and social relationships made as they are in the street cafés of continental Europe . . .

The shopping centre and its practices – a new indoor *flânerie* (strolling), the habit of window shopping as much as 'hanging out' or being an onlooker enjoying the crowds – have become established features of contemporary urban life . . . *Flânerie* has acquired a less gendered character to become an almost universal diversion amongst new middle-mass consumers . . . Thus, in the United States, for example, the most frequented public spaces are shopping malls . . . The large regional centres . . . have a sizeable group of users who visit simply for enjoyment and to observe the world . . .

[But] unlike the street-life of the European tradition, the surrounding environment in the centre is carefully and consciously managed . . . To congregate in such spaces . . . requires that one observe bourgeois norms of social docility and conservatism both in dress and action. The displays of 'peacock clothing ' or 'punk spectacle' common in the United Kingdom or USA are relegated to just outside the doors of . . . [the] mall. Instead, the clothing adopted is a fashion industry edition of punk or gothic style . . . One must always look as if one has bought something or is about to buy. Hence their uniform, classless appearance. Also, there are thus no hangouts. The greatest rebellion is the act of sitting down on the floor, ignoring benches . . . Movement after a few minutes is often essential to avoid the security guards patrolling for loiterers . . .

Why patronize such a place? The key to the success of the mall appears to lie in the manner of appropriation by its users, the *flâneurs* . . . The carnivalization of the mall by its users provides the only means at hand to balance its 'commercial terror' . . . Like Urry's 'post-tourist' who knows that mass tourism is a game played for status, the mall has its 'post-shoppers' who, as *flâneurs*, play at being consumers in complex, self-conscious mockery . . . This tinges the complicitous self-implication and apathy of the *flâneurs* . . . with parody. The effect is to disrupt the pretentiousness of [the] mall . . . The users, both young and old, are not just resigned victims, but actively subvert the ambitions of the mall developers by developing the insulation value of the stance of the jaded, world-

> weary *flâneurs*; asserting their independence in a multitude of ways apart from consuming ... It is this practice ... which is the heart of the postmodern experience of the mall.
>
> **Extracted from: Rob Shields (1989): Social spatialization and the built environment: the West Edmonton Mall. *Environment and Planning D: Society and Space*, 7, 147–64.**

Shields' work discussed above places emphasis on the mall as part and parcel of 'post-modernism' (Harvey, 1989a) but, as other authors have recognized, the 'post-modern retail environment' (Goss, 1992) comprises far more than the suburban mega-mall or regional shopping centre. Indeed, as we noted in our 1996 review of the new retail geography, 'the differentiation of shopping malls is far more advanced than a reading of the literature to date would suggest' (Lowe and Wrigley, 1996: 26). And there are many examples of 'malls' which have not been covered so far in this chapter. It is to these alternative consumption spaces that we will now turn.

The malling of retail space: new landscapes of consumption

From the 1980s onwards malls in North America approached saturation point. The serial reproduction of mall formats across the United States – 'the malling of America' – had proceeded apace throughout the previous quarter of a century and in this climate the concept of differentiation became important. However, as Crawford (1992: 10) suggests, 'the system demonstrated a surprising adaptability [and] in spite of its history of rigidly programmed uniformity, new economic and locational opportunities prompted new prototypes'. Speciality centres – an anchorless collection of upmarket shops and restaurants pursuing a specific retail and architectural theme – were joined by downtown 'mega-structures' and 'festival marketplaces'.

Reading 12.7 by John Goss, captures the range of these new mall types in North America. Such new landscapes of consumption often became important vehicles of 'civic boosterism' in conditions of heightened interregional competition in the 1980s (Harvey, 1989b).

> ### Reading 12.7 – The post-modern retail environment
>
> The essential forms of the post-modern retail environment – the speciality centre and the downtown 'megastructure' – reflect the vernacular and high forms of post-modernism respectively, while a hybrid form – the festival marketplace – combines elements of both. The speciality centre is an 'anchorless' collection of upmarket shops pursuing a specific retail and architectural theme. It is prone to quaintification. Typical designs in North America include New England villages (Pickering Wharf, Salem, Massachusetts); French provincial towns (The Continent, Columbus, Ohio); Spanish–American haciendas (The

Pruneyard, San Jose); Mediterranean villages (Atrium Court, Newport Beach, California); and timber mining camps (Jack London Village, Oakland, California). Pride of place must, however, go to the Borgota in Scottsdale, Arizona, a mock thirteenth-century walled Italian village, with bricks imported from Rome and shop signs in Italian . . . and to the Mercado in nearby Phoenix, Arizona, modelled on traditional hillside villages of Mexico, with original components imported from Guadalajara and buildings given Hispanic names . . .

The downtown megastructure, on the other hand, is a self-contained complex including retail functions, hotel, offices, restaurants, entertainment, health centres and luxury apartments. Typical examples include Water Tower Place in Chicago, the Tower City Centre in Cleveland and Town Square in St Paul . . . These small worlds ensure that the needs of affluent residents, office workers, conference attenders and tourists can be met entirely within a single hermetically sealed space. Several features distinguish the downtown megastructure from the suburban shopping mall, although by now many of these have been extensively 'retrofitted' in the post-modern style . . . Daylight [has been returned] in glazed malls reminiscent of nineteenth-century European arcades . . . natural light allows the planting of ficus, bamboos and 'interiorised' palms to simulate the tropical environments of tourism . . . elaborate watercourses and waterfalls simulate nature, rather than urban fountains . . . Pure and perfected nature . . . is ironically found indoors within the city, and no longer in the deteriorating environments of the suburbs . . .

The downtown malls are also no longer primarily 'machines for shopping', although the aesthetics of movement are retained in the sweep of huge escalators and the trajectory of 'bubble' elevators. Now passage through the mall is an interactive experience, an adventure in winding alleys resembling the Arabian Souk or medieval town, with the unpredictability of 'pop-out' shop fronts – glass display cases which jut out into the mall – and mobile vendors . . .

The festival market combines [shopping and entertainment] with an idealised version of historical urban community and the street market, typically in a restored waterfront district after the model of Faneuil Hall Marketplace in Boston. These environments reflect a nostalgia for manual labour, public gatherings and the age of commerce. Buildings and vessels are restored, and there is usually a historic museum on site. The marketplace is typically decorated with antique signage and props which casually suggest an authentic stage upon which the modern consumer can act out a little bit of history. The street entertainers, barrow vendors and costume staff often support this image.

Extracted from: John Goss (1992): Modernity and postmodernity in the retail landscape. In Kay Anderson and Fay Gale (eds) *Inventing places: studies in cultural geography*. Melbourne: Longman Cheshire, 158–77.

Meanwhile, in the UK, different economic and political imperatives – specifically the shifting nature of government policy towards out-of-town retail developments (see Chapter 7; also Guy, 1994b, 1998b, 1998c, 1998d; Wrigley, 1998a) – has led to a similar duplication of new retail formats. Indeed, by the 1990s a 'fourth wave' of innovative retail developments comprising 'more specialised, up market formats, such as warehouse clubs, airport retailing and factory outlets centres' was identified (Fernie, 1995, 1998).

Table 12.1: *Existing and proposed factory outlet centres in the UK, December 1995*

Operator	Trading name	Location of factory outlet	Date of opening	Square footage
Existing				
BAA/MacArthur Glenn	Cheshire Oaks	Chester	Open	110,000
Brighton Marina Co.	Brighton Marina	Brighton	Open	40,000
C & J Clark	Clarks Village	Street, Somerset	Open	72,000
	K Village	Kendal, Cumbria	Open	19,000
Freeport Leisure	Hornsea Freeport	Hornsea, East Yorkshire	Open	40,000
	Freeport Village	Fleetwood	Open	60,000
Schroders Property (formerly the Guinea Group)	Jacksons Landing	Hartlepool	Open	60,000
Lightwater Holdings	Lightwater Valley	Yorkshire	Open	19,000
Value Retail	Bicester Village	Bicester	Open	107,000
Proposed				
BAA/MacArthur Glenn-Tarmac/Richardson	Great Western	Swindon	1997	180,000
	Bridgend	Bridgend	1997	175,000
	Junction 28	Mansfield	1998	200,000
BAA/MacArthur Glenn	Western Links	Weston-super-Mare	NA	150,000
	NA	York	1997	300,000
	NA	Bathgate	1998	110,000
Bisley Properties	Tobacco Dock	London	NA	160,000
European Outlet Markets Ltd	Killarney Company Stores	Killarney	1996	79,000
Fairclough Homes	NA	Dover	NA	130,000
Freeport Leisure	NA	Westwood, Scotland	NA	50,000
The Guinea Group	Clacton Common	Clacton-on-Sea	NA	116,000
	Leven Fields	Kinross	1997	80,000
Lansfastigheter Prop	The Galleria	Hatfield	NA	300,000
Lister & Co.	Manningham Mills	Bradford	1996	100,000
Morrison Construction Group	Royal Quays	North Shields	1996	100,000
Oldway Property Group	New House Village	Chepstow	NA	50,000
Oldrids	NA	Grantham	1996	55,000
PSIT	Rolling Stock	Haydock	1996	160,000
Ram Euro-Centres	Cotswold Village	Tewkesbury	NA	155,000
Ram Euro-Centres/C & J Clark	Yorkshire Outlet	Doncaster	1996	155,000
Rockeagle Developments	Festival Shopping	Ebbw Vale	1996	90,000
Schroder Property	Chorley Mills	Chorley	1997	46,000
Sea Containers/Ellison Harte	Harwich Europort F/O Centre	Harwich	1996	150,000
Value Retail	Sandbach Village	Sandbach	NA	100,000
Walton Commercial	NA	Liverpool	NA	150,000

Source: Adapted from Fernie (1996), 17–18.

Warehouse clubs, which sell food and non-food items to club members at low prices in 'no frills' surroundings, and factory outlet centres, which sell cut-price merchandise direct from manufacturers in purpose-built centres, both have their origins in the US where they have enjoyed considerable success (Guy, 1994b). In the UK, warehouse clubs such as the US chain Costco met with more limited fortunes in the wider context of the severe mid-1990s shake-out in the discount food retail sector (Wrigley and Clarke, 1999). In contrast, factory outlet centres initially performed more strongly. In 1993 there were two such centres – at Hornsea and Street; by 1995 there were nine and proposals for a further 25 (Fernie, 1998). Table 12.1 illustrates existing and proposed factory outlet centres in the UK in December 1995. Several of the factory outlet centres, particularly the designer outlet mall at Bicester Village in Oxfordshire, are modelled on the New England village design which Goss in Reading 12.7 describes as 'quaintification'. More recently, due to tightening land use planning regulation, there is evidence of a slowing of this trend. For example, Fernie (1998) shows that by mid-1997 only 15 such centres were trading, that the scale of some of the developments planned may be downgraded, and that their format is likely to be more 'downmarket' than in the USA.

Airport retailing has expanded from its original duty/tax-free orientation (wines and spirits, perfumes and tobacco) to encompass a broad range of merchandise tailored to the lifestyles of various passenger groups. As a result, many high-street retailers are now trading successfully in airports (Freathy and O'Connell, 1998) and the interiors of airport terminals present the traveller with an array of consumption possibilities in mall-type settings. In 1994 Chesterton calculated that Britain's 15 largest airports had a total of 248 speciality shops occupying 323,792 square feet of retail floorspace (Chesterton, 1994) and there has been considerable expansion since that date.

New landscapes of consumption such as the factory outlet centre or airport mall have developed rapidly in the past decade and have joined new purchasing opportunities at the railway station, petrol/service station and hospital in widening the retail offering. This is not to suggest, of course, that other kinds of retail development present in North America have been absent from the UK. In particular, speciality centres and festival marketplaces – such as those found in London (Covent Garden or Whiteleys, a former department store), Glasgow (Princes Square), Leeds (Corn Exchange) or Southampton (Canute's Pavilion) (Guy, 1994b) have become familiar sites in many of Britain's towns and cities. But it is the 'home' the traditional site of reproduction/consumption – which has been recast as a new consumption landscape of the most considerable potential. Our next (and final) chapter addresses this alternative consumption space in detail.

13
The home

This chapter will examine the history of home shopping from its earliest origins in late nineteenth-century USA to its current status within the profitable and developing sphere of niche retailing. In the contemporary period the home as the traditional site of reproduction/consumption has been recast as a new consumption landscape of considerable potential, with other non-place routes to market such as television shopping and the Internet complementing catalogue retailing and enabling the substantial growth of this sector. But the home is not only an important site into which the ever expanding consumer market can penetrate; it also, and very obviously, acts as a space within which household provisioning takes place. Here we centre our analysis on the home as a specific and unique focus of contemporary consumption and demonstrate how various types of home consumption are equally entrenched in notions of identity construction and social formation as are more formal (and more documented) consumption activities, for example those that take place in the department store or mall.

Catalogue retailing and the home

Retailing through the mail developed in the USA under the leadership of Montgomery Ward (launched in 1872) and Sears Roebuck (started in 1886). These new mail-order houses sold the American consumer revolution to an isolated rural population. Consumers could purchase virtually everything from the pages of catalogues (from baby chicks to gravestones), and by the 1930s Sears Roebuck were even merchandising off-the-shelf houses which were ordered by mail and arrived in rural communities complete with suitably adorned domestic interiors. Aided by the expanding network of railroads and a series of changes in the US Post Office – such as rural parcel post – the new mail-order catalogues thus offered a realm of consumer choices to populations relatively isolated from America's emerging cities and were a central force in the cultural transformations of the period.

Reading 13.1 by American cultural historian Thomas J. Schlereth, captures the dramatic effect that mail-order merchandising had on nineteenth-century rural life in the USA. In the broader essay from which this reading is extracted Schlereth documents the various ways in which the 'Farmers' Bibles' or 'Wishbooks' doubled as readers, textbooks and encyclopaedias in many rural schoolhouses, and served as almanacs for adults since they usually contained inspirational readings, epigrams, poetry and farming and household tips (Schlereth, 1989: 365). Here, though, the focus is on the catalogue providing rural dwellers with 'a department store in a book', as well as on the novel means of distribution and payment adopted by the mail-order catalogues.

Reading 13.1 – The birth of mail-order merchandising

Often called 'Farmers' Bibles', the mail-order catalogues of the Chicago giants – Montgomery Ward and Sears – often expressed a secular hope for salvation from want. There was, for example, the off-quoted tale of the little boy from rural Idaho who, upon being asked by his Sunday School teacher where the Ten Commandments came from, unhesitatingly replied, 'From Sears, Roebuck, where else?'

Mail-order catalogues were called 'a department store in a book', 'the nation's largest supply house', 'a consumer guide', 'a city shopping district at your fingertips' and 'the world's largest country store'. Large-scale retailing through the mails took off under the nationwide marketing skills of Aaron Montgomery Ward. Ward claimed to have launched in 1872 the modern mail-order industry. Catalogues were distributed to people who then placed their orders by mail. The ordered goods were delivered to customers' homes by some established shipping service such as freight, express or post . . .

Richard Sears, a former Minnesota railway agent who got into the mail-order business in his spare time selling watches, launched his company in 1886. Two decades later Sears claimed his catalogue was 'the largest supply house in the world', selling 10,000 other items in addition to a watch every minute. The key medium for displaying his merchandise was the semi-annual *Consumer's Guide*, as his mail-order catalogue began to be labelled in the 1894 edition . . .

Mail-order merchandising required constant attention to maintain credibility . . . Sears and Ward jammed their catalogues with beguiling illustrations and detailed descriptions written in a folksy vernacular . . . Both guaranteed customer satisfaction or an immediate cash refund [and] used their catalogue covers to depict their stores as beehives of industrious clerks filling thousands of orders daily from a seemingly unlimited abundance of products . . .

In 1905 Sears improvised a plan – called his 'Iowaization' scheme – to distribute catalogues and recruit customers. The company wrote to all its current customers in Iowa and asked each to pass on catalogues to friends and neighbours. The customers in turn sent the names of people given the catalogue to the company. The company kept track of who ordered what and gave each 'distributor' premiums on the basis of the number of incoming orders from friends and relatives. Sales from Iowa soon outstripped all other states, and Sears went on to apply the concept throughout the country . . .

Instalment buying for many rural dwellers grew out of their mail-order merchandising experience. Many products selling for $25 or more could be had for a minimal down-payment and a check or money order dispatched monthly to Chicago.

Extracted from: Thomas J. Schlereth (1989): Country stores, country fairs, and mail-order catalogues: consumption in rural America. In Simon J. Bronner (ed.) *Consuming visions: accumulation and display of goods in America 1880–1920*. New York: W. W. Norton, 339–75.

By the mid-twentieth century America's mail-order giants had been joined by British equivalents such as Littlewoods, Freemans, Kays, Grattan and Empire. Based primarily in the north of England, the UK's mail-order industry was dominated by women who sold goods to their local communities through catalogues. An elaborate system of buying and selling developed around the 'big books' with consumers mostly purchasing on weekly credit and agents being paid commission on their sales. Mail-order retailing via these networks of local agents grew steadily from the 1950s, and between 1950 and 1970 its share of retail sales increased from 0.9 to 4.2 per cent, with mail-order penetration being particularly high in certain product areas such as soft furnishings and womenswear/infantswear (McGoldrick, 1991).

By the late 1980s/early 1990s, however, traditional 'big book' agency catalogues were experiencing more difficult times and mail-order's share of retail sales in the UK had declined to around 3 per cent. In the USA, Sears, after suffering major losses, was forced to close down its mail-order catalogue business in 1993, and in the UK traditional agency catalogues had experienced similar declining fortunes. Notwithstanding this, however, new entrants into the catalogue market, together with shifts in the style of established firms, began in the mid- to late 1990s to give the sector a new lift. Indeed, direct mail or direct marketing (as opposed to agency) catalogues which cater for specific lifestyle groups have become an important force in contemporary retailing and have positioned themselves carefully within the profitable and developing niche retailing sector. In the UK, Next pioneered the new approach to catalogue retailing with its Next Directory. Launched in the late 1980s, the Directory built on the previous success of the Next concept and was utilized as a means of gaining important feedback on best-selling lines for subsequent in-store sales. Next was followed by similar new mail-order merchandisers, many of which have developed out of, or have subsequently built up, a retail presence in conventional high street or mall locations.

Unlike their agency counterparts these new mail-order moguls in both the US and UK such as L.L. Bean, Lands' End, Patagonia and Boden target specific consumer segments – often those with little time for traditional modes of consumption such as busy professionals or dual income families – and promise such wealthy consumers high-quality products and good service. Lands' End, for example, the US direct-mail company has expanded its business in the UK since 1991 and targets middle-class shoppers, specifically women aged 35–55, keen on travel and golf, in a professional and managerial position and with a degree. In the USA, Lands' End 'maintains a mailing list of 9 million people, 45 per cent of whom have purchased merchandise from the firm in the previous 36 months. In 1996 it mailed out 150 million catalogues which generated sales of 925 million' (Levy and Weitz, 1998: 72). Great Universal Stores (GUS) the largest home shopping retailer in the UK, and operator of some of the traditional catalogues, has also entered the direct mail business with sector-specific catalogues such as Disney Selection, Sports Elite and Style Plus. Even Sears has re-entered catalogue retailing in the USA, setting up joint ventures with other firms, and many other operators such as Talbots and Spiegel print and mail 'specialogs' catering to specific customer segments – e.g. petite women – as well as or instead of annual catalogues that contain all their offerings. In this way traditional mail-order business has been revamped and repositioned. Indeed, what Christopherson (1996) calls 'non-store' retailing has expanded dramatically in recent years. Visa credit card purchases at home in the USA doubled between 1988 and 1992 to $17.8 billion, leading some commentators to suggest – somewhat ambitiously perhaps – that 'stores are becoming places where people kick the tires, lift the lid on a washing machine or listen to the sound of a stereo speaker – and then go home and call an 800 number to order the same item at a 40 per cent discount' (Morgenson, 1993, quoted in Christopherson, 1996: 169). What is clear, however, is that non-store formats – albeit small – are growing faster than store sales (in the USA at 7 per cent compared to 4 per cent in the late 1990s).

Of course these figures not only include shopping by phone and mail, but also take into account the small but expanding arena of TV home shopping and Internet sales. It is to these new forms of home consumption that we now turn.

Television home shopping and the Internet

In the USA television home shopping retailing is dominated by two giant networks, QVC and the Home Shopping Network (HSN). Customers watch a dedicated TV channel on which various types of merchandise are demonstrated, and then place orders by telephone. In the ten-year period from 1988 to 1998 the percentage of US consumers buying goods via home shopping programmes increased from 5 to 10 per cent. However, most sales are to lower-income consumers (specifically older women) and are concentrated heavily on inexpensive jewellery, women's clothing and personal care products, with little emphasis on nationally recognized brands. QVC, the leading US operator, established a home shopping network in the UK in 1993 and it was predicted that TV retailing could capture 5 per cent of the home shopping market – a total market estimated at around £8 billion a year in 1998. Interactive shopping technology (which lets customers browse through channels and ask for data and advice) is expected to significantly increase the popularity of the TV home shopping concept. Notwithstanding the novelty of television home shopping, however, and specifically its obvious advantage over catalogue retailing – customers can actually see merchandise on the TV screen – it is the Internet which many have argued has the potential to generate the biggest revolution in home consumption.

Reading 13.2 by Jonathan Reynolds highlights the way in which new electronic channels to market (specifically the Internet) allow careful targeting of specific market niches. Based on research carried out in the mid-1990s – when the rhetoric concerning the potential of the Internet ran considerably ahead of the reality – Reynolds draws attention to the challenges of what he terms 'a virtual geography of demand and supply' and, in the paper from which this reading is extracted, emphasizes the fact that Internet retail sites demand as much care in presentation and market research as conventional stores.

Reading 13.2 – Internet channels to market: rhetoric versus reality

Market conditions are [now] more favourably disposed towards electronic channels than even before. Trends towards 'market fragmentation' suggest that channels to markets which permit the more careful targeting of identifiable market segments (as electronic channels do) represent . . . efficient business development strategies for retailers . . . Key to understanding the potential attraction of an Internet channel to market for retailers is the emerging user profile: hitherto hard to reach but very desirable market segments . . . In terms of lifestyle groups active on the Internet, actualizers ('individuals characterized by higher incomes and educational levels') and experiencers ('innovative stimulation-seeking and fashionable young people') are most prominent. However, these segments are among the most demanding and sophisticated in their use of conventional retail channels and therefore potentially provide a significant marketing challenge to retailers on the Internet.

Outside the USA the country which has experienced the largest amount of rhetoric about the Internet and its commercial potential has been the UK . . . At the end of 1995 there were over 300 retail sites on the Web based within the UK . . . These varied from independent operators, such as Lossie Seafoods, to The Body Shop, Austin Reed and Toys R Us. Sixteen virtual shopping centres opened, including Barclays' BarclaySquare, Highland Trail (offering net access to Scottish products) and the London Mall [and in addition] most UK food retailers have an Internet presence. However, after rhetoric we must address the reality . . . After a year of trading J. Sainsbury reports that, while it is satisfied with the experience, sales of wine, flowers and . . . chocolate, are 'not like hot cakes'. The finance director of Argos revealed in March 1996 that, after nine months on-line, the Argos web site had sold just 22 items, with the biggest selling product being wine racks: 'biggest selling product in terms of hello, we have sold more than one . . . we have sold about three . . .'

The most thoughtful retailers regard their web sites as experiments from which they will learn. Looking forward, some UK retailers are entirely pragmatic: 'By the end of the century I would be surprised if the home sector (including all electronic channels to market) was delivering for us the equivalent in sales of one average store . . . That could mean sales of up to £3 million so it could still be worth our while doing it, depending on the cost. But it will not have much impact on our High Street operations' (Dixons).

Extracted from: Jonathan Reynolds (1997): Retailing in computer-mediated environments: electronic commerce across Europe. *International Journal of Retail and Distribution Management*, 25, 29–37.

Like Reynolds, other commentators similarly questioned the potential of the Internet's role in home consumption. Indeed, by early 1998 at the Merrill Lynch 'Retailing Leaders' Conference in New York, Carl Steidtmann, director of research for Management Horizons, referred to the 'enormous amount of hype' about the Internet, suggesting that Bill Gates' well-known view in his book *The road ahead* that 30 per cent of retail sales would eventually go through the Internet was fundamentally flawed – 'My feeling about that was that he must get access to much better hallucinogens than the rest of us do to make that kind of forecast.' Nevertheless, Steidtmann does concede that Management Horizons see the Internet as becoming a significant channel of distribution particularly for certain product categories such as books and consumer electronics – 'High end products and products that require a lot of information will do quite well across the Internet. Ultimately we see it as being probably about 5 per cent to 6 per cent of total retail sales, which is about the size of the catalogue industry today' (Merrill Lynch, 1998b: 11). We now examine this contention in detail by focusing on one of Steidtmann's identified product categories – the retailing of books.

Following hard on the heels of Amazon.com one of the most successful US Internet retailers and the largest retailer of books and music on-line with sales which had reached $219 million per annum by the late 1990s, the existing major book retailers in the USA, Barnes and Noble and Borders launched their Internet sites BarnesandNoble.com and Borders.com in 1997 and 1998 respectively. After one year on-line BarnesandNoble.com had attracted over 500,000 customers in over 158 countries. In the same way that bookstore/coffee shop formats have become community centres where people go to buy books, read magazines, drink coffee and socialize (see Chapter 11), Barnes and Noble is trying to create a similar sense of community on-line. Live author chats are held every day similar to the in-store events held in their book superstores, and community members are encouraged to write reviews of book they have read – reviews which are made available to other consumers searching for books. Notwithstanding these advances, however, Internet sales represented by the late 1990s less than 2 per cent of total book sales in the USA and less than 1 per cent of global book sales, with international customers, those living in regions without a bookstore and consumers who dislike traditional shopping providing the main market. Reading 13.3 by Robert Di Romualdo, chairman and chief executive of the Borders Group, the US bookstore chain which in 1997 acquired UK chain Books etc., highlights the importance of Borders' Internet operations but also suggests how this new electronic channel not only acts as an important vehicle for Borders' international expansion but also, and critically, improves Borders' in-store service via its potential to facilitate on-line customer searches – in this way enhancing Borders' core bookstore business (see also Chapter 4).

Reading 13.3 – Borders.com: expanding the brand and the potential of the Internet

Our objective on the Internet is to extend our brand globally, provide an on-line alternative for Borders' customers and build a profitable business ... Our Internet site will be all media including books, music and video. We cannot deliver a hot cup of coffee on line but we will sell you the bean. The selection, service and community content on line will basically leverage our stores. We are integrating our Internet operations with our fulfilment center in Nashville [which adds] capacity to our main distribution center. This fulfilment centre has 53 miles of shelving and is capable of stocking 750,000 skus (stock-keeping units) of books, music and video ...

Our goal is to achieve zero wasted trips when people come to shop at our stores. When you come to our stores you will see 200,000 to 250,000 skus of books, music and video that we carry. There you can sample it, open the book, etc. If we do not have what you want we will have 50,000 skus available for immediate shipment to your home from the fulfilment center. If we do not have it there we will search through 'all else in print' through Ingram, Baker and Taylor or anybody else who has it. If they do not have what you want ... We can go to our chase team who can go to Borders stores where very obscure titles may be sitting on a store shelf ... If all else fails we will go to the publisher. We also have an investment in Harvest, which is the premier out of print search service who is our partner in our Internet venture ...

We are forecasting sales of about $25 million internally in 1998 ... [in the longer term we] do not know how big a business [on-line sales] will be. There have been catalogues around for a long time. Certain people are catalogue shoppers for certain commodity type products. Customers are time pressed but no one has to go out and get a cup of coffee and yet Starbucks is booming. So people do not want to just stay at home. There is a limit to cocooning. We will have to watch how big a business the Internet evolves into.

Extracted from: Robert Di Romualdo (1998): Presentation to Merrill Lynch & Co., Inc. 'Retailing Leaders' Conference, New York, 11 March 1998. Transcript available in *'Retailing Leaders' Conference management presentations*, 18 May 1998, New York: Merrill Lynch, 13–22.

The above extract is, we believe, instructive concerning the potential role of the Internet in the growing home consumption market. Whilst book retailers like Barnes and Noble and Borders have been identified as potential success stories of on-line retailing (after all this sector has no 'quality' differentiation built into the products it sells – consumers recognize that a particular book purchased is the same whether acquired via the Internet or via a visit to their local bookstore), Di Romualdo, chief executive of one of the largest book retailers in the US in the late 1990s remained fairly sceptical of the long-term potential of Internet sales, and his company saw on-line possibilities as *complementary* to, rather then a replacement for, conventional main street or mall book sales. Importantly, it is the cost of home delivery (in this case for single book titles rather than bulk orders) which is viewed as a brake on the expansion of such a means of home consumption. Indeed, limitation on growth due to fulfilment cost became a critically important factor affecting the growth of Internet retailing after the dot com bubble burst during 2000/2001 (see Wrigley and Lowe, 2001). In addition, the above extract also points very forcefully to a further limitation of the Internet as a basis for household consumption – that is, the fact that such consumption from home often denies consumers the essential sociality associated with the other landscapes of consumption – the street, the store and the mall that we have considered in Chapters 10, 11 and 12. Whilst making a direct comparison between on-line sales and more traditional forms of catalogue retailing, Di Romualdo highlights the fact that 'people do not want to just stay at home', and as a result draws attention to a fundamental oversight of those who predicted during the late 1990s the massive expansion of the Internet as the latest electronic channel to the home consumption market. Indeed, his concern regarding the lack of sociality of Internet-based consumption has been similarly voiced by others commenting on the relative lack of commercial success so far of other non-place routes to market such as mail-order merchandising or television home shopping. Notwithstanding this, however, it is arguable that certain forms of home consumption promote unique forms of sociality, at least in certain communities and it is to these aspects of contemporary home consumption that we will now turn.

Shopping at home and sociality

In our account of mail-order merchandising above we rather skimmed over the means of distribution of mail-order catalogues, preferring instead to focus our attention on the history of mail-order shopping and on the various ways in which catalogue retailing made consumer products available to those without geographical or financial access to the department stores and other spaces of more formal consumption. However, in the early part of the twentieth century, and to a certain extent in the present day, the unique means of distribution of mail-order catalogues promoted in both the USA and the UK led to an important form of consumer sociability. Richard Sears' 'Iowaization' scheme discussed in Reading 13.1 involved communication between friends and neighbours in order to distribute his catalogues, whilst the agent network established by mail-order merchandisers in the north of England provided a distinct means by which women in particular could meet. Although such networks are considerably diluted in the present day, mail order – at least via the traditional 'big book' catalogues – is clearly not as individualistic and privatized a form of consumption as TV home shopping and the Internet are proving to be. But all of the above forms of home consumption pale into insignificance regarding their 'social' aspects when set against 'shopping at home' which we have so far failed to consider in this chapter. It is to direct selling – shopping *at* home as distinct from shopping *from* home that we now turn our attention.

The concept of direct selling includes direct contact with consumers in their homes or offices as well as phone solicitations initiated by retailers. Cosmetics, fragrances, jewellery, cooking and kitchenware, encyclopaedias and cleaning products are among the items most often sold in this manner, with independent agents (as in the case of catalogues) acting as the retail workforce. In 1995, direct selling comprised $18 billion in US sales and employed 7.2 million people mostly on a part-time basis, with 60 per cent of all direct sales taking place in the home (Berman and Evans, 1998). Retail leaders in this field – many with an international presence include Avon (skin care products and cosmetics), Mary Kay (cosmetics), Amway (household supplies) and Fuller Brush Company (small household products). Whilst the concept of direct selling is now considered relatively outmoded, the unique 'party plan' system utilized by a number of direct sellers, whereby salespeople encourage customers to act as hosts and invite friends and neighbours to a 'party' where products are demonstrated, has been adopted by a number of more contemporary retailers such as The Body Shop (with their 'direct' range) and Ann Summers. Of all direct selling retailers, however, it is Tupperware which in both the USA and the UK is most emblematic of this sales technique. Reading 13.4 by Alison Clarke, a design historian, captures the essential sociality inherent in the Tupperware product as it swept through the suburbs of America in the 1950s (and subsequently Britain in the 1960s). In the broader article from which this reading is extracted, Clarke situates Tupperware as an 'icon of suburbia' and considers its role in the active formation of post-war feminine identity.

Reading 13.4 – Suburbia, sociality and the Tupperware Party

Tupperware initially failed to sell in sufficient amounts from hardware or department store displays and its mail-order business was minimal. Inspired by direct sales concerns such as Stanley Home Products, and Fuller Brush Co., the Tupper Corporation adopted the 'Party Plan' system in 1951 with the employment of sales woman Brownie Wise. Wise attracted Earl Tupper, the Corporation's President's attention with enormous sales orders placed by her small direct sales company, Patio Parties. Whilst Tupperware formed only a proportion of her door-to-door sales, Wise found it a particularly suitable product for demonstration to groups of enthusiastic homemakers. Earl Tupper was persuaded of the virtues of the 'Party Plan' system in furthering the 'Tupperization' of American homes. By 1954 over twenty thousand women belonged to the Tupperware Party network as dealers, distributors and managers.

The Tupperware Party elaborated established door-to-door selling techniques by incorporating party games, refreshments and sophisticated product demonstrations. As well as serving as a highly rarefied sales forum it acted as a ritual interface between maker, buyer and user . . . The 'hostess' offered the intimacy of her home and the range of her social relations with other women (relatives, friends, neighbours) to the Tupperware dealer in exchange for a non-monetary gift. The dealer, supplied by an area distributor used the space to set up a display of products and recruit further parties from amongst the hostess's guests. The dealer benefited from commission accrued on sales and the potential for further party reservations.

Suburban homemaking provided Tupperware with its most comfortable niche: the corporate slogan read, 'the modern way to shop for your houseware needs; conveniently, leisurely, economically'. Concepts of thrift and excess co-existed happily in the Tupperware ethos. Whilst items such as the 'Econo-canister' and 'Wonderlier Bowl' made possible bulk storage, and the ability to turn 'Left-Overs into Plan-Overs', spindly legged condiment holders provided the perfect centrepiece at any impromptu buffet party. Suburban rituals such as the barbecue, flourished with the Tupperware range . . .

Tupperware Parties provided sanctioned all-female gatherings outside the family. Loyalty to fellow neighbours and friends was the linchpin for attendance to many parties . . . but for numerous women it was an opportunity to socialize outside the home at little expense. Whilst the pretext of the gatherings was domestic this did not preclude women from directing the conversation and interaction towards other concerns . . . Tupperware, then, embodied the contradictions of a growing postwar consumer culture. Whilst substantiating predominantly conservative and traditional feminine roles, it provided a pragmatic, pro-active alternative to domestic subordination . . .

In 1961 Tupperware was appropriated and re-invented by a British public. *Which?* consumer magazine criticized Tupperware (despite rating its functional performance) for its elaborate and immoral American sales technique, which brought commerce directly into the sanctity of the home. Numerous newspaper reports condemned the divisive Tupperware Party; a typical headline read 'As Soft Sell Steals into Suburbia'. But British women defied all marketing surveys and embraced the network enthusiastically. The National Housewives' Register integrated Tupperware Parties as part of their 'getting to know you schemes' on new suburban estates and by the mid 1960s the product and its party, featured on the front cover of the fashionable and aspirational *Queen* magazine, both were symbols of a newly perceived social mobility.

Extracted from: Alison J. Clarke (1997): Tupperware: suburbia sociality and mass consumption. Chapter 5 of Roger Silverstone (ed.) *Visions of suburbia*. London: Routledge, 132–60.

Of course not all commentary on Tupperware is as positive as that revealed in Reading 13.4, and it is, we feel, instructive at this point to reflect on what Clarke terms 'the ambiguity of the Tupperware Party, the intrusion of the 'market' into the sanctity of domesticity'. This ambiguity – here ascribed to Tupperware – can, of course equally be applied to the various other forms of home consumption from mail-order merchandising to TV and Internet sales discussed earlier in this chapter. Retail methods like Tupperware Parties can also be seen as 'anti-feminist capitalist sales devices' (Clarke 1997: 142), allowing the further penetration of the consumer market into the private sphere of reproduction/consumption. In this context, it is perhaps surprising that despite the very obvious role of the home as the primary territory of reproduction/consumption, the home as an explicit consumption space has been relatively ignored, with most accounts of historical and contemporary consumption (as we have seen in Chapters 10, 11 and 12) being dominated by assessments of the public sphere. Likewise, commentaries on the domestic sphere have often trivialized consumption. And yet, increasingly, the boundaries of the home are being further eroded, with new non-place routes to market taking consumption directly to the heart of the consumption landscape – the living rooms and kitchens of the middle classes. But the home is not only an important site into which the ever expanding consumer market can penetrate, but also, and very obviously, acts as a space within which household provisioning takes place and home consumption takes shape. It is to the role of the home as a specific consumption space that we now turn.

The home as a consumption space

In assessing the place of the home as a specific and unique focus of contemporary consumption it is useful to revisit the commodity chains literature which we examined in Chapter 3. In their wide-ranging critique of these literatures (see Reading 3.5), Leslie and Reimer (1999) suggest that 'relatively few commentators have extended their analysis to final consumption'. In a similar vein, Jackson and Thrift's (1995: 205) view of consumption as a social process which merely begins with a single isolated act of purchase, suggests the necessity of extending analyses of consumption 'forward into cycles of use and re-use'.

Specific studies of the final spaces of contemporary consumption began to emerge in the late 1990s with the parallel work of Miller et al. (1998) and Clarke (1998, 2000) on the dynamics of household provisioning in north London. In their ethnographic study of the residents of a single street, Miller focused on the formal shopping habits of informants (specifically their interactions with Brent Cross and Wood Green shopping centres – see Reading 12.5), whilst Clarke dealt with home-based and informal modes of acquisition such as stolen goods, Tupperware Parties, Colour-Me-Beautiful sessions, jumble and rummage sales, door-to-door sales and so on, as well as, and more specifically, the households' uses of *Loot*, a local classified paper, and *Argos*, a catalogue linked to a nationwide retail distribution outlet. This latter work demonstrates how various types of home consumption are equally entrenched in notions of identity construction and social formation as are more formal (and more documented) consumption activities such as those taking place in the department store or mall.

Within geography, and in a much more explicitly spatial framework, Robert Sack's (1992) *Place, modernity and the consumer's world* emphasizes the role of commodities in creating environments, contexts or places – 'Each time we bring a product home our home is transformed physically – something is there which was not there before – and transformed in terms of values and meaning' (Sack, 1992: 3). Sack specifically mentions the purchase of furnishings and the role of these commodities in the creation of particular environments, for example living rooms or dining rooms within the home. Drawing on the work of Sack, and linking this with the commodity chains literature (see Reading 3.5), Leslie and Reimer have studied commodity chains within the home furnishings industries of Canada and the UK – chains which end with the 'placing' of female consumers in the home. Reading 13.5 from Leslie and Reimer is concerned with the role of home consumption in the construction of identity and has several parallels with the work of Alison Clarke discussed above.

Reading 13.5 – Creating space through home consumption

Having argued for the geographical contingency of commodity chains at a national level, we turn to focus upon the creation of space through home consumption. Our argument emphasises both the creation of space by consumers and the *creation* of (the) *space* of the home. The commodity chain literature often ignores what happens when consumers place commodities in the context of their everyday lives. In part this reflects a tendency to emphasise the power that retailers and advertisers exert over consumers ... Such arguments have a distinctly gendered inflection ... the 'consumer as dupe' ... is generally seen as 'feminine' and 'other' to rational man.

In highlighting the constitutive nature of home consumption, we argue that consumers do not straightforwardly draw upon meanings prescribed by retailers and advertisers, but rather that commodity meanings are often contested and reworked by consumers. For example, the provision of room set displays which began in shops such as Habitat substitutes consumer participation for dealer expertise. Consumers are positively encouraged to imagine the spaces of the home while shopping and to construct identity through the creation of space ...

'Home making has become very closely related to identity formation' (Löfgren, 1990). The home has become an important arena of playfulness and creativity ... the project of 'home' is never complete and through constant visits to furniture stores such as Ikea, notions of 'family' and 'gender' are continuously renovated.

However, one can proceed too far in celebrating consumer autonomy, reflexivity and resistance. Producers and marketers exploit consumer interests in playing with identity in order to speed up fashion cycles in furniture and reduce its durability. Furniture retailers increasingly promote seasonal colour changes ... A consideration of the neglected space of the home reveals the complex process which transpires when goods move from the store to the home and are situated within the home ...

In focusing upon the creation of the particular space of the home, we support suggestions that geographers' accounts of the spatiality of consumption have tended to focus on a narrow range of sites, particularly the department store, shopping mall and theme park. Other spaces of conviviality such as streets, restaurants and bars are often neglected. The consumption literature has also tended to ignore places of final consumption ... Further investigation of the relationship between home consumption practices and identities is needed.

Extracted from: Deborah Leslie and Suzanne Reimer (1999): Spatializing commodity chains. *Progress in Human Geography*, 22, 401–20.

Leslie and Reimer's account of home consumption and what they term 'furnishing identity' seeks to begin to illuminate the home as a focus of final consumption. But, as Reading 13.5 indicates, Leslie and Reimer are not only concerned with the creation of the particular space of the home and its relationship to identity construction (an important and valuable task in itself), but are also keen to add weight to the arguments of others concerning the fact that consumption geographies have thus far focused on too narrow a range of consumption sites. It is our hope that this volume – which offers a distinctly geographical reading of retail and consumption spaces – has gone at least some way towards addressing this research lacuna.

Bibliography

Abelson, E.S. 1989: *When ladies go-a-thieving: middle class shoplifters in the Victorian department store*. Oxford: Oxford University Press.

ABN-AMRO, 1999: *European food retail: short cut to consolidation*. London: ABN-AMRO Equities, 12 March.

ABN-AMRO, 2001: *Carrefour – Globalization: from victor to victim*. London: ABN-AMRO Equities, 11 January.

Adelman, M.A. 1959: *A&P: a study in price–cost behaviour and public policy*. Cambridge, MA: Harvard University Press.

Adburgham, A. 1989: *Shops and shopping 1800–1914*. London: Barrie and Jenkins.

Alexander, A. 1997: Strategy and strategists: evidence from an early retail revolution in Britain. *International Review of Retail Distribution and Consumer Research*, 7, 61–78.

Alexander, A. and **Pollard, J.** 2000: Banks, grocers and the changing retailing of financial services in Britain. *Journal of Retailing and Consumer Services*, 7, 137–47.

Alexander, N. 1995: Internationalization: interpreting the motives. In McGoldrick, P.J. and Davies, G. (eds) *International retailing: trends and strategies*. London: Pitman, 77–98.

Alexander, N. 1997: *International retailing*. Oxford: Blackwell.

Appadurai, A. 1986: Introduction: commodities and the politics of value. In Appadurai, A. (ed.) *The social life of things: commodities in cultural perspective*. Cambridge: Cambridge University Press, 1–51.

Arce, A. and **Marsden, T.K.** 1993: The social construction of international food: a new research agenda. *Economic Geography*, 69, 293–311.

Baker, R.G.V. 1994: The impact of trading hour deregulation on the retail sector and the Australia community. *Urban Policy and Research*, 12, 104–17.

Baker, R.G.V. 1995: What hours will we trade Mr Superstore? A review of the 1994 Australian experience. *Urban Policy and Research*,13, 28–36.

Balto, D.A. 2001: Supermarket merger enforcement. *Journal of Public Policy and Marketing*, 20, 38–50.

Baret, C., Lehndorff, S. and **Sparks, L.** 2000: *Flexible working in food retailing*. London: Routledge.

Barnes, T. 1995: Political economy 1: 'the culture' stupid. *Progress in Human Geography*, 19, 423–31.

Barrett, H.R., Ilbery, B.W., Browne, A.W. and **Binns, T.** 1999: Globalization and the changing networks of food supply: the importation of fresh horticultural produce from Kenya into the UK. *Transactions of the Institute of British Geographers*, NS24, 159–74.

Beckett, A. 1994: Take a walk on the safeside. *Independent on Sunday*, 27 February.

Benson, J. and **Shaw, G.** 1992: *The evolution of retail systems c1880–1914*. Leicester: Leicester University Press.

Benson, S.P. 1986: *Counter cultures: saleswomen, managers and customers in American department stores 1890–1940*. Urbana and Chicago: University of Illinois Press.

Berle, A. and **Means, G.** 1932: *The modern corporation and private property*. New York: Macmillan.

Berman, B. and **Evans, J.R.** 1998: *Retail management: a strategic approach*, seventh edition. Upper Saddle River, NJ: Prentice-Hall.

Berry, B.J.L. 1963: *Commercial structure and commercial blight*. University of Chicago, Department of Geography, Research Paper 85.

Berry, B.J.L. 1967: *Geography of market centers and retail distribution*. Englewood Cliffs, NJ: Prentice Hall.

Berry, B.J.L. and **Tenant, R.J.** 1963: *Chicago commercial reference handbook*. University of Chicago, Department of Geography, Research Paper 86.

Berry, B.J.L., Cutler, I., Draine, E.H., Kiang, Y.-C., Tocalis, T.R. and **de Vise, P.** 1976: *Chicago: transformations of an urban system*. Cambridge, MA: Ballinger.

Best, M.H. 1990: *The new competition: institutions of industrial restructuring*. Oxford: Blackwell.

Birkin, M., Clarke, G.P., Clarke, M. and **Wilson, A.G.** 1996: *Intelligent GIS*. Cambridge: GeoInformation International.

Blair, M. 1995: *Ownership and control: rethinking corporate governance for the 21st century*. Washington DC: Brookings Institute.

Blomley, N.K. 1985: The Shops Act (1950): the politics and the policing. *Area*, 17, 25–33.

Blomley, N.K. 1986a: Retail law at the urban and national levels: geographical aspects of the operation and possible amendment of the Shops Act (1950). Unpublished Ph.D. thesis, Department of Geography, University of Bristol.

Blomley, N.K. 1986b: Regulatory legislation and the legitimation crisis of the state: the enforcement of the Shops Act (1950). *Environment and Planning D: Society and Space*, 4, 183–200.

Blomley, N.K. 1994: Retailing, geography of. In Johnston, R.J., Gregory, D. and Smith, D.M. (eds) *The Dictionary of Human Geography*. Oxford: Blackwell, 533–5.

Blomley, N.K. 1996: 'I'd like to dress her all over': masculinity, power and retail capital. In Wrigley, N. and Lowe, M.S. (eds) *Retailing, consumption and capital: towards the new retail geography*. Harlow: Addison Wesley Longman, 238–56.

Bowlby, R. 1997: Supermarket futures. In Falk, P. and Campbell, C. (eds) *The shopping experience*. London: Sage, 92–110.

Bowlby, S.R. and **Foord, J.** 1995: Relational contracting between UK retailers and manufacturers. *International Review of Retail, Distribution and Consumer Research*, 5, 333–61.

Brooks, D. 2000: Exiles on mainstreet. *New York Times*, 9 April 2000, Section 6, 64–8.

Burt, S. 1992: Retail brands in British grocery retailing: a review. WP-9204, Institute for Retail Studies, University of Stirling.

Burt, S. and **Sparks, L.** 1994: Structural change in grocery retailing in Great Britain: a discount reorientation? *International Review of Retail, Distribution and Consumer Research*, 4, 195–217.

Burt, S. and **Sparks, L.** 2001: The implications of Wal-Mart's takeover of Asda. *Environment and Planning A*, 33, 1463–87.

Caves, R. and **Porter, M.** 1976: Barriers to exit. In Masson, R. and Qualls, D. (eds) *Essays on industrial organization in honor of Joe S. Bain*. Cambridge, MA: Ballinger, 36–69.

Caves, R. and **Porter, M.** 1977: From entry barriers to mobility barriers: conjectural decisions and contrived deterrence to new competition. *Quarterly Journal of Economics*, 91, 247–61.

Chaney, D. 1990: Subtopia in Gateshead: The Metrocentre as a cultural form. *Theory, Culture and Society*, 7, 49–68.

de Chernatony, L. 1989: Branding in an era of retailer dominance. *International Journal of Advertising*, 8, 245–60.

Chesterton 1994: *Airport retailing: the growth of a new high street*. Chesterton International Property Consultants, London.

Chevalier, J.A. 1995a: Capital structure and product – market competition: empirical evidence from the supermarket industry. *American Economic Review*, 85, 415–35.

Chevalier, J.A. 1995b: Do LBO supermarkets charge more? An empirical analysis of the effects of LBOs on supermarket pricing. *Journal of Finance*, 50, 1095–112.

Christopher, M. 1986: *The strategy of distribution management*. London: Heinemann.

Christopherson, S. 1989: Flexibility in the US service economy and emerging spatial divisions of labour. *Transactions of the Institute of British Geographers*, NS14, 131–43.

Christopherson, S. 1993: Market rules and territorial outcomes: the case of the United States. *International Journal of Urban and Regional Research*, 17, 274–88.

Christopherson, S. 1994: The fortress city: privatized spaces, consumer citizenship. In Amin, A. (ed.) *Post Fordism: a reader*. Oxford: Blackwell, 409–27.

Christopherson, S. 1996: The production of consumption: retail restructuring and labour demand in the USA. In Wrigley, N. and Lowe, M.S. (eds) *Retailing, consumption and capital: towards the new retail geography*. Harlow: Addison Wesley Longman, 159–77.

City of Chicago Planning Department 1987: *Englewood concourse revitalization strategy*. Chicago.

Clark, G.L. 1989: Remaking the map of corporate capitalism: the arbitrage economy of the 1990s. *Environment and Planning A*, 21, 997–1000.

Clark, G.L. 1992: Real regulation reconsidered. *Environment and Planning A*, 24, 615–27.

Clark, G.L. 1998: Stylized facts and close dialogue: methodology in economic geography. *Annals of the Association of American Geographers*, 88, 73–87.

Clark, G.L. and **Wrigley, N.** 1995: Sunk costs: a framework for economic geography. *Transactions of the Institute of British Geographers*, NS20, 204–23.

Clark, G.L. and **Wrigley, N.** 1997a: The spatial configuration of the firm and the management of sunk costs. *Economic Geography*, 73, 285–304.

Clark, G.L. and **Wrigley, N.** 1997b: Exit, the firm, and sunk costs: reconceptualizing the corporate geography of disinvestment and plant closure. *Progress in Human Geography*, 21, 338–58.

Clarke, A.J. 1997: Tupperware: suburbia, sociality and mass consumption. In Silverstone, R. (ed.) *Visions of suburbia*. London: Routledge, 132–60.

Clarke, A.J. 1998: Window shopping at home: classified, catalogues and new consumer skills. In Miller, D. (ed.) *Material cultures: why some things matter*. London: UCL Press, 73–99.

Clarke, A.J. 2000: Mother swapping: the trafficking of nearly new children's wear. In Jackson, P., Lowe, M.S., Miller, D. and Mort, F. (eds) *Commercial Cultures*. Oxford: Berg, 85–100.

Clarke, D.B. 1996: The limits to retail capital. In Wrigley, N. and Lowe, M.S. (eds) *Retailing, consumption and capital: towards the new retail geography*. Harlow: Addison Wesley Longman, 284–301.

Clarke, I. 2000: Retail power, competition and local consumer choice in the UK grocery sector. *European Journal of Marketing*, 34, 975–1002.

Clarke, I. and **Rimmer, P.** 1997: The anatomy of retail internationalisation: Daimaru's decision to invest in Melbourne, Australia. *Service Industries Journal*, 17, 361–82.

Cockburn, C. 1977: *The local state*. London: Pluto Press.

Comeau, E. 1995: *Food and drug retailers industry viewpoint*. Donaldson, Lufkin and Jenrette Securities Corp, New York, 9 October.

Competition Commission, 2000: *Supermarkets: a report on the supply of groceries from multiple stores in the United Kingdom*, 3 volumes. Cm 4842. London: The Stationery Office.

Connor, J.M. and **Schiek, W.A.** 1997: *Food processing: an industrial powerhouse in transition*, second edition. New York: John Wiley.

Cook, I.J. 1993: Constructing the exotic: the case of tropical fruit. Paper presented at the Annual Conference of the Institute of British Geographers, Royal Holloway and Bedford New College, University of London, 5–8 January.

Cook, I.J. 1994: New fruits and vanity: symbolic production in the global food economy. In Bonanno, A., Busch, L., Friedland, W.H., Gouveig, L. and Mingiovie, E. (eds) *From Columbus to ConAgra: the globalisation of agriculture and food*. Lawrence, KS: University Press of Kansas, 232–48.

Cook, I.J. and **Crang, P.** 1996: The world on a plate: culinary culture, displacement and geographical knowledges. *Journal of Material Culture* 1, 131–53.

Cook, I.J., Crang, P. and **Thorpe, M.** 1998: Biographies and geographies: consumer understandings of the origins of foods. *British Food Journal*, 100, 162–7.

Cooper, J., Browne, M. and **Peters, M.** 1991: *European logistics*. Oxford: Blackwell.

Cotter, J. and **Hutchinson, R.W.** 1998: The use and abuse of accounting information in the UK retailing sector. Working paper, School of Commerce, University of Ulster, Coleraine, Northern Ireland.

Cotterill, R.W. 1993: Food retailing: mergers, leveraged buyouts and performance. In Deutsch, L. (ed.) *Industry studies*. Englewood Cliffs: Prentice Hall, 157–81.

Cotterill, R.W. 1997: The food distribution system of the future: convergence towards the US or UK model. *Agribusiness* 13, 121–35.

Cotterill, R.W. 1999: An antitrust economic analysis of the proposed acquisition of Supermarkets General Holding Corporation by Ahold Acquisition Inc. Food Marketing Policy Centre Research Report 46, Dept. of Agriculture and Resource Economics, University of Connecticut, USA.

Crang, P. 1994: It's showtime: on the workplace geographies of display in a restaurant in south-east England. *Environment and Planning D: Society and Space*, 12, 675–704.

Crang, P. 1996: Displacement, consumption and identity. *Environment and Planning A*, 28, 47–67.

Crang, P. 1997: Cultural turns and the (re)constitution of economic geography. In Lee, R. and Wills, J. (eds) *Geographies of economies*. London: Arnold, 3–15.

Crang, P. and **Jackson, P.** 1998: Consuming geographies: localizing the global and globalizing the local. In Robins, K. and Morley, D. (eds) *British cultural studies*. Oxford: Oxford University Press.

Crawford, M. 1992: The world in a shopping mall. In Sorkin, M. (ed.) *Variations on a theme park: the new American city and the end of public space*. New York: Noonday Press, 3–30.

Credit Suisse First Boston 1999: *UK food retail sector: struggling for growth in a low inflation world*. Credit Suisse First Boston (Europe) Ltd, London, 26 February.

Crewe, L. 2000: Geographies of retailing and consumption. *Progress in Human Geography*, 24, 275–90.

Crewe, L. and **Davenport, E.** 1992: The puppet show: changing buyer–supplier relationships within clothing retailing. *Transactions of the Institute of British Geographers*, NS17, 183–97.

Crewe, L. and **Gregson, N.** 1998: Tales of the unexpected: exploring car boot sales as marginal spaces of contemporary consumption. *Transactions of the Institute of British Geographers*, NS 23, 39–53.

Crewe, L. and **Lowe, M.** 1995: Gap on the map? Towards a geography of consumption and identity. *Environment and Planning A*, 27, 1877–98.

Davies, K., Gilligan, C. and **Sutton, C.** 1985: Structural changes in grocery retailing: the implications for concentration. *International Journal of Physical Distribution and Materials Management*, 15, 3–48.

Dawson, J.A. 1993: The internationalization of retailing. In Bromley, R.D.F. and Thomas, C.J. (eds) *Retail change: contemporary issues*. London: UCL Press, 15–40.

Dawson, J.A. and **Shaw, S.A.** 1990: The changing character of retailer–supplier relationships. In Fernie, J. (ed.) *Retail distribution management*. London: Kogan Page.

Denis, D.J. 1994: Organizational form and the consequences of highly leveraged transactions: Kroger's recapitalization and Safeway's LBO. *Journal of Financial Economics*, 36, 193–224.

Deutsche Bank, 2000a: *Delhaize – Le Lion*. Deutsche Bank AG European Equity Research, London, 21 June.

Deutsche Bank, 2000b: *J. Sainsbury – quantifying the restructuring programme*. Deutsche Bank AG European Equity Research, London, 16 November.

Doel, C. 1995: Market development, organizational change and the food industry. Unpublished Ph.D. thesis, University of Cambridge.

Doel, C. 1996: Market development and organizational change: the case of the food industry. In Wrigley, N. and Lowe, M.S. (eds) *Retailing, consumption and capital: towards the new retail geography*. Harlow: Addison Wesley Longman, 48–67.

Doel, C. 1999: Towards a supply chain community? Insights from governance processes in the food industry. *Environment and Planning A*, 31, 69–85.

Doeringer, P. and **Piore, M.** 1971: *Internal labour markets and manpower analysis*. Lexington, MA: D.C. Heath.

Domosh, M. 1996a: The feminized retail landscape: gender ideology, and consumer culture in nineteenth century New York City. In Wrigley, N. and Lowe, M.S. (eds) *Retailing, consumption and capital: towards the new retail geography*. Harlow: Addison Wesley Longman, 257–70.

Domosh, M. 1996b: *Imagined cities*. New Haven, CT: Yale University Press.

Domosh, M. 1998: Gender, consumer culture and the moral geography of 19th century New York City. Paper presented at 94th Annual Meeting of Association of American Geographers, Boston, 26 March.

Dowling, R. 1991: Shopping and the construction of femininity in the Woodwards department store, Vancouver, 1945–1960. Unpublished MA thesis, Department of Geography, University of British Columbia.

Dowling, R. 1993: Femininity, place and commodities: a retail case study. *Antipode* 25, 295–319.

Ducatel, K. and **Blomley, N.K.** 1990: Rethinking retail capital. *International Journal of Urban and Regional Research*, 14, 207–27.

du Gay, P. 1996: *Consumption and identity at work.* London: Sage.

Eaton, B.C. and **Lipsey, R.G.** 1980: Exit barriers are entry barriers: the durability of capital as a barrier to entry. *Bell Journal of Economics*, 11, 721–9.

Eaton, B.C. and **Lipsey, R.G.** 1981: Capital, commitment and entry equilibrium. *Bell Journal of Economics*, 12, 593–604.

Fama, E. 1980, Agency problems and the theory of the firm. *Journal of Political Economy*, 88, 288–307.

Fama, E. and **Jensen, M.C.** 1983a: Separation of ownership and control. *Journal of Law and Economics*, 26, 301–25.

Fama, E. and **Jensen, M.C.** 1983b: Agency problems and residual claims. *Journal of Law and Economics*, 26, 327–49.

Featherstone, M. 1998: The flâneur, the city and virtual public life. *Urban Studies*, 35, 909–25.

Fernie, J. 1992: Distribution strategies of European retailers. *European Journal of Marketing*, 26 (8/9), 35–47.

Fernie, J. 1994: Quick response: an international perspective. *International Journal of Physical Distribution and Logistics Management*, 24(6), 38–46.

Fernie, J. 1995: International comparisons of supply chain management in grocery retailing. *Service Industries Journal*, 15, 134–47.

Fernie, J. 1997: Retail change and retail logistics in the United Kingdom: past trends and future prospects. *Service Industries Journal*, 17, 383–96.

Fernie, J. 1998: The breaking of the fourth wave: recent out-of-town retail developments in Britain. *International Journal of Retail, Distribution and Consumer Research*, 8, 303–17.

Fernie, J. and **Arnold, S.** 2002: Wal-Mart in Europe: prospects for Germany, the UK and France. *International Journal of Retail and Distribution Management*, 30.

Fernie, J., Moore, C.M., Lawrie, A. and **Hallsworth, A.** 1997: The internationalization of the high fashion brand: the case of central London. *Journal of Product and Brand Management* 6, 151–62.

Fernie, S. 1995: The coming of the fourth wave. *International Journal of Retail and Distribution Management*, 23, 3–10.

Fernie, S. 1996: The future for factory outlet centres in the UK: the impact of changes in planning policy guidance on the growth of a new retail format. *International Journal of Retail and Distribution Management*, 24, 11–21.

Ferry, J. 1960: *A history of the department store*. New York: Macmillan.

Fine, B. and **Leopold, E.** 1993: *The world of consumption*. London: Routledge.

Florida, R. and **Kenney, M.** 1991: Transplanted organizations: the transfer of Japanese industrial organization to the US. *American Sociological Review*, 56, 381–98.

Flynn, A. and **Marsden, T.K.** 1992: Food regulation in a period of agricultural retreat: the British experience. *Geoforum*, 23, 85–93.

Flynn, A., Harrison, M. and **Marsden, T.K.** 1998: Regulation, rights and the structuring of food choices. In Murcott, A. (ed.) *The nation's diet: the social science of food choice*. Harlow: Addison Wesley Longman, 152–67.

Foord, J., Bowlby, R.R. and **Tillsley, C.** 1996: The changing place of retailer–supplier relations in British retailing. In Wrigley, N. and Lowe, M.S. (eds) *Retailing, consumption and capital: towards the new retail geography*. Harlow: Addison Wesley Longman, 68–89.

Freathy, P. and **O'Connell, F.** 1998: *European airport retailing: growth strategies for the new millennium*. Basingstoke: Macmillan.

Freathy, P. and **Sparks, L.** 1996: Understanding retail employment relations. In Wrigley, N. and Lowe, M.S. (eds) *Retailing, consumption and capital: towards the new retail geography*. Harlow: Addison Wesley Longman, 178–95.

Freathy, P. and **Sparks, L.** 2000: The organization of working time in large UK food retail firms. In Baret, C., Lehndorff, S. and Sparks, L. (eds) *Flexible working in food retailing*. London: Routledge, 83–113.

Frieden, B.J. and **Sagalyn, L.B.** 1989: *Downtown Inc.: how America rebuilds cities*. Cambridge, MA: MIT Press.

Gadrey, J. and **Lehndorff, S.** 2000: A societal interpretation of the differences and similarities in working time practices. In Baret, C., Lehndorff, S. and Sparks, L. (eds) *Flexible working in food retailing*. London: Routledge, 143–69.

Galbraith, J.K. 1980: *American capitalism: the concept of countervailing power*, revised edition. Oxford: Blackwell.

Gap Inc. 1991: *Annual report*. San Francisco: Gap Inc.

Gardner, C. and **Sheppard, J.** 1989: *Consuming passion: the rise of retail culture*. London: Unwin Hyman.

Garreau, J. 1991: *Edge City: life on the new frontier*. New York: Anchor Books.

Gertler, M.S. 1988: The limits to flexibility: comments on the post-Fordist vision of production and its geography. *Transactions of the Institute of British Geographers*, NS 13, 419–32.

Gertler, M.S. 1995: 'Being there': proximity, organization and culture in the development and adoption of advanced manufacturing technologies. *Economic Geography*, 71, 1–26.

Gertler, M.S. 1997: The invention of regional culture. In Lee, R. and Wills, J. (eds) *Geographies of economies*. London: Arnold, 47–58.

Glennie, P.D. 1998: Consumption, consumerism and urban form: historical perspectives. *Urban Studies*, 35, 927–51.

Glennie, P.D. and **Thrift, N.J.** 1992: Modernity, urbanism and modern consumption. *Environment and Planning D: Society and Space*, 10, 423–43.

Glennie, P.D. and **Thrift, N.J.** 1996a: Consumers, identities, and consumption spaces in early-modern England. *Environment and Planning A*, 28, 25–45.

Glennie, P.D. and **Thrift, N.J.** 1996b: Consumption, shopping and gender. In Wrigley, N. and Lowe, M.S. (eds) *Retailing, consumption and capital: towards the new retail geography*. Harlow: Addison Wesley Longman, 221–37.

Goldberger, P. 1997: The store strikes back. *New York Times Magazine*, 6 April, 45–9.

Goss, J. 1992: Modernity and postmodernity in the retail landscape. In Anderson, K. and Gale, F. (eds) *Inventing places: studies in cultural geography*. Melbourne: Longman Cheshire, 158–77.

Goss, J. 1993: The 'magic of the mall': an analysis of form, function and meaning in the retail built environment. *Annals of the Association of American Geographers*, 83, 18–47.

Goss, J. 1996: Disquiet on the waterfront: reflections on nostalgia and utopia in the urban archetypes of festival marketplaces. *Urban Geography*, 17, 221–47.

Goss, J. 1999: Once-upon-a-time in the commodity world: an unofficial guide to the Mall of America. *Annals of the Association of American Geographers*, 89, 45–75.

Gottdiener, M. 1998: The semiotics of consumer spaces: the growing importance of themed environments. In Sherry, J.F. (ed.) *Servicescapes: the concept of space in contemporary markets*. Lincolnwood, IL: NTC Business Books.

Graff, T.O. and **Ashton, D.** 1994: Spatial diffusion of Wal-Mart. *Professional Geographer*, 46, 19–29.

Granovetter, M. 1985: Economic action and social structure: the problem of 'embeddedness'. *American Journal of Sociology*, 91, 491–510.

Grant, R.M. 1987: Manufacturer–retailer relations: the shifting balance of power. In Johnson, G. (ed.) *Business strategy and retailing*. London: Wiley, 43–58.

Gregson, N. and **Crewe, L.** 1997: The bargain, the knowledge, and the spectacle: making sense of consumption in the space of the car boot sale. *Environment and Planning D: Society and Space*, 15, 87–112.

Gregson, N. and **Crewe, L.** 1998: Dusting down 'Second Hand Rose': gendered identities and the world of second-hand goods in the space of the car boot sale. *Gender, Place and Culture* 5, 77–100.

Gregson, N. and **Crewe, L.** 2001: Discourse, displacement and the production of (charity) retail space. *Environment and Planning A*, 33.

Gregson, N., Crewe, L. and **Longstaff, B.** 1997: Excluded spaces of regulation: car boot sales as an enterprise culture out of control. *Environment and Planning A*, 29, 1717–37.

Gregson, N., Brooks, K. and **Crewe, L.** 2000: Narratives of consumption and the body in the space of charity/shops. In Jackson, P., Lowe, M.S., Miller, D. and Mort F. (eds) *Commercial cultures: economies, practices, spaces.* Oxford: Berg, 101–21.

Gruen, V. and **Smith, L.** 1960: *Shopping towns USA: the planning of shopping centres.* New York: Reinhold Publishing Corporation.

Guy, C.M. 1994a: *The retail development process.* London: Routledge.

Guy, C.M. 1994b: Whatever happened to regional shopping centres? *Geography*, 79, 293–312.

Guy, C.M. 1996: Corporate strategies in food retailing and their local impacts: a case study of Cardiff. *Environment and Planning A*, 28, 1575–602.

Guy, C.M. 1997: Fixed assets or sunk costs? An examination of retailers' land and property investment in the UK. *Environment and Planning A*, 29, 1449–64.

Guy, C.M. 1998a: Exit strategies and sunk costs: the implications for multiple retailers. Paper presented at 5th International Conference on Retailing and Consumer Services, Baveno, Italy. Subsequently published in *International Journal of Retail and Distribution Management*, 27, 237–45.

Guy, C.M. 1998b: High Street retailing in off-centre retail parks: a review of the effectiveness of land use planning policies. *Town Planning Review*, 69, 291–313.

Guy, C.M. 1998c: Alternative-use valuation, open A1 planning consent, and the development of retail parks. *Environment and Planning A*, 30, 37–47.

Guy, C.M. 1998d: Off-centre retailing in the UK: prospects for the future and the implications for town centres. *Built Environment*, 24, 16–30.

Guy, C.M. 1998e: Controlling new retail spaces: the impress of planning policies in western Europe. *Urban Studies*, 35, 953–79.

Guy, C.M. and **Lord, J.D.** 1993: Transformation and the city centre. In Bromley, R.D.F. and Thomas, C.J. (eds) *Retail change: contemporary issues.* London: UCL Press, 88–108.

Hallsworth, A.G. 1991: The Campeau takeovers – the arbitrage economy in action. *Environment and Planning A*, 23, 1217–23.

Hallsworth, A.G. and **McClatchey, J.** 1994: Interpreting the growth of superstore retailing in Britain. *International Review of Retail, Distribution and Consumer Research*, 4, 317–28.

Hancher, L. and **Moran, M.** 1989: Organising regulatory space. In Hancher, L. and Moran, M. (eds) *Capitalism, culture and economic regulation.* Oxford: Clarendon Press, 271–99.

Handelman, D. 1997: The billboards of Madison Avenue. *New York Times Magazine*, 6 April, 50–103.

Hankins, K. 2002: The restructuring of retail capital and the street. *Tijdschrift voor Economische en Sociale Geographie*, 93(1).

Hannigan, J. 1998: *Fantasy city: pleasure and profit in the postmodern metropolis.* London and New York: Routledge.

Harrison, M., Flynn, A. and **Marsden, T.K.** 1997: Contested regulatory practice and the implementation of food policy: exploring the local and national interface. *Transactions of the Institute of British Geographers*, NS22, 473–87.

Harvey, D. 1985: *The urbanization of capital*. Baltimore: Johns Hopkins University Press.

Harvey, D. 1989a: *The condition of postmodernity*. Oxford: Blackwell.

Harvey, D. 1989b: From managerialism to entrepreneurialism: the transformation in urban governance in late capitalism. *Geografiska Annaler*, 71B, 3–17.

Harvey, D. 1990: Between space and time: reflections on the geographical imagination. *Annals of the Association of American Geographers*, 80, 418–34.

Henderson Crosthwaite, 1992: Three plus one. Food retail report No. 10. London: Henderson Crosthwaite Institutional Brokers Ltd.

Hillier Parker, 1996: *Fashion designers in central London*. Hillier Parker, 77 Grosvenor Street, London W1A 2BT.

Hochschild, A.R. 1983: *The managed heart: commercialization of human feeling*. Berkeley, CA: University of California Press.

Hodgson, G. 1988: *Economics and institutions: a manifesto for a modern institutional economics*. Cambridge: Polity Press.

Honeycombe, G. 1984: *Selfridges: seventy five years. The story of the store*. London: Park Lane Press.

Hopkins, J. 1990: West Edmonton Mall: landscape of myths and elsewhereness. *Canadian Geographer*, 34, 2–17.

Hughes, A.L. 1996a: Retail restructuring and the strategic significance of food retailers' own-labels: a UK–USA comparison. *Environment and Planning A*, 28, 2201–26.

Hughes, A.L. 1996b: Changing food retailer–manufacturer power relations within national economies: a UK–USA comparison. Unpublished Ph.D. thesis, Department of Geography, University of Southampton.

Hughes, A.L. 1999: Constructing competitive spaces: the corporate practice of British retailer–supplier relationships. *Environment and Planning A*, 31, 819–39.

Hughes, A.L. 2000: Retailers, knowledges and changing commodity networks: the case of the cut flower trade, *Geoforum*, 31, 175–90.

Hughes, A.L. 2001: Multi-stakeholder approaches to ethical trade: towards a reorganization of UK retailers' global supply chains. *Journal of Economic Geography*, 1, 421–37.

Hutchinson, R.W. and **Hunter, R.L.** 1995: Determinants of capital structure in the retailing sector in the UK. *International Review of Retail, Distribution and Consumer Research*, 5, 63–78.

J.P. Morgan Securities 1997: *Food retail industry analysis*. J.P. Morgan Securities Inc., New York, 27 May.

J.P. Morgan Securities 1998: *Industry analysis – department stores*. J.P. Morgan Securities Inc., New York, 18 June.

Jackson, P. 1991: The cultural politics of masculinity: towards a social geography. *Transactions of the Institute of British Geographers*, NS16, 199–213.

Jackson, P. 1994: Black male: advertising and the cultural politics of masculinity. *Gender, Place and Culture*, 1, 49–59.

Jackson, P. and **Taylor, J.** 1996: Geography and the cultural politics of advertising. *Progress in Human Geography*, 20, 356–71.

Jackson, P. and **Thrift, N.J.** 1995: Geographies of consumption. In Miller, D. (ed.) *Acknowledging consumption*. London: Routledge, 204–37.

Jackson, P., Lowe, M.S., Miller, D. and **Mort, F.** 2000: *Commercial cultures: economies, practices, spaces*. Oxford: Berg.

Jacobs, J. 1961: *The death and life of great American cities*. New York: Random House.

Jensen, M.C. 1986: Agency costs of free cash flow, corporate finance and takeovers. *American Economic Review*, 76, 323–9.

Jensen, M.C. 1993: The modern industrial revolution, exit and the failure of internal control systems. *Journal of Finance*, 48, 831–80.

Jensen, M.C. and **Meckling, W.** 1976: Theory of the firm: managerial behaviour, agency costs and ownership structure. *Journal of Financial Economics*, 3, 305–60.

Jerde, J. 1998: Capturing the leisure zeitgeist: creating places to be. *Architectural Design*, Profile No. 131, 69–71.

Johnson, D.B. 1991: Structural features of West Edmonton Mall. *Canadian Geographer*, 35, 249–61.

Johnston, R.J., Gregory, D. and **Smith, D.M.** (eds) 1994: *The dictionary of human geography*, third edition. Oxford: Blackwell.

Jones, K. and **Biasiotto, M.** 1999: Internet retailing: current hype or future reality? *International Review of Retail, Distribution and Consumer Research*, 9, 69–79.

Jones, K. and **Simmons, J.** 1990: *The retail environment*. London: Routledge.

Kahn, A. 1992: Filling every Gap. *New York Times*, 23 August.

Karmel, R.S. 1993: Implications of the stakeholder model. *George Washington Law Review*, 61, 1156–76.

Katz, D. 1994: *Just do it: the Nike spirit in the corporate world*. New York, Random House.

King, C.L. 1913: Can the cost of distributing food products be reduced? *Annals of the American Academy of Political and Social Science*, 48, 199–224.

Kowinski, W.S. 1985: *The malling of America: an inside look at the great consumer paradise*. New York: Morrow.

Kurt Salmon 1993: *Efficient consumer response: enhancing consumer value in the supply chain*. Washington DC: Kurt Salmon.

Laaksonen, H. 1994: *Own brands in food retailing across Europe*. Oxford Institute of Retail Management, Templeton College, University of Oxford.

Lancaster, B. 1995: *The department store: a social history*. London: Leicester University Press.

Langman, L. 1992: Neon cages: shopping for subjectivity. In Shields, R. (ed.) *Lifestyle shopping: the subject of consumption*. London: Routledge, 40–82.

Langston, P., Clarke, G.P. and **Clarke, D.B.** 1997: Retail saturation, retail location, and retail competition: an analysis of British grocery retailing. *Environment and Planning A*, 29, 77–104.

Langston, P., Clarke, G.P. and **Clarke, D.B.** 1998: Retail saturation: the debate in the mid-1990s. *Environment and Planning A*, 30, 49–66.

Laulajainen, R. 1987: *Spatial strategies in retailing*. Dordrecht: Reidel.

Laulajainen, R. 1988: The spatial dimension of an acquisition. *Economic Geography*, 64, 171–87.

Laulajainen, R. 1990: Defense by expansion. *Professional Geographer*, 42, 277–88.

Lavin, M. 2000: Problems and opportunities of retailing in the US inner city. *Journal of Retailing and Consumer Studies*, 7, 47–57.

Leach, W. 1984: Transformations in a culture of consumption: women and department stores, 1890–1925. *Journal of American History*, 71, 319–42.

Leach, W. 1993: *Land of desire: merchants, power and the rise of a new American culture*. New York: Pantheon Books.

Leahy, T. 1987: Branding – the retailer's viewpoint. In Murphy, J.M. (ed.) *Branding: a key marketing tool*. London: Macmillan, 116–24.

Lee, R. and **Wills, J.** (eds) 1997: *Geographies of economies*. London: Arnold.

Leidner, R. 1993: *Fast food, fast talk: service work and the routinization of everyday life*. Berkeley, CA: University of California Press.

Leslie, D. and **Reimer, S.** 1999: Spatialising commodity chains. *Progress in Human Geography*, 22, 401–20.

Levy, M. and **Weitz, B.A.** 1998: *Retailing management,* third edition. Boston, MA: Irwin McGraw-Hill.

Longley, P.A. and **Clarke, G.P.** (eds) 1995: *GIS for business and service planning*. Cambridge: GeoInformation International.

Longstreth, R. 1999: *The drive-in, the supermarket, and the transformation of commercial space in Los Angeles, 1914–1941*. Cambridge, MA: MIT Press.

Lowe, M.S. 1993: Local hero! An examination of the role of the regional entrepreneur in the regeneration of Britain's regions. In Kearns, G. and Philo, C. (eds) *Selling places: the city as cultural capital, past and present*. Oxford: Pergamon Press, 211–30.

Lowe, M.S. 1998: The Merry Hill regional shopping centre controversy: PPG6 and the new urban geographies. *Built Environment*, 24, 57–69.

Lowe, M.S. 2000a: From Victor Gruen to Merry Hill: Reflections on regional shopping centres and urban development in the US and UK. In Jackson, P., Lowe, M.S., Miller, D. and Mort, F. (eds) *Commercial cultures: economies, practices, spaces*. Oxford: Berg, 245–59.

Lowe, M.S. 2000b: Britain's regional shopping centres: new urban forms? *Urban Studies*, 37, 261–74.

Lowe, M.S. and **Crewe, L.** 1996: Shopwork: image, customer care and the restructuring of retail employment. In Wrigley, N. and Lowe, M.S. (eds) *Retailing, consumption and capital: towards the new retail geography*. Harlow: Addison Wesley Longman, 196–207.

Lowe, M.S. and **Wrigley, N.** 1996: Towards the new retail geography. In Wrigley, N. and Lowe, M. (eds) *Retailing, consumption and capital: towards the new retail geography*. Harlow: Addison Wesley Longman, 3–30.

Magowen, P.A. 1989: The case for LBOs: the Safeway experience. *California Management Review*, 32, 9–18.

Marden, P. 1992: Real regulation reconsidered. *Environment and Planning A*, 24, 751–67.

Marion, B.W. 1995: Changing power relationships in the US food industry: brokerage arrangements for private label products. Paper presented at conference on Food Retailer–Manufacturer Competitive Relationships in the EU and USA, University of Reading, 17–19 July.

Marsden, T.K. and **Wrigley, N.** 1995: Regulation, retailing and consumption. *Environment and Planning A*, 27, 1899–912.

Marsden, T.K. and **Wrigley, N.** 1996: Retailing, the food system and the regulatory state. In Wrigley, N. and Lowe, M.S. (eds) *Retailing, consumption and capital: towards the new retail geography*. Harlow: Addison Wesley Longman, 33–47.

Marsden, T.K., Flynn, A. and **Harrison, M.** 1997: Retailing, regulation and food consumption: the public interest in a privatized world? *Agribusiness*, 13, 211–26.

Marsden, T.K., Harrison, M. and **Flynn, A.** 1998: Creating competitive space: exploring the social and political maintenance of retail power. *Environment and Planning A*, 30, 481–98.

Marshall, N. and **Wood, P.A.** 1995: *Services and space: Key aspects of urban and regional development*. Harlow: Longman.

Marston, S. and **Modarres, A.** 2002: Flexible retailing: Gap Inc. and the new spaces of retail in the United States. *Tijdschrift voor Economische en Sociale Geographie*, 93(1).

Mata, J. 1991: Sunk costs and entry by small and large plants. In Geroski, P.A. and Schwalbach, J. (eds) *Entry and market contestability*. Oxford: Blackwell, 49–62.

Mathias, P. 1967: *Retailing revolution: a history of multiple retailing in the food trades based upon the Allied Suppliers Group of companies*. London: Longman.

McCaffery, K., Hutchinson, R.W. and **Jackson, R.** 1997: Aspects of the finance function: a review and survey into the UK retailing sector. *International Review of Retail, Distribution and Consumer Research*, 7, 125–44.

McClelland, W.G. 1962: The supermarket and society. *Sociological Review*, 10 133–44.

McGoldrick, P. 1991: *Retail marketing*. Maidenhead: McGraw-Hill.

McGoldrick, P. and **Davies, G.** 1995: *International retailing: trends and strategies*. London: Pitman.

McKinnon, A.C. 1989: *Physical distribution systems*. London: Routledge.

McMillan, J. 1990: Managing suppliers; incentive systems in Japanese and US industry. *California Management Review*, 38–55.

Merrill Lynch 1997a: *Food and drug merchandising: investment snapshots*. New York: Merrill Lynch, 5 June.

Merrill Lynch, 1997b: *J.C. Penney & Co*. New York: Merrill Lynch, 21 August.

Merrill Lynch, 1998a: *Supercenter forecaster*. New York: Merrill Lynch, 4 May.

Merrill Lynch, 1998b: *'Retailing Leaders' Conference management presentations*. New York: Merrill Lynch, 18 May.

Merrill Lynch, 1999a: *e-commerce: virtually here*. New York: Merrill Lynch, April.

Merrill Lynch, 1999b: *Leaders emerging: trends in retail consolidation*. New York: Merrill Lynch, 29 April.

Merrill Lynch, 2000: *Wal-Mart Stores*. New York: Merrill Lynch, 18 May.

Merrill Lynch, 2001: *Retailing industry: the globalization of retailing*. New York: Merrill Lynch, 11 June.

Messinger, P.R. and **Narasimham, C.** 1995: Has power shifted in the grocery channel? *Marketing Science*, 14, 189–223.

Miller, D. (ed.) 1995: *Acknowledging consumption: a review of new studies*. London: Routledge.

Miller, D. 1998a: Coca-Cola: a black sweet drink from Trinidad. In Miller, D. (ed.) *Material cultures: why some things matter*. London: UCL Press, 169–87.

Miller, D. 1998b: *A theory of shopping*. Cambridge: Polity Press.

Miller, D. 2001: *The dialectics of shopping*. Chicago: University of Chicago Press.

Miller, D., Jackson, P., Thrift, N.J., Holbrook, B. and **Rowlands, M.** 1998: *Shopping, place and identity*. London: Routledge.

MMC, 1981: *Discounts to retailers*. Monopolies and Mergers Commission HC 311. London: HMSO.

Modigliani, F. and **Miller, M.** 1958: The cost of capital, corporate finance and the theory of investment. *American Economic Review*, 48, 261–97.

Morgenson, G. 1993: The death of a salesman. *Forbes*, 1511, 106–12.

Morris, J. and **Imrie, R.** 1992: *Transforming buyer–supplier relations: Japanese-style industrial practices in a Western context*. Basingstoke: Macmillan.

Moir, C. 1990: Competition in the UK grocery trades. In Moir, C. and Dawson, J.A. (eds) *Competition and markets: essays in honour of Margaret Hall*. London: Macmillan, 91–118.

Monks, R. and **Minow, N.** 1995: *Corporate governance*. New York: Blackwell.

Moore, C.M. 2000: Streets of style: fashion designer retailing within London and New York. In Jackson, P., Lowe, M.S., Miller, D. and Mort, F. (eds) *Commercial cultures: economics, practices, spaces*. Oxford: Berg, 261–77.

Mort, F. 1995: Archaeologies of city life: commercial culture, masculinity, and spatial relations in 1980s London. *Environment and Planning D: Society and Space*, 13, 573–90.

Mort, F. 1996: *Cultures of consumption: masculinities and social space in late twentieth-century Britain*. London: Routledge.

Mort, F. 1998: Cityscapes: consumption, identity and the mapping of London since 1950. *Urban Studies*, 35, 889–907.

Mueller, W.F. and **Paterson, T.W.** 1986: Policies to promote competition: In Marion, B.W. (ed.) *The organization and performance of the US food industry*, Lexington, MA: Lexington Books, 371–412.

Murray, R. 1989: Fordism and post-Fordism. In Hall, S. and Jaques, M. (eds) *New times: the changing face of politics in the 1990s*. London: Lawrence and Wishart, 38–53.

Nava, M. 1997: Modernity's disavowal: women, the city and the department store. In Falk, P. and Campbell, C. (eds) *The shopping experience*. London: Sage, 56–91.

Nava, M. 1998: The cosmopolitanism of commerce and the allure of difference: Selfridges, the Russian ballet and the tango 1911–1914. *International Journal of Cultural Studies*, 1, 163–96.

Nixon, S. 1996: *Hard looks: masculinities, spectatorship and contemporary consumption*. London: UCL Press.

Noyelle, T. 1987: *Beyond industrial dualism: market and job segmentation of the new economy*. Boulder, CO: Westview.

OFT 1985: *Competition and retailing*. Office of Fair Trading. Field House, Bream Buildings, London EC4A IPR.

Ogbonna, E. and **Wilkinson, B.** 1998: Power relations in the UK grocery supply chain. Developments in the 1990s. *Journal of Retailing and Consumer Services*, 5, 77–86.

Omar, O.E. 1995: Retail influence on food technology and innovation. *International Journal of Retail and Distribution Management*, 23 (3), 11–16.

OXIRM 1994: *Sheffield retail study*. Oxford: Oxford Institute of Retail Management, Templeton College.

Panek, R. 1997: Superstore inflation. *New York Times Magazine*, 6 April, 66–8.

Pellegrini, L. 1993: Retailer brands: a state of the art review. In Burt, S.L. and Sparks, L. (eds) *Proceedings of the 7th International Conference on Research in the Distributive Trades*, University of Stirling, 348–58.

Pine, B.J. and **Gilmore, J.H.** 1999: *The experience economy: work is theatre and every business is a stage*. Boston, MA: Harvard Business School Press.

Piore, M. 1975: Notes for a theory of labour market stratification. In Edwards, R., Reich, M. and Garden, D. (eds) *Labour market segmentation*. Lexington, MA: Lexington Books, 125–50.

Polhemus, T. 1994: *Street style: from sidewalk to catwalk*. London: Thames and Hudson.

Poole, R., Clarke, G.P. and **Clarke, D.B.** 2002: Grocery retailers and regional monopolies. *Regional Studies.*

Porter, M.E. 1980: *Competitive strategy: techniques for analysing industries and competition*. New York: The Free Press.

Posner, R.A. 1979: The Chicago school of antitrust. *University of Pennsylvania Law Review*, 63, 925–48.

Powell, D. 1991: *Counter revolution: the Tesco story*. London: Grafton.

Pred, A. 1996: Interfusions: consumption, identity and the practices and power relations of everyday life. *Environment and Planning A*, 28, 11–24.

Prentice, D.D. and **Holland, P.R.J.** (eds) 1993: *Contemporary issues in corporate governance*. Oxford: Oxford University Press.

Reekie, G. 1992: Changes in the Adamless Eden: the spatial and sexual transformation of a Brisbane department store 1930–90. In Shields, R. (ed.) *Lifestyle shopping: the subject of consumption*. London: Routledge, 170–94.

Reve, T. 1990: The firm as a nexus of internal and external contracts. In Aoki, M. et al. (eds) *The firm as a nexus of treaties*. London: Sage, 136–88.

Reynolds, J. 1990: Food courts: tasty! *Stores*, August, 52–54.

Reynolds, J. 1993: The proliferation of the planned shopping centre. In Bromley, R.D.F. and Thomas, C.J. (eds) *Retail change: contemporary issues*. London: UCL Press, 70–87.

Reynolds, J. 1997: Retailing in computer-mediated environments: electronic commerce across Europe. *International Journal of Retail and Distribution Management*, 25, 29–37.

Richards, J. and **MacNeary, A.** 1991: A city view of retailing: challenge and opportunity in the nineties. In Treadgold, A. (ed.) *The city view of retailing*. Harlow: Longman.

Ritzer, G. 1993: *The McDonaldization of society: an investigation into the changing character of contemporary social life*. Thousand Oaks, CA: Pine Forge Press.

Rowley, G. 1993: Prospects for the central business district. In Bromley, R.D.F. and Thomas, C.J. *Retail change: contemporary issues*. London: UCL Press, 110–25.

Sack, R. 1992: *Place, modernity and the consumer's world*. Baltimore, MD: Johns Hopkins University Press.

Sayer, A. 1989: Post-Fordism in question. *International Journal of Urban and Regional Research*, 13, 666–95.

Sayer, A. 1997: The dialectic of culture and economy. In Lee, R. and Wills, J. (eds) *Geographies of economies*. London: Arnold, 16–26.

Sayer, A. and **Walker, R.** 1992: *The new social economy: reworking the division of labour*. Oxford: Blackwell.

Schiller, R. 1986: Retail decentralisation: the coming of the third wave. *The Planner*, July, 13–15.

Schleifer, A. and **Vishny, R.W.** 1997: A survey of corporate governance. *Journal of Finance*, 52, 737–83.

Schlereth, T.J. 1989: Country stores, country fairs, and mail-order catalogues: consumption in rural America. In Bronner, S.J. (ed.) *Consuming visions: accumulation and display of goods in America 1880–1920*. New York: W.W. Norton, 339–75.

Schoenberger, E. 1988: From Fordism to flexible accumulation: technology, competitive strategies and international location. *Environment and Planning D: Society and Space*, 6, 245–62.

Schoenberger, E. 1994: Corporate strategy and corporate strategists: power, identity and knowledge within the firm. *Environment and Planning A*, 26, 435–51.

Schoenberger, E. 1997: *The cultural crisis of the firm*. Oxford: Blackwell.

Schumpeter, J. 1942: *Capitalism, socialism and democracy*. New York: Harper and Row.

Scola, R. 1982: Retailing in the nineteenth century town: some problems and possibilities. In Johnson, J. and Pooley, C. (eds) *The structure of nineteenth century cities*. London: Croom Helm, 153–70.

Scott, A.J. and **Soja, E.W.** (eds) 1996: *The city: Los Angeles and urban theory at the end of the twentieth century*. Berkeley: University of California Press.

Senker, J. 1986: Technological cooperation between manufacturers and retailers to meet market demand. *Food Marketing*, 2 (3), 88–100.

Shackleton, R. 1998a: Exploring corporate culture and strategy: Sainsbury at home and abroad during the early to mid 1990s. *Environment and Planning A*, 30, 921–40.

Shackleton, R. 1998b: Part-time working in the 'super-service' era: labour force restructuring in the UK food retailing industry during the late 1980s and early 1990s. *Journal of Retailing and Consumer Services*, 5, 223–34.

Shaw, G. 1985: Change in consumer demand and food supply in nineteenth century British cities. *Journal of Historical Geography*, 11, 280–96.

Shaw, G., Alexander, A., Benson, J. and **Hodson, D.** 2000: The evolving culture of retailer regulation and the failure of the 'Balfour Bill' in interwar Britain. *Environment and Planning A*, 32, 1977–89.

Shaw, S.A., Nisbet, D.J. and **Dawson, J.A.** 1989: Economies of scale in UK supermarkets: some preliminary findings. *International Journal of Retailing*, 4 (5), 12–16.

Sherry, John F., Jr. 1998: The soul of the company store, Nike Town Chicago and the emplaced brandscape. In John. F. Sherry, Jr. (ed.) *Servicescapes: the concept of place in contemporary markets*. Lincolnwood, IL: NTC Business Books, 109–46.

Shields, R. 1989: Social spatialization and the built environment: the West Edmonton Mall. *Environment and Planning D: Society and Space*, 7, 147–64.

Shields, R. (ed.) 1992: *Lifestyle shopping: the subject of consumption*. London: Routledge.

Smith, D.G. 1996: Corporate governance and managerial incompetence: lessons from Kmart. *North Carolina Law Review*, 74, 101–203.

Smith, D.L.G. and **Sparks, L.** 1993: The transformation of physical distribution in retailing: the example of Tesco plc. *International Review of Retail, Distribution and Consumer Research*, 3, 35–64.

Smith, M.D. 1996: The empire filters back: consumption, production and the politics of Starbucks coffee. *Urban Geography*, 17, 502–24.

Smith, S. 1988: How much change at the store? The impact of new technologies and labour processes on managers and staff in retail distribution. In Knights, D. and Wilmott, H. (eds) *New technology and the labour process*. Basingstoke: Macmillan, 143–62.

Soja, E.W. 1996: Los Angeles 1965–1992: from crisis generated restructuring to restructuring-generated crisis. In Scott, A J. and Soja, E.W. (eds) *The city: Los Angeles and urban theory at the end of the twentieth century*. Berkeley: University of California Press.

Sorkin, M. 1992: See you in Disneyland. In Sorkin, M. (ed.) *Variations on a theme park: the new American city and the end of public space*. New York: Noorday Press.

Sparks, L. 1990: Spatial–structural relationships in retail corporate growth: a case study of Kwik Save Group plc. *Service Industries Journal*, 10, 25–84.

Sparks, L. 1992: Customer service in retailing: the next leap forward. *Service Industries Journal*, 12, 165–84.

Sparks, L. 1994: Delivering quality: the role of logistics in the post-war transformation of British food retailing. In Jones, G. and Morgan, N.J. (eds) *Adding value: brands and marketing in food and drink*. London: Routledge, 310–35.

Sparks, L. 1996a: Challenge and change: Shoprite and the restructuring of grocery retailing in Scotland. *Environment and Planning A*, 28, 261–84.

Sparks, L. 1996b: Space wars: Wm Low and the 'auld' enemy. *Environment and Planning A*, 28, 1465–84.

Sparks, L. 1996c: Investment recommendations and commercial reality in Scottish grocery retailing. *Service Industries Journal*, 16, 165–90.

Sparks, L. 1997: From coca-colonization to copy-cotting: the Cott Corporation and retailer brand soft drinks in the UK and the US. *Agribusiness*, 13, 153–67.

Sternquist, B. 1998: *International retailing*. New York: Fairchild Publications.

Storper, M. and **Walker, R.** 1989: *The capitalist imperative: territory, technology and industrial growth*. Oxford: Blackwell.

Thomas, C.J. and **Bromley, R.D.F.** 1993: The impact of out-of-centre retailing. In Bromley, R.D.F. and Thomas, C.J. (eds) *Retail change: contemporary issues*. London: UCL Press, 126–52.

Thrift, N.J. and **Olds, K.** 1996: Refiguring the economic in economic geography. *Progress in Human Geography*, 20, 311–37.

Tokatli, N. and **Boyaci, Y.** 2001: Globalization and the changing political economy of distribution channels in Turkey, *Environment and Planning A*, 33.

Treadgold, A. 1990: The developing internationalization of retailing. *International Journal of Retail and Distribution Management*, 18, 4–11.

Treadgold, A. and **Davies, R.** 1988: *The internationalization of retailing*. Harlow: Longman.

Walker, M. 1994: Quick response: the road to lean logistics. In Cooper, J. (ed.) *Logistics and distribution planning*. London: Kogan Page.

Weatherhead, M. 1997: *Retail estate in corporate strategy*. Basingstoke: Macmillan.

Weber, R.N. 1997: Governance and obligation: a legal–institutional approach to the restructuring of the US defense industry. Working Paper, Department of City and Regional Planning, Cornell University, Ithaca, New York.

Whitehead, M.B. 1992: Internationalization of retailing: developing new perspectives. *European Journal of Marketing*, 26 (8/9), 74–9.

Williams, D.E. 1992: Motives for retailers internationalization: their impact, structure and implications. *Journal of Marketing Management*, 8, 269–85.

Williamson, O.E. 1985: *The economic institutions of capitalism: firms, markets, relational contracting*. New York: The Free Press.

Wood, J. 1989: The world stage in retailing. *Retail and Distribution Management*, 17 (6) 14–16.

Wood, S. 2001a: Corporate restructuring, regulation and competitive space: the US department store in the 1990s. Unpublished Ph.D. thesis, Department of Geography, University of Southampton.

Wood, S. 2001b: Regulatory constrained portfolio restructuring: the US department store industry in the 1990s. *Environment and Planning A*, 33, 1279–304.

Wood, S. 2001c: Organizational restructuring, knowledge and spatial scale: the case of the US department store industry. *Tijdschrift voor Economische en Sociale Geographie*, 92.

Wood, S. 2002: The limits to portfolio restructuring: lessons from regional consolidation in the US department store industry. *Regional Studies*.

Wrigley, N. 1987: The concentration of capital in UK grocery retailing. *Environment and Planning A*, 19, 1283–8.

Wrigley, N. 1991: Is the 'golden age' of British grocery retailing at a watershed? *Environment and Planning A*, 23, 1537–44.

Wrigley, N. 1992: Antitrust regulation and the restructuring of grocery retailing in Britain and the USA. *Environment and Planning A*, 24, 727–49.

Wrigley, N. 1993a: Retail concentration and the internationalization of British grocery retailing. In Bromley, R.D.F. and Thomas, C.J. (eds) *Retail change: contemporary issues*. London: UCL Press, 41–68.

Wrigley, N. 1993b: Abuses of market power? Further reflections on UK food retailing and the regulatory state. *Environment and Planning A*, 25, 1545–57.

Wrigley, N. 1994: After the store wars: towards a new era of competition in UK food retailing. *Journal of Retailing and Consumer Services*, 1, 5–20.

Wrigley, N. 1996: Sunk costs and corporate restructuring: British food retailing and the property crisis. In Wrigley, N. and Lowe, M.S. (eds) *Retailing, consumption and capital: towards the new retail geography*. Harlow: Addison Wesley Longman, 116–36.

Wrigley, N. 1997a: British food retail capital in the USA. Part 1: Sainsbury and the Shaw's experience. *International Journal of Retail and Distribution Management*, 25, 7–21. (Reprinted in *British Food Journal*, 99, 412–26.)

Wrigley, N. 1997b: British food retail capital in the USA. Part 2: Giant prospects?, *International Journal of Retail and Distribution Management*, 25, 48–58. (Reprinted in *British Food Journal*, 99, 427–37.)

Wrigley, N. 1997c: Foreign retail capital on the battlefields of Connecticut: competition regulation at the local scale and its implications. *Environment and Planning A*, 29, 1141–52.

Wrigley, N. 1998a: Understanding store development programmes in post-property-crisis UK food retailing. *Environment and Planning A*, 30, 15–35.

Wrigley, N. 1998b: European retail giants and the post-LBO reconfiguration of US food retailing, *International Review of Retail, Distribution and Consumer Research*, 8, 127–46.

Wrigley, N. 1998c: How British retailers have shaped food choice. In Murcott, A. (ed.) *The nation's diet: the social science of food choice*. Harlow: Addison Wesley Longman, 112–28.

Wrigley, N. 1999a: Corporate finance, leveraged restructuring and the economic landscape: the LBO wave in US food retailing. In Martin, R.L. (ed.) *Money and the space economy*, Chichester: John Wiley, 185–205.

Wrigley, N. 1999b: Market rules and spatial outcomes: insights from the corporate restructuring of US food retailing. *Geographical Analysis*, 31, 288–309.

Wrigley, N. 2000a: Strategic market behaviour in the internationalization of food retailing: interpreting the third wave of Sainsbury's US diversification. *European Journal of Marketing*, 34, 891–918.

Wrigley, N. 2000b: The globalization of retail capital: themes for economic geography. In Clark, G.L., Feldman, M.P. and Gertler, M.S. (eds) *The Oxford handbook of economic geography*. Oxford: Oxford University Press, 292–313.

Wrigley, N. 2001a (subsequently published 2002): Transforming the corporate landscape of US food retailing: market power, financial re-engineering, and regulation. *Tijdschrift voor Economische en Sociale Geographie*, 93(1), 63–83.

Wrigley, N. 2001b: Local spatial monopoly and competition regulation: reflections on recent UK and US rulings. *Environment and Planning A*, 33, 189–94.

Wrigley, N. 2002a: Food deserts in British cities: policy context and research priorities. *Urban Studies*, 39.

Wrigley, N. 2002b: The landscape of pan-European food retail consolidation. *International Journal of Retail and Distribution Management*, 30.

Wrigley, N. and **Clarke, G.P.** 1999: Discount shake-out: the transformation of UK discount food retailing, 1993–98. Unpublished manuscript, Department of Geography, University of Southampton.

Wrigley, N. and **Lowe, M.S.** (eds) 1996: *Retailing, consumption and capital: towards the new retail geography*. Harlow: Addison Wesley Longman.

Wrigley, N. and **Lowe, M.S.** 2001: Progress report 2: Retailing and e-tailing. *Urban Geography*, 22.

Yago, G. 1991: *Junk bonds: how high yield securities restructured corporate America*. New York: Oxford University Press.

Zimmerman, M.M. 1937: *Super market: spectacular exponent of mass distribution*. New York: Super Market Publishing Company.

Zola, É. 1992: *Au bonheur des dames* ('The ladies' paradise'). Berkeley, CA: University of California Press (originally published 1883).

Zukin, S. 1991: *Landscapes of power: from Detroit to Disney World*. Berkeley, CA: University of California Press.

Zukin, S. 1995: *The culture of cities*. Oxford: Blackwell.

Zukin, S. 1998: Urban lifestyles: diversity and standardisation in spaces of consumption. *Urban Studies*, 35, 825–39.

Author index

Subject index